KEITH JOHNSON

PHYSICS FOR YOU

National Curriculum Edition for GCSE

Stanley Thornes (Publishers) Ltd

First published in 1980 by Hutchinson Education
Reprinted 1980, 1981, 1982, 1983 (twice), 1984, 1985, 1986
Revised edition 1986, 1987
Reprinted 1988, 1989
Major new edition with revisions published in 1991 by:
Stanley Thornes (Publishers) Ltd
Old Station Drive
Leckhampton
CHELTENHAM GL53 0DN
England

Reprinted 1992

British Library Cataloguing in Publication Data

Johnson, Keith
    Physics for you.
    I. Title
    530

    ISBN 0-7487-0565-1

Typeset by Oxprint Ltd OX2 6TR
Printed and bound in Hong Kong

'What is the use of a book,' thought Alice, 'without pictures or conversations?'

Lewis Carroll, *Alice in Wonderland*

Everything should be made as simple as possible, but not simpler.

Albert Einstein

There is no higher or lower knowledge, but one only, flowing out of experimentation.

Leonardo da Vinci

I do not know what I may appear to the world, but to myself
I seem to have been only a boy playing on the seashore, and
diverting myself in now and then finding a smoother pebble or
a prettier shell than ordinary, while the great ocean of truth
lay all undiscovered before me.

Sir Isaac Newton

Introduction

Physics For You is designed to introduce you to the basic ideas of Physics, and show you how these ideas can help to explain the world in which we live.

This book is based on successful earlier editions of the same name, but new pages and extra questions have been added to cover the requirements of the GCSE Examination and the National Curriculum.

Physics For You has been designed to be interesting and helpful to you, whether you are using it for a pure Physics course or as part of an Integrated or Coordinated Science course.

The book is carefully laid out so that each new idea is introduced and developed on a single page or on two facing pages. Words have been kept to a minimum and as straightforward as possible. Pages with a red band in the top corner are the more difficult pages and may be left out at first.

Throughout the book there are many simple experiments for you to do. Plenty of guidance is given on the results of these experiments, in case you find the work difficult.

Each important fact or new formula is printed in **heavy type** or is in a box. There is a summary of important facts at the end of each chapter.

At the back of the book there is advice for you on practical work, the nature of science, careers, revision and examination techniques, as well as some help with mathematics.

Questions at the end of a chapter range from simple fill-in-a-missing-word sentences (useful for writing notes in your notebook) to more difficult questions that will need some more thought. In calculations, simple numbers have been used to keep the arithmetic as straightforward as possible.

At the end of each main topic you will find a section of further questions taken from actual GCSE examination papers.

Throughout the book, cartoons and rhymes are used to explain ideas and ask questions for you to answer. In many of the cartoons, Professor Messer makes a mistake because he does not understand Physics very well. Professor Messer does not think very clearly, but I expect you will be able to see his mistakes and explain where he has gone wrong.

Here I would like to thank my wife, Ann, for her constant encouragement and help with the many diagrams and cartoons.

I hope you will find Physics interesting as well as useful. Above all, I hope you will enjoy **Physics For You**.

Keith Johnson

Professor Messer gets in messes,
Things go wrong when he makes guesses.
As you will see, he's not too bright,
It's up to you to put him right.

Contents

The small . . .

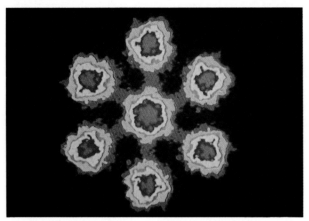

Seven atoms in a uranyl microcrystal, photographed with an electron microscope and false colour added. They are magnified 100 million times.

. . . and the large.

The Andromeda galaxy. It contains about 100 000 million stars, and the distance across it is over 100 000 light-years (10^{21} metres).

Waves: Light and Sound

Electricity and Magnetism

Nuclear Physics

Extra sections

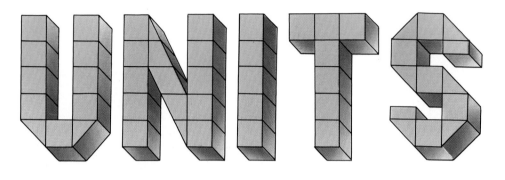

chapter 1

UNITS

▷ Length

Length is measured in a unit called the **metre** (often shortened to m).
A door knob is usually about 1 metre from the ground; doorways are about 2 m high.

For shorter distances we often use centimetres (100 cm = 1 metre) or millimetres (1000 mm = 1 metre).

Experiment 1.1
a) Look at a metre rule. Which marks are centimetres and which are millimetres?
b) It is useful to know the length of your handspan. Mine is 22 cm (0.22 m); what is yours?
c) Use the metre rule to measure
    – the length of your foot
    – the length of a long stride
    – your height.
    In each case write down the answer in millimetres and also in metres.

When measuring in Physics we try to do it as accurately as we can.
Professor Messer is trying to measure the length of a block of wood with a metre rule but he has made at least six mistakes.

How many mistakes can you find?

Experiment 1.2
Measure the length of a block of wood taking care not to make any of the Professor's mistakes.

▷ Mass

If you buy a bag of sugar in a shop, you will find the **mass** of sugar marked on the bag. It is written in **grams** (g) or in **kilograms** (kg). 'Kilo' always means a thousand, so 1 kilogram = 1000 grams.

The mass of this book is about 1 kilogram.
People often get confused between mass and weight, but they are **not** the same (see pages 81 and 84).

Experiment 1.3
Lift some masses labelled 1 kg, 2 kg, 5 kg and 1 g.

▷ Time

In Physics, time is always measured in **seconds** (sometimes shortened to s).
You can count seconds very roughly, without a watch, by saying at a steady rate: ONE (thousand) TWO (thousand) THREE (thousand) FOUR . . .

Experiment 1.4
Use a stopclock or stopwatch to measure the time for a complete swing of a pendulum or the beating of your heart.
What is the time for 100 of your heartbeats?
What is the time for one heartbeat?
By how much does it change if you run upstairs?

All the other units you will meet in this book are based on the metre, the kilogram and the second. They are called **SI units**.

Very large and small numbers

For very large or very small numbers, we sometimes use a shorthand way of writing them, by counting the number of zeros (see also page 375).

For example:

a) 1 million = 1 000 000 (6 zeros) = 10^6

b) 0.000 001 = $\frac{1}{1\,000\,000}$ (1 millionth) = 10^{-6}

In this shorthand way, write down:
one thousand, one thousandth, 10 million, one hundredth.

In Maths and in Physics, a 'k'
Means a thousand of whatever you say
For grams and for metres
And even, for teachers,
The size of their annual pay.

'kilo' is not the only prefix:

Mega (M) = 1 million	= 1 000 000
kilo (k) = 1 thousand	= 1 000
centi (c) = 1 hundredth	= $\frac{1}{100}$
milli (m) = 1 thousandth	= $\frac{1}{1000}$
micro (μ) = 1 millionth	= $\frac{1}{1\,000\,000}$
nano (n) = 1 thousand-millionth	= $\frac{1}{1\,000\,000\,000}$

Approximate length of time in seconds	Events
10^{18}	Expected lifetime of the Sun
10^{17}	Age of the Earth
10^{15}	Time since the dinosaurs lived
10^{13}	Time since the earliest human
10^{10}	Time since Isaac Newton lived
10^{9}	Average human life span
10^{7}	A school term
10^{5}	One day
10^{0}	One second
10^{-2}	Time for sound to cross a room
10^{-7}	Time for an electron to travel down a TV tube
10^{-8}	Time for light to cross a room
10^{-11}	Time for light to pass through spectacles
10^{-22}	Time for some events inside atoms

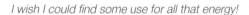

chapter 2

I wish I could find some use for all that energy!

Energy can exist in different forms, as you can see in the cartoon.

People get their energy from the **chemical energy** in their food.

Cars run on the chemical energy in petrol. A firework in the cartoon has chemical energy which it converts into **thermal energy (heat)** and **light** and **sound** energy when it explodes.

Chemical energy is one kind of stored energy or *potential* energy. Another kind of potential energy is the **strain energy** stored in the stretched elastic of the catapult.
The bucket over the door also has some stored potential energy, called **gravitational energy**. When the bucket falls down, this gravitational energy is converted into movement energy.

The moving pellets from the catapult and the moving people all have movement energy, called **kinetic energy**.

The television set is taking in **electrical energy** and converting it to thermal energy and light and sound energy.

Another form of energy is **nuclear energy**, which is used in nuclear power stations.

These forms of energy are shown in the diagram on the opposite page (see also page 114).

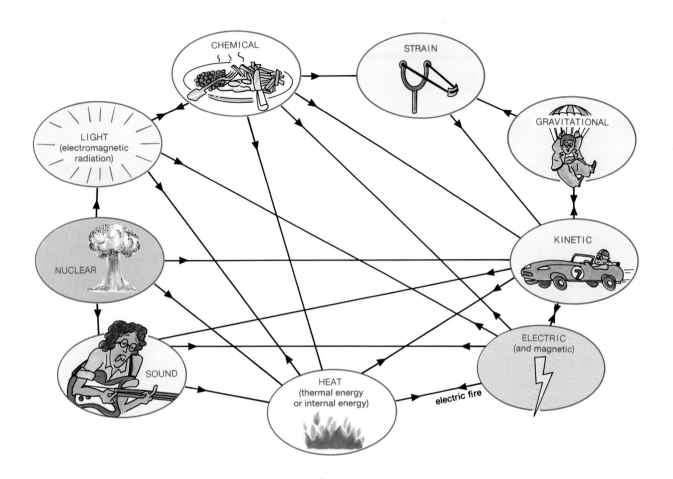

The connecting lines on the diagram show the different ways
that energy can be converted from one form to another.

See if you can decide what the energy changes are in the following objects.

For example:
A firework converts **chemical** energy to **thermal** energy, **light** and **sound** energy.
Copy and complete these sentences.

1. A TV set converts . . . . energy to . . . . energy.
2. A match converts . . . . energy to . . . . energy.
3. A light-bulb converts . . . . energy to . . . . energy.
4. A catapult converts . . . . energy to . . . . energy.
5. A falling bucket converts . . . . energy to . . . . energy.
6. An electric fire converts . . . . energy to . . . . energy.
7. A human body converts . . . . energy to . . . . energy.
8. A microphone converts . . . . energy to . . . . energy.
9. An atomic bomb converts . . . . energy to . . . . energy.
10. A car engine converts . . . . energy to . . . . energy.

You can learn more about energy changes in chapter 16.

Energy changes

In the diagram on the previous page (page 9), one energy change has been labelled 'electric fire'. Copy out the diagram into your book and then add the correct label to every arrow. Use words from the following list: coal fire, electric fire, steam engine, atom bomb (on four arrows), car engine, battery, loudspeaker, dynamo, very hot object, friction, bow and arrow, falling parachutist, cricket ball rising in the air, vibrations, microphone, thermocouple, photographer's light-meter, plants, glow-worm, fluorescent lamp, girl landing on a trampoline, hanging a weight on a spring, an arm muscle tightening, sound-absorbing material, light-absorbing material, electroplating.

Taking in chemical energy

Saving money

Energy costs money and we use large amounts of energy/money each day.

There are several ways of saving energy/money in your home (and making your home more comfortable at the same time). The table shows the time taken before they have paid for themselves and start to show a 'profit'.

How well is your home insulated?
How well is your school insulated? Write a list of recommendations.

Energy is measured in units called **joules.**
The joule is a small unit. To lift this book through a height of 10 cm needs about 1 joule.
When you walk upstairs you use over 1000 joules.

The diagram below shows the energy (in joules) involved in different events.
(Remember: 10^5 = number with 5 zeros = 100 000 and $10^{-5} = \frac{1}{100\,000}$)

Method	'Profit' times*
Lagging the hot-water cylinder	Less than a month
Draught excluders	A few weeks
Lagging the loft (see page 49)	About 3 years
Wall cavity insulation	4–7 years
Double glazing in windows	About 10 years

*all these times are shorter if you get a government grant

SAVE IT

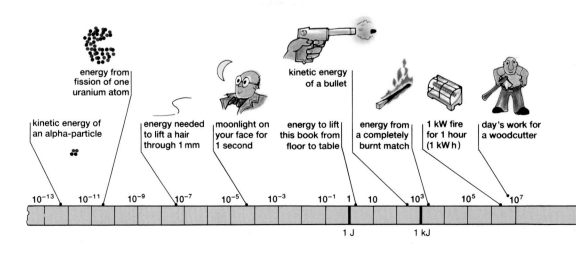

energy from fission of one uranium atom

kinetic energy of a bullet

kinetic energy of an alpha-particle

energy needed to lift a hair through 1 mm

moonlight on your face for 1 second

energy to lift this book from floor to table

energy from a completely burnt match

1 kW fire for 1 hour (1 kW h)

day's work for a woodcutter

10^{-13} 10^{-11} 10^{-9} 10^{-7} 10^{-5} 10^{-3} 10^{-1} 1 10 10^3 10^5 10^7

1 J   1 kJ

▷ The energy crisis

We are only just beginning to realise fully that our planet Earth is a spaceship with **limited** food and fuel (and we are taking on more passengers each year as the population increases).

Our supplies of energy cannot last for ever.

Oil and **natural gas** will be the first to disappear. If the whole world used oil at the rate it is used in America and Europe, our oil supplies would end in about 4 years!
As it is, the world's oil supplies might last for about 30 years.

How old will you be then? In what ways will your life be different without oil (and therefore without petrol and plastics)?

Coal will last longer – perhaps 300 years with careful mining. We might see (improved) steam engines on the railways again!

Nuclear energy might help for a while – but it causes problems due to the very dangerous radioactive waste that is produced. Also, each power station lasts only about 30 years and is difficult to dismantle because of the radioactivity.

We **waste** huge amounts of energy. It takes over 5 million joules of energy to make one fizzy-drink-can and we throw away 700 million of them each year! Making paper and steel uses particularly large amounts of energy, but very little is re-cycled.

We **must** find new ways of obtaining energy. The Sun's energy is free but it is not easy to capture it. Scientists are experimenting with new sources of energy (see next page). They will probably not give enough energy for the future. Our main hope is that the H-bomb **fusion** process (page 351) will eventually be controlled.

*Power stations are wasteful.
The overall efficiency (from power station to your home) is about 25%. Three-quarters of the energy is entirely wasted! The wasted energy could be used to heat nearby homes.*

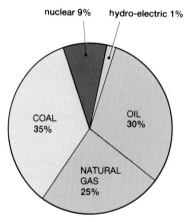

Energy supplies of Britain at present. What will happen when the oil and natural gas run out?

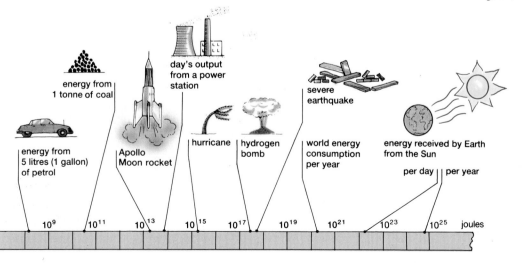

▷ New sources of energy

Some sources of energy are **renewable.** They are not used up like coal or oil.
Physicists and engineers are working hard to develop machines to use these sources of energy.

Source of energy	Original source
Solar	Sun
Biomass	Sun
Wind	Sun
Waves	Sun
Hydro-electric	Sun
Tides	Moon
Geothermal	Earth

Solar energy

The Earth receives an enormous amount of energy directly from the Sun each day, but we use very little of it. Some homes have solar panels on the roof (see page 56). In hot countries solar ovens can be used for cooking (see page 54).

Space ships and satellites use **solar cells** to convert sunlight into electricity. You may have seen calculators powered by solar cells.

Covering part of the Sahara Desert with solar cells would produce energy but it would be very expensive. To equal the power of one modern power station you would need 40 square kilometres of solar cells.

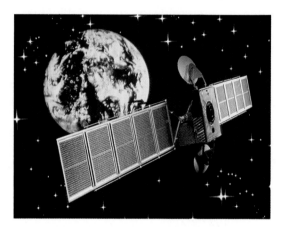

Biomass

Some of the sunlight shining on the Earth is trapped by plants, as they grow. We use this **biomass** when we eat plants or when we burn wood.

In Brazil they grow sugar cane and then use the sugar to make alcohol. The alcohol is then used in cars, instead of petrol.

Rotting plants can produce a gas called methane which is the same as the 'natural gas' we use for cooking. If the plants rot in a closed tank, called a **digester**, the gas can be piped away and used as fuel for cooking.
This is often used in China and India.

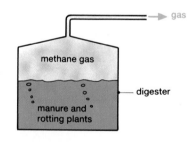

Wind energy

This energy also comes from the Sun, because winds are caused by the Sun heating different parts of the Earth unequally.

Modern windmills are very efficient but we would need about 2000 very large windmills to provide as much power as one modern power station.

Wave energy

Waves are caused by the winds blowing across the sea. They contain a lot of free energy.

One method of getting this energy might be to use large floats which move up and down with the waves. The movement energy could be converted to electricity. However, we would need about 20 kilometres of floats to produce as much energy as one power station.

Hydro-electric energy

Dams can be used to store rain-water, and then the falling water can be used to make electricity (see experiment 16.5 on page 117).

This is a very useful and clean source of energy for mountainous countries like Norway and China.

The same idea can be used to store energy from power stations that cannot easily shut down. At night, when demand is low, spare electricity can be used to pump water up to a high lake. During the day, the water can be allowed to fall back down to produce electricity when it is needed.

Tidal energy

As the Moon goes round the Earth it pulls on the seas so that the height of the tide varies.

If a dam is built across an estuary, it can have gates which trap the water at high tide. Then at low tide, the water can be allowed to fall back through the dam and make electricity (see experiment 16.5).

Geothermal energy

The inside of the Earth is hot (due to radioactivity, see chapter 39). In some parts of the world (like New Zealand) hot water comes to the surface naturally. In other countries cold water is pumped down very deep holes and hot water comes back to the surface.

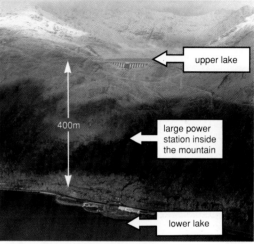

A pumped storage scheme in Scotland

A tidal power station in France

Summary

Energy exists in several different forms: chemical, electric, magnetic, kinetic, potential, sound, nuclear, electromagnetic radiation (including light) and heat (also called internal energy or thermal energy).

Energy can be changed from one form to another.

It is measured in joules.

Some sources of energy are non-renewable and will be used up: coal, oil, gas, nuclear.
Other sources are renewable: solar, biomass, wind, wave, hydro-electric, tidal and geothermal. See also chapter 16.

chapter 3

MOLECULES

Do you know that as you are reading this sentence you are punched on the nose more than 100 billion billion times ! ! ?

This is because the air is made up of billions and billions of tiny particles called *molecules* – rather like the title at the top of this page is made up of tiny dots. But the molecules are so small that we cannot see them even through a powerful microscope. Since each molecule is very small and light, it does not hurt as much as when a big fist hits your nose.

▷ High speed gas

The molecules hitting your face are moving very fast. Their average speed is about 1000 m.p.h. (1600 km/h), but with some moving much faster and some slower.

The billions and billions of air molecules around you are bouncing and colliding with each other at this very high speed. It's just as though the room is full of tiny balls bouncing everywhere and filling the whole room.

If the air is cooled down, the molecules have less energy and move more slowly.
If the air is heated up, the molecules have more energy and so move faster (and hit your nose harder).

This idea about molecules is called the *kinetic theory.*

▶ Three states of matter

Every substance can exist in three states: *gas, liquid* and *solid.*

The air in this room is a gas. But if it is cooled down it can become liquid or even solid.

In the same way, water can be a solid (ice), a liquid, or a gas (steam):

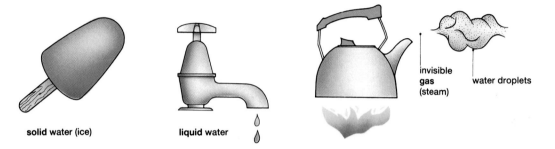

solid water (ice)　　　　liquid water　　　　invisible gas (steam)　　　water droplets

In a **gas**, the molecules are moving very quickly.

In a **liquid**, the molecules are moving more slowly and are held together more closely by forces between the molecules. This is why a large volume of gas *condenses* into a smaller volume of liquid.

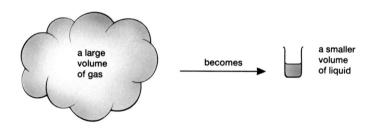

a large volume of gas　　becomes　　a smaller volume of liquid

These forces between molecules make the liquid have a definite size, although it can still change its shape and flow.

In a **solid**, the forces between molecules are so strong that the solid has a definite shape as well as a definite size.

There are two kinds of forces between molecules:
– *attractive* forces if molecules try to move apart
– *repulsive* forces if molecules try to move closer.
Normally these forces balance.
If you try to stretch a solid (and so pull the molecules apart) you can feel the attractive forces.
If you squeeze a solid (and so push its molecules closer together) you can feel the repulsive forces.

These forces are *electric* forces (see page 239).

Molecular forces are strong:

▷ Experiments

You will need some marbles (to act as 'molecules')
and a flat box or tray with vertical sides.

Experiment 3.1   A solid
Place the marbles in the tray. Tilt the tray slightly
so that the 'molecules' collect at one end.

You can see that the group of 'molecules' has a
definite size and shape. This is imitating a **solid**
(a very cold solid). Sketch the pattern of the
molecules (a regular **crystalline** pattern).

The molecules in a solid are continually vibrating.
If the solid gets hotter, its molecules vibrate more.

Experiment 3.2   A liquid
Now rock the tray gently from side to side.
You can see that the group of 'molecules' now has more energy
and does not have a definite shape. This is imitating a **liquid**.

Experiment 3.3   A gas
Next, with the tray flat on the table, shake the tray to and fro
and from side to side to give the 'molecules' a lot more energy.
You can see that the 'molecules' fill all the available space in the tray.
This is imitating a **gas**.

Heat up the gas by shaking the tray faster.

Experiment 3.4   Brownian movement
No-one has ever seen an air molecule, but here is an experiment to
help us believe that they exist.

A microscope is used to look into a small glass box which contains
some smoke as well as air molecules.

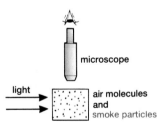

In a bright light, the smoke particles show up as bright specks
which move in a continuous and jerky random movement. This is
called **Brownian movement**.

It happens because the smoke particles (like your nose) are being
punched by molecules. Because the smoke particles are so light,
they twist and jerk as you can see through the microscope.

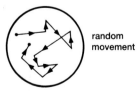

Experiment 3.5
While shaking your marbles tray to imitate a gas, drop into it a
larger marble (to act as a particle of smoke).
Watch the random movement of the 'smoke particle' as it is
punched by the 'molecules'. This is Brownian movement.

▷ Diffusion

When food is being cooked in the kitchen, it can be smelt in other rooms in the house, even when there are no draughts to move the air.

This is because molecules from the food are moving around at high speed and after billions of collisions eventually reach all parts of the house. This is called *diffusion*.

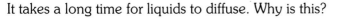

Experiment 3.6   Diffusion
While shaking your marbles tray to imitate a gas, drop into one corner a coloured marble (acting as a 'smell molecule'). Watch it while you keep shaking. You will see that eventually the 'smell molecule' will diffuse to any part of the tray.

Draw a rough sketch of the random movement of the 'molecule'.

Experiment 3.7   Diffusion in a liquid
Fill a tall beaker with water and leave it for a while so that it becomes still and at room temperature.
Carefully drop in a single crystal of potassium permanganate. When this dissolves it makes the water purple.
Leave the beaker where it will not be touched or knocked for several weeks and observe it.

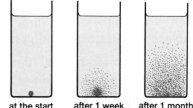

at the start   after 1 week   after 1 month

It takes a long time for liquids to diffuse. Why is this?

Experiment 3.8   Diffusion in a gas
Your teacher may be able to show you the diffusion of some bromine. This is a gas that you can see (it is brown), but it is very poisonous so special apparatus and care must be used. When the tap is opened, the bromine runs down and starts diffusing.

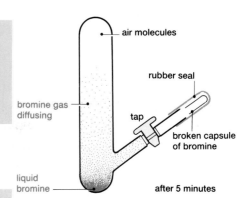

Why is the diffusion in this experiment faster than in the previous experiment?

Experiment 3.9   Diffusion into a vacuum
Your teacher may be able to repeat the bromine experiment but with the main tube completely empty of air molecules (by using a vacuum pump to take out all the molecules).
What happens this time as the tap is opened to let in the bromine? (Don't blink or you might miss it.)

Because there are no air molecules to get in the way, this shows you the speed of bromine molecules – it is about 500 m.p.h. (800 km/h) at room temperature!

▷ Molecular forces in liquids

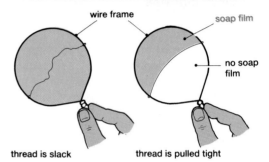

wire frame   soap film

no soap film

thread is slack   thread is pulled tight

These experiments all show that there is a surface force or **surface tension** on the surface of liquids. This is due to the forces attracting each surface molecule to its neighbouring molecules.

There is a tension because the molecules on the surface are slightly farther apart (like the molecules in a stretched elastic band).
The tension tries to reduce the surface area. This is why bubbles are **spherical** (as this shape has the smallest area for a given volume).

This force of attraction between molecules of the **same** substance is called **cohesion**. This is the force that holds a raindrop together.

The force of attraction between molecules of **different** substances is called **adhesion**. This is the force that holds glue or paint to a wall.

section through a needle
floating without getting wet

Some insects can walk on water because of this. The photograph shows a 'pond-skater' and its reflection in the water.

Can you see where the water surface is pushed down by its legs?

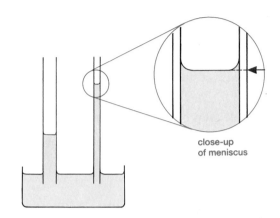

close-up
of meniscus

In which tube does it rise higher?
The water rises because the adhesion (between
water and glass molecules) is greater than the
cohesion (between water molecules).

This also explains why the **meniscus** shape at the
top of the water is curved as shown.

If you repeated this experiment with mercury, you
would find there is a capillary **fall**. This is
because the adhesion (between mercury and glass
molecules) is **less** than the cohesion (between
mercury molecules).
Which way does a mercury meniscus curve? (See the
diagram on page 95.)

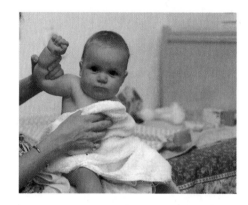

Using capillary rise

What happens when you use a towel to dry your face?

The water soaks up into the tiny tubes in the cloth
of the towel.
The same thing happens with nappies, paper tissues,
blotting paper and kitchen cloths.

The wick of a candle or a paraffin heater also uses
capillary action.
Rain-water soaks through soil in the same way.

Can you invent a device to feed your house plants
from a tank of water so they are kept watered when
you are on holiday?

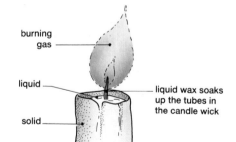

burning
gas

liquid

liquid wax soaks
up the tubes in
the candle wick

solid

Preventing capillary rise

Water will soak up through the floor of a house
unless we prevent it with a sheet of plastic.

Bricks are **porous** and will allow 'rising damp'
unless a 'damp course' is included just above
ground level.

Laying a damp-proof course

Summary

All substances can exist in 3 states: *solid*, *liquid* and *gas*.

All substances are made of **molecules**.
If the substance is heated, the molecules have more energy and move faster.

Molecules can **diffuse** through other molecules.

Brownian movement of smoke particles helps us to believe in molecules.

In a *gas*, the molecules are moving freely at high speed.
In a *liquid*, the molecules are loosely held together by forces.
A *solid* has a definite shape because its molecular forces are very strong.

In liquids, molecular forces cause a 'skin' or **surface tension** and cause liquids to rise up narrow tubes.

▷ Questions

1. Copy out and fill in the missing words:
   a) The three states of matter are . . . . , ... and . . . . . All matter consists of tiny . . . . travelling at . . . . speed.
   b) A solid has a definite size and a definite . . . . because the forces between its . . . . are very . . . . . A liquid has weaker forces between its . . . . and so does not have a definite . . . . .
   c) If a substance is cooled down, its molecules move at a . . . . speed. When some molecules pass through some other molecules, we call it . . . . . The twisting and jerking of smoke particles when they are hit by air . . . . is called . . . . .

2. Copy out and fill in the missing words:
   The forces between the . . . . in liquids cause a kind of 'skin' or . . . . The forces between . . . . also cause water to . . . . inside narrow tubes. This is called . . . . The curved surface at the top of the water is called a . . . . . This occurs because the force of adhesion is . . . . than the force of . . . . .

3. Explain why
   a) solids have a definite shape but liquids flow
   b) solids and liquids have a fixed size but gases fill whatever container they are in.

4. Explain why perfume can be smelled some distance away from the person wearing it.

5. a) Explain what is meant by Brownian movement and how it helps us to believe in molecules.
   b) What change would you expect to see in the movement if the air was cooled down?

6. Copy and complete the rhyme and explain why you do not see this happening:
   A rather small student called Brown,
   Was asked why he danced up and down,
   He said "Look, you fools,
   It's the air . . . . ,
   They constantly knock me around."

7. A tin can containing air is sealed. If it is then heated, what can you say about:
   a) the average speed of the molecules,
   b) how often the molecules hit the walls of the can,
   c) how hard the molecules hit the walls,
   d) the air pressure inside the can?

8. Explain the following:
   a) Raindrops are almost spherical.
   b) A needle can be floated on water.
   c) The needle sinks if detergent is added.
   d) Hairs (on your head or on a paint brush) cling together when they are wet but not when they are dry.
   e) Small insects can stand on the surface of a pond without sinking.
   f) You can dry your hands with a towel but not with a sheet of polythene.
   g) Paint sticks to a wall.

Further questions on page 74.

chapter 4

Expansion

Experiment 4.1
This metal bar will **just** fit into the gap when they are both cold. Heat the bar with a Bunsen burner. Does it fit the gap now?

What happens when it cools down?

Experiment 4.2
This metal ball will **just** slip through the metal ring when they are both cold.

Does it go through when the ball is heated?

When objects get hotter they grow bigger. We say they **expand**.
When objects cool down they get smaller. We say they **contract**.
It is difficult to see this change in size because it is so small.

Experiment 4.3
This thick bar is heated with a Bunsen burner and then the big nut is tightened.

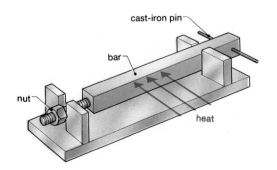

What happens to the cast-iron pin when the bar cools down? (Do not stand too near.)

This shows that:
– when bars **contract** you get big **pulling** forces.
– when bars **expand** you get big **pushing** forces.

▷ Expansion can cause trouble

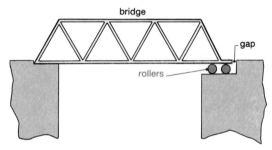

Bridges are often made of big steel bars. When they get hotter, they get longer.
The Forth Railway Bridge is more than 1 metre longer in summer than in winter.

Bridges are usually put on rollers so that they can expand and contract without causing damage.

There must be an *expansion gap* in the road at the end of a bridge.
Have you seen one?

What would happen if there was no gap?

Roads are often made of large concrete slabs. They also expand. There are *expansion gaps* between the slabs, filled with a soft substance which can be squeezed easily in hot weather.

If you look carefully, you may find similar expansion gaps in the floor of your school.

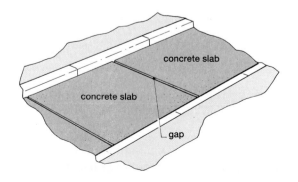

Telephone wires look like this in summer.
What will they look like in winter?

What would happen if the wires were made tight in summer and then it went cold?

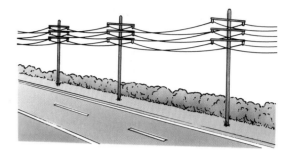

Thick glass beakers and milk-bottles sometimes crack if boiling water is put in them. This is because the inside tries to expand while the outside stays the same size.

To wash out a bottle I tried,
By pouring hot water inside,
But the glass was too thick,
(The expansion too quick)
The result was a crack very wide.

'Pyrex' glass beakers do not crack because 'pyrex' does not expand as much as ordinary glass.

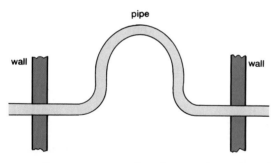

Steam and water pipes often have a large bend in them. The bend lets the pipe expand without breaking anything.

Steel railway lines need expansion gaps. Sometimes the gaps are at the end of each length of rail (you can hear the clicks as the wheels go over them). Nowadays many rails are welded together and the gaps are much farther apart.

A railway mechanic was sent
To lay down some track across Kent.
He forgot to design
Some gaps in the line,
And when it got hot, it went bent.

▶ Expansion and contraction can be useful

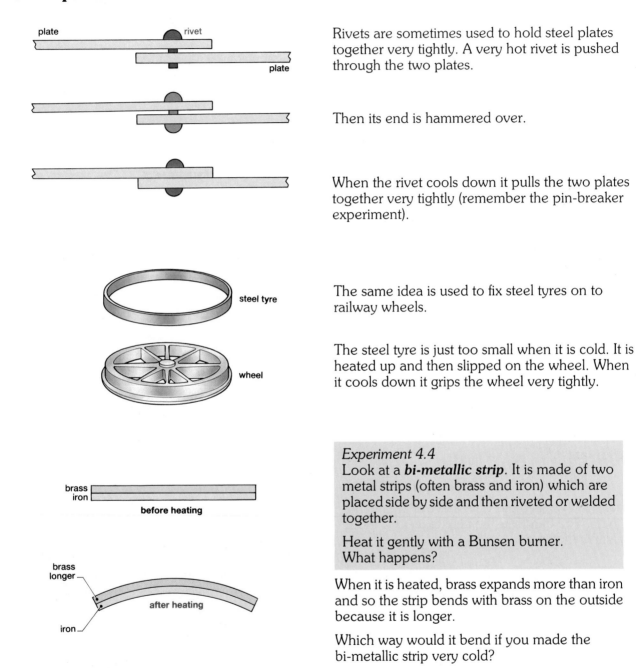

Rivets are sometimes used to hold steel plates together very tightly. A very hot rivet is pushed through the two plates.

Then its end is hammered over.

When the rivet cools down it pulls the two plates together very tightly (remember the pin-breaker experiment).

The same idea is used to fix steel tyres on to railway wheels.

The steel tyre is just too small when it is cold. It is heated up and then slipped on the wheel. When it cools down it grips the wheel very tightly.

Experiment 4.4
Look at a **bi-metallic strip**. It is made of two metal strips (often brass and iron) which are placed side by side and then riveted or welded together.

Heat it gently with a Bunsen burner.
What happens?

When it is heated, brass expands more than iron and so the strip bends with brass on the outside because it is longer.

Which way would it bend if you made the bi-metallic strip very cold?

▶ Using bi-metallic strips

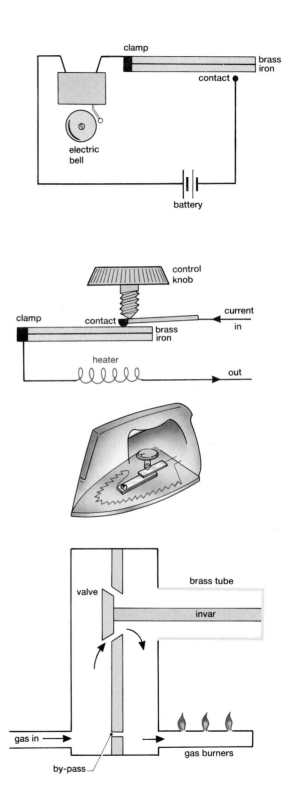

Electric thermostat

A *thermostat* is used to keep something at the same temperature, without getting too hot or too cold.
In the diagram, the electric current is flowing through the contact and the bi-metallic strip to the heater.

What happens if it gets too hot?
What happens later when it cools?

This is used in electric irons, to keep the iron at the correct temperature.

Which way would you turn the control knob if you wanted a cooler iron, for ironing nylon?

Gas thermostat for a gas oven

This uses a rod made of a special metal called 'invar' which does not expand. When the oven gets hot, the brass tube expands but the invar does not – and so the valve is pulled to the right. This closes the gap so that the gas cannot get through.

What happens when the oven cools down?

For safety, there should be a small 'by-pass' to let some gas through even when the valve is closed. What would happen if there was no 'by-pass'?

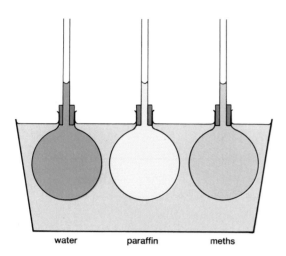

water   paraffin   meths

Experiment 4.6
We need three glass bottles or flasks which are all the same size and have narrow tubes coming out of the top.

Fill them with three different liquids up to the same level as in the diagram. Put them in a dish and then fill the dish with hot water.

What happens?
Which liquid expands the most?
Do liquids expand more or less than solids?

Liquids expand more than solids

Most liquids expand when they get hotter. But *water is different* from other liquids:

If you heat water from freezing point (0 °C) up to 4 °C, it gets *smaller* even though you are making it hotter.
Above 4 °C, it behaves like other liquids and expands as it gets hotter.
This is shown on the graph:

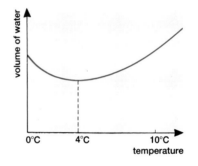

You can see that water is smallest at 4 °C. This means that water is *densest* at 4 °C.

Now think what this means to a fish in a pond in winter. Because the densest water falls to the bottom of the pond, the bottom stays at 4 °C even when the top freezes over.

So the bottom is warmer than the top. The fish can still swim around and stay alive underneath the ice.

▶ Expansion of gases

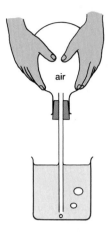

Experiment 4.7
Get a glass flask with a rubber bung and a tube as in the diagram.
Dip the end of the tube in a beaker of water and then warm the flask with your hands.

What do you see?
What is expanding?
Do gases expand more or less than liquids?

Now take your hands off and wait for the flask to cool down.

What happens?   Why?

Gases expand more than liquids.
We will see how this is used in rockets and other engines when we come to chapter 10 (page 68).
There is more about expanding air on page 36.

▶ Explaining expansion

If a group of people are just standing still then they can stay close together and they don't take up much space.
But if the people start to dance then they take up more space – the group expands.

In the same way, if a substance is heated, the molecules start moving more and so they take up more space – the substance expands.

Notice that the molecules themselves do not get any bigger – but as a group they take up more space.

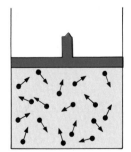

Summary

When objects are heated, they expand and push with large forces. Expansion gaps are needed in bridges, roads, etc.

Liquids expand more than solids.

Water is different from other liquids – it contracts as it is heated from 0 °C up to 4 °C and then it expands above 4 °C.

A bi-metallic strip is made of two different metals fixed together. When heated, it bends (with the metal with the bigger expansion on the outside). It is used in thermostats and fire-alarms.

Gases expand more than liquids.

Substances expand because the molecules are moving more and take up more space.

Can you explain this cartoon?

▶ Questions

1. Copy out and fill in the missing words:
   a) When an object is heated it . . . . and when it cools it . . . . . If we try to prevent it from contracting, it exerts a very large . . . . . If a bi-metallic strip made of brass and iron is heated, it . . . . because . . . . expands more than . . . . and the . . . . goes on the outside of the bend.
   Liquids expand more than . . . . but less than . . . . .
   b) When water is heated from 0 °C up to 4 °C it . . . . . When it is heated above 4 °C it . . . . . When substances are heated, they get bigger because the . . . . are moving more.

2. Explain how expansion affects the design of
   a) bridges
   b) railway lines and
   c) steam pipes.

3. Which contracts the most when it is cooled – iron, water or air?

4. The diagram shows a bi-metallic strip. Draw diagrams to show what it looks like
   a) if it is heated up
   b) if it is cooled down.

brass
iron

5. a) Draw a labelled diagram of a fire alarm and explain how it works.
   b) How could you change it to a frost alarm?

6. a) Draw a labelled diagram of a thermostat for an electric iron. Explain how it works.
   b) How could you change it to make a thermostat for a refrigerator?

7. Draw a labelled diagram of a thermostat for a gas oven and explain how it works.

8. Why are the doors and shelves of ovens made loosely fitting?

9. a) What happens to your ruler when the room gets warmer?
   b) What does this mean about the lines you measure?

10. What will happen to the pendulum of a clock when the room gets warmer? What effect will this have on the clock's timekeeping?

11. In a motor-car engine:
   a) why are the cylinder liners cooled before they are pushed into the cylinder block?
   b) why are piston rings needed and why do they have a gap in them?

12. A steel bridge over a motorway is 20 m long at 0 °C. How much longer is it at 20 °C? (Steel expands by 0.000 01 of its length, for each °C rise in temperature.)

   If you were the engineer designing the bridge, what expansion gap would you plan for (in the British climate)?

13. Can you explain this cartoon?

14. Here is a circuit often used in cars:

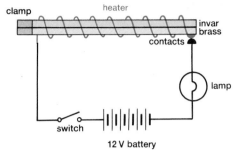

a) When the switch is connected, what happens to the light and the heating coil?
b) What happens to the bi-metallic strip?
c) What happens then to the light and the heater?
d) What happens then to the bi-metal strip?
e) What happens next?
f) What would you see?
g) Where is this used on a car?

15. a) If a bottle was filled with a liquid, tightly sealed and then heated, what might happen?
   b) Why does a bottle of lemonade always have space between the top of the liquid and the cap?
   c) Why should you never heat an un-opened can of baked beans in an oven?
   d) Why might a beach-ball burst if left in the sun?
   e) Why can a dented table-tennis ball often be mended by warming it in hot water?
   f) Why must you never put sealed bottles or 'aerosol' spray cans on a fire?

Further questions on page 75.

chapter 5

Thermometers are used to measure **temperature**.
Temperature is not the same thing as heat or thermal energy.

To understand this, let's compare a white-hot 'sparkler' firework with a bath-full of warm water:

The tiny 'sparkles' are at a very high temperature but contain little energy because they are very small

The water is at a lower temperature but it contains more internal energy because it has many more molecules

Each sparkle contains a few molecules vibrating at very high temperature but with little total energy

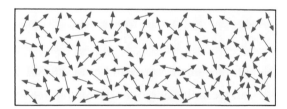

The bath has many molecules, each vibrating at low temperature but with more total energy

You can see that the bath has more **thermal energy**.
Its molecules have more **internal energy**.
To raise the temperature of an object you must give more energy to its molecules.

A laboratory thermometer

This uses the expansion of a liquid to measure temperatures accurately. Look at the diagram opposite. If the liquid gets warmer, it expands along the tube, where there is a scale of numbers marked °C or **degrees Celsius** (or degrees centigrade). There is a vacuum (no air) above the liquid so it can move easily along the tube.

To make the thermometer **sensitive**, with a long scale, it should have a narrow-bore tube and a large bulb.

To make the thermometer **quick-acting**, it should have a bulb made of thin glass so that the heat can get through easily.

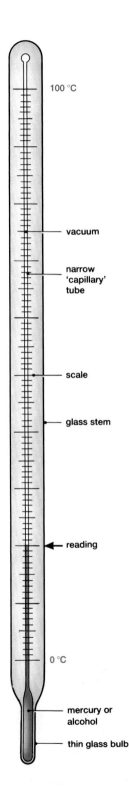

100 °C

vacuum

narrow 'capillary' tube

scale

glass stem

reading

0 °C

mercury or alcohol

thin glass bulb

What is the reading on the thermometer in the diagram? (This is the temperature of a warm room.)

Experiment 5.1
Practise reading the scale of a thermometer by measuring the temperature of the room; of the cold-water supply; of the hot-water supply; of your armpit (take care – thermometers break easily).

Calibration

How are the marks put on the thermometer in the correct places?

Two *fixed points* are marked first of all:

The *upper fixed point*, marked 100 °C, is the temperature of the steam over pure boiling water at standard atmospheric pressure (see page 95).

The *lower fixed point*, marked 0 °C, is the temperature of pure melting ice at standard atmospheric pressure.

When the two fixed points are marked, the scale between them is divided into 100 equal divisions.

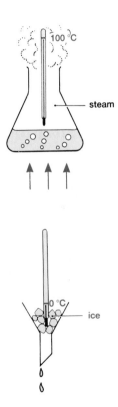

100 °C

steam

0 °C

ice

Experiment 5.2
Check the fixed points on a thermometer (or 'calibrate' an unmarked thermometer) by using the apparatus shown.

Choice of liquids

Mercury is used because a) it expands evenly as the temperature rises and b) it is a good conductor of heat. Unfortunately it freezes solid if it is used in very cold places.

Mercury is poisonous and care must be taken if the glass breaks.

Alcohol is often used instead, because a) it can be used at very low temperatures and b) its expansion is six times greater than mercury. Unfortunately it cannot be used in very hot places because it boils at a lower temperature than mercury.

Clinical thermometer

A doctor or nurse needs a thermometer that will continue to show its maximum temperature after it is taken out of your mouth.

A clinical thermometer has a very narrow constriction in the tube just above the bulb. When the mercury expands it pushes past the constriction, but when it is taken out of your mouth it does not go back to the bulb – and so the doctor can read your temperature.

The mercury can be shaken back into the bulb after it has been read.

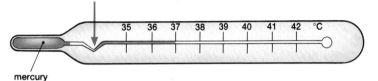

Doctor Shoddy, old and frail,
Can't see the numbers on this scale,
Take the reading – tell old Shoddy,
This is the temperature of your body!

Another maximum thermometer

This contains a little iron *index* which is pushed along by the mercury.

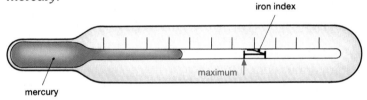

How could you use a magnet to reset this thermometer?

A minimum thermometer

This can be used to find the coldest temperature during the night. It contains an iron index placed inside the liquid which is alcohol. As it cools down, the surface meniscus of the alcohol pulls the index back.

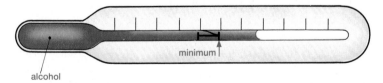

How would you reset this thermometer?

A liquid-crystal thermometer

The numbers on the scale are made from different chemicals, and they show up at different temperatures:

It is easy to read but not very accurate.

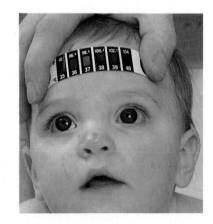

Other thermometers

1. Since gases expand by a greater amount and more evenly than liquids, **gas thermometers** are more accurate than liquid thermometers and can work at higher and lower temperatures. (See also page 38.)

2. A **resistance thermometer** uses the fact that electricity does not flow so easily through a wire when the wire gets hot (see also page 252).

3. A **thermistor thermometer** uses the fact that electricity flows through it more easily when it is hot (see also page 317).

4. A **thermocouple thermometer** consists of two wires made of different metals and connected to a sensitive ammeter. Different temperatures produce different electric currents.
   High and low temperatures can be measured and the end of the thermometer can be made very small.

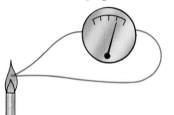

5. The temperature of a very hot object like a furnace or the Sun can be measured by a **pyrometer** that measures the amount of radiation given off by the object.

Absolute zero

As substances get colder, gases condense into liquids and liquids freeze into solids. If the temperature continues to get colder and colder, the molecules vibrate less and less until eventually they have their lowest possible energy. This happens at the coldest possible temperature, at $-273\,°C$, called **absolute zero**. (See also page 37.)

Scientists often measure temperatures on the **Kelvin scale**, which begins at absolute zero but increases just like the Celsius scale. This means $0\,°C$ becomes 273 kelvin (written 273 K) and $100\,°C = 373$ K.

On a hot day, the temperature is $27\,°C$ – what is this in kelvin?

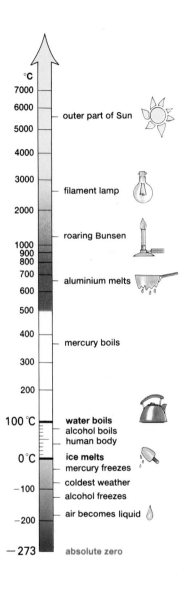

°C
7000
6000 — outer part of Sun
5000
4000
3000 — filament lamp
2000
— roaring Bunsen
1000
900
800
700 — aluminium melts
600
500
400 — mercury boils
300
200
100 °C — **water boils**
      alcohol boils
      human body
0 °C — **ice melts**
      mercury freezes
-100 — coldest weather
      alcohol freezes
-200 — air becomes liquid
-273 — absolute zero

Summary

The temperature of an object depends upon how much its molecules are vibrating.
Internal energy (heat or thermal energy) is the total amount of energy in an object.

A clinical thermometer has a constriction to prevent the mercury from returning to the bulb (until it is shaken).

The upper fixed point, $100\,°C$, is the temperature of the steam over pure boiling water at standard atmospheric pressure.
The lower fixed point, $0\,°C$, is the temperature of pure melting ice at standard atmospheric pressure.
Absolute zero is the coldest possible temperature at $-273\,°C$. Temperature in kelvin = °C + 273.

▶ Questions

1. Copy out and fill in the missing words:
   a) Thermometers are used to measure . . . . which depends on how much the . . . . are vibrating. Internal energy is the amount of . . . . in an object.
   b) The Upper Fixed Point of the Celsius scale is marked . . °C and is the temperature of . . . . over pure . . . . water at normal atmospheric . . . . . The Lower Fixed Point of the Celsius scale is marked . . °C and is the . . . . of pure . . . . ice at . . . . atmospheric . . . . . The scale between the two fixed points is marked into . . equal divisions.
   c) The coldest possible temperature is . . °C, called . . . .

2. Copy out and complete the rhyme:
   Thermometers were often made
   With marks described as c . . . . .
   But modern times have seen a fuss
   To change the name to C . . . . .

3. Describe and explain the features of a thermometer which will make it
   a) sensitive,
   b) quick-acting.

4. a) Convert 10 °C to kelvin.
   b) Convert 300 K to °C.
   c) What is your body temperature in kelvin?

5. *Professor Messer blows his top, Tell him why he's such a flop:*

6. Copy out and place ticks in the table to show which liquid is better in each case:

	Mercury	Alcohol
Expands more evenly		
Expands more		
A better conductor of heat		
Useful at higher temperatures		
Useful at lower temperatures		

7. Professor Messer was taken ill. The table shows how his temperature varied:

Time (hours)	0	1	2	3	4	5	6	7	8	9	10
Temperature (°C)	37	39	40	40	40	39.5	38.3	37.7	37.3	37	37

   a) Plot a graph of his temperature (for the range 36–40 °C) against time (in hours).
   b) What was his highest temperature?
   c) When did this happen?
   d) When do you think an ice-bath was used to cool him?
   e) For how long was his temperature above normal?
   f) At what times do you think his temperature was 38 °C?
   g) When was he cooling fastest?

8. Explain the difference between heat and temperature.

Further questions on page 75.

chapter 6

THE  LAWS

When investigating the behaviour of gases, we must consider *three* varying quantities: **pressure**
volume
temperature.
In the following experiments we always keep one of these variables constant and investigate the other two variables.

Experiment 6.1
Keeping the temperature constant – **Boyle's Law**
Use the apparatus in the diagram to investigate how the **volume** of air depends on the **pressure** applied (with the temperature kept constant).

Use the vertical scale to find the **volume** of the trapped air.
What is the volume of air in the diagram?
What is the volume of air in your apparatus?

The pressure on the trapped air is measured using a Bourdon pressure gauge (see also page 96).
What is the pressure on the gauge in the diagram?
What is the pressure in your apparatus?
Why is the pressure not zero? (See page 95.)

Now use the pump to exert more pressure and squeeze the air. This will make the trapped air slightly warmer, so wait a minute to let it cool back to room temperature.

Now measure the new pressure and volume and put your results in a table.
Use the pump again, to get more results.

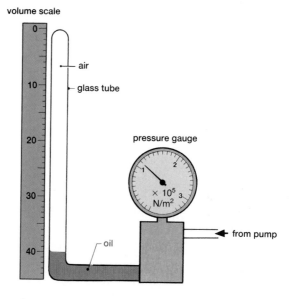

Some possible results:

Pressure p	Volume V	$p \times V$
1.1	40	44
1.7	26	44
2.2	20	
2.6	17	

Then calculate pV (pressure × volume) for each set of results.   What do you notice?

This is called **Boyle's Law** after its discoverer, Robert Boyle:

> For a fixed mass of gas, at constant temperature, $pV = $ constant

Did you notice that if p is doubled, V is halved?
If p increases to 3 times as much, V decreases to $\frac{1}{3}$rd. This means:

> **volume is *inversely proportional* to pressure,   or $V \propto \dfrac{1}{p}$**

▶ Charles' Law

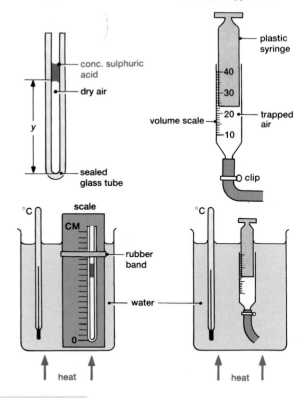

usual apparatus   *alternative apparatus*

Experiment 6.2
Keeping the pressure constant – **Charles' Law**
In this experiment you will investigate how the
volume of air changes as the **temperature** changes
(with the pressure kept constant).

In the usual apparatus, the air is held inside a glass
capillary tube by a short length of concentrated
sulphuric acid (mercury can be used but the acid
dries the air to give better results).

The length **y** of the trapped air is a measure of the
volume of the air.

Hold the glass tube on to a ruler with rubber
bands. Move the tube until the bottom of the
trapped air is opposite the zero mark of the ruler.

An alternative is to use a plastic syringe which is
lubricated with silicone oil and sealed by a rubber
tube and a clip so that it is half-full of air.

Put your apparatus in a tall beaker of cold water with a thermometer.
Wait until the trapped air is the same temperature as the water and
then measure the **volume** and the **temperature**.

Then heat the water until it is about 20 °C hotter. Again, wait before
taking the readings of volume and temperature.
Repeat this at other temperatures until the water boils.
Put your results in a table.

At constant pressure:

Volume	Temperature (°C)

Then plot a graph to show how the **volume** varies with **temperature**
(at constant pressure).

You used air in your experiment but other gases give the same result.

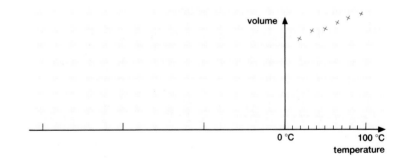

Through the points on your graph, draw the **best** straight line (it helps to hold the paper near your eye and look along the crosses to see the best line).

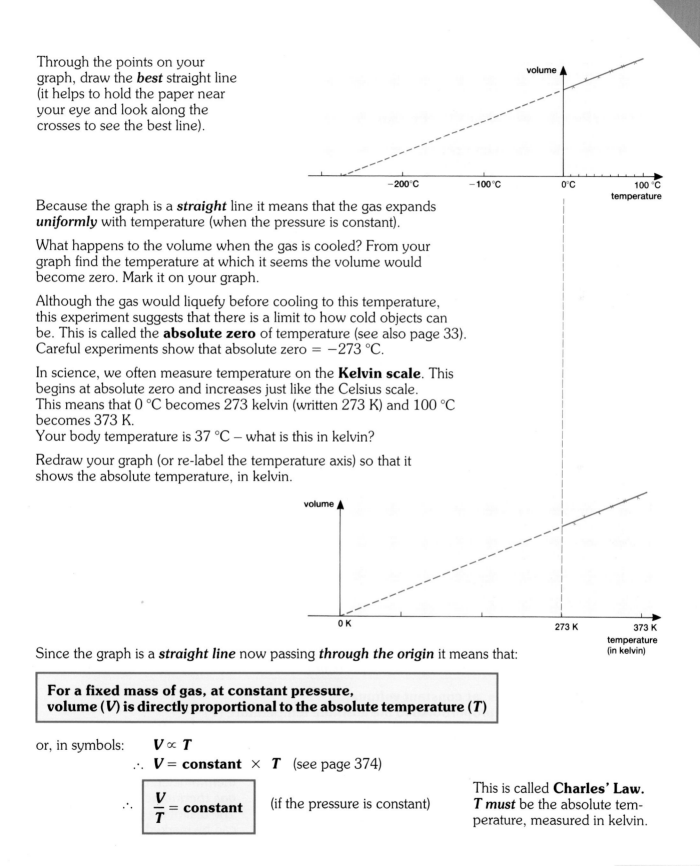

Because the graph is a **straight** line it means that the gas expands **uniformly** with temperature (when the pressure is constant).

What happens to the volume when the gas is cooled? From your graph find the temperature at which it seems the volume would become zero. Mark it on your graph.

Although the gas would liquefy before cooling to this temperature, this experiment suggests that there is a limit to how cold objects can be. This is called the **absolute zero** of temperature (see also page 33). Careful experiments show that absolute zero = −273 °C.

In science, we often measure temperature on the **Kelvin scale**. This begins at absolute zero and increases just like the Celsius scale. This means that 0 °C becomes 273 kelvin (written 273 K) and 100 °C becomes 373 K.
Your body temperature is 37 °C – what is this in kelvin?

Redraw your graph (or re-label the temperature axis) so that it shows the absolute temperature, in kelvin.

Since the graph is a **straight line** now passing **through the origin** it means that:

> **For a fixed mass of gas, at constant pressure,**
> **volume (V) is directly proportional to the absolute temperature (T)**

or, in symbols: $V \propto T$

∴ $V = \text{constant} \times T$ (see page 374)

∴ $\boxed{\dfrac{V}{T} = \text{constant}}$ (if the pressure is constant)

This is called **Charles' Law.** T **must** be the absolute temperature, measured in kelvin.

▶ Pressure Law

Use the apparatus shown in the diagram to investigate how the **pressure** of air changes as the **temperature** changes (with the volume kept constant).

The **pressure** is measured using a Bourdon pressure gauge (see also page 96). What is the pressure on the gauge in the diagram?

What is the **pressure** in your apparatus?
What is the **temperature** of your apparatus?

Now heat the water by about 20 °C. Then wait for several minutes to allow the air in the flask to reach the temperature of the water.
Take the readings of pressure and temperature.
Repeat this at several temperatures until the water boils. Put your results in a table.

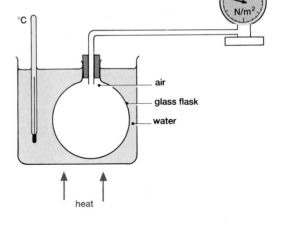

Bourdon pressure gauge

°C · air · glass flask · water · heat

Plot a graph of pressure : temperature in °C.
Compare it with the graph on the previous page.
According to your graph, at what temperature would the pressure become zero? How does this compare with the accepted value?
What is this temperature called?

What happens to the molecules of a gas as it cools? What could you say about the molecules if the gas cooled to absolute zero?

Redraw your graph (or re-label the temperature axis) so that it shows
pressure : *absolute* temperature.

At constant volume:

Pressure × 10⁵ (N/m²)	Temperature (°C)	Absolute temperature (K)

Pressure $\times\ 10^5$ (N/m²)	Temperature (°C)	Absolute temperature (K)

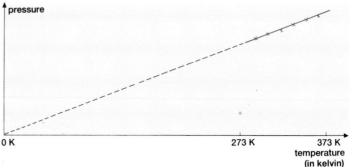

pressure

0 K   273 K   373 K
temperature (in kelvin)

Since the graph is a *straight line* passing through the *origin*, it means:

For a fixed mass of gas, at constant volume,
pressure (p) is directly proportional to the absolute temperature (T)

---

or, in symbols: $p \propto T$

∴ $p = \text{constant} \times T$

∴ $\boxed{\dfrac{p}{T} = \text{constant}}$ (if the volume is constant)

As before, **T** must be in kelvin.

This apparatus can be used as a thermometer: *a constant volume gas thermometer* (see page 33). The dial of the pressure gauge can be marked out in °C.

▶ The gas equation

From the last three experiments, we have 3 equations:

pV = constant (Boyle's Law) $\dfrac{V}{T}$ = constant (Charles' Law) $\dfrac{p}{T}$ = constant (Pressure Law)

These 3 equations can be combined into one, called the **ideal gas equation**:

$$\boxed{\dfrac{pV}{T} = \text{constant}}$$

If a fixed mass of gas has values p_1, V_1, and T_1, and then some time later has values p_2, V_2 and T_2, then the equation becomes:

$$\boxed{\dfrac{p_1 V_1}{T_1} = \dfrac{p_2 V_2}{T_2}}$$

Example
A bicycle pump contains 70 cm³ of air at a pressure of 1.0 atmosphere and a temperature of 7 °C.
When the air is compressed to 30 cm³ at a temperature of 27 °C, what is the pressure?

What do we know? p_1 = 1.0 atmosphere p_2 = ?
                   V_1 = 70 cm³ V_2 = 30 cm³
                   T_1 = 273 + 7 = 280 K T_2 = 273 + 27 = 300 K

Then the formula: $\dfrac{p_1 V_1}{T_1} = \dfrac{p_2 V_2}{T_2}$

Put in the numbers: $\therefore \dfrac{1 \times 70}{280} = \dfrac{p_2 \times 30}{300}$

$\therefore \quad p_2 = \dfrac{70 \times 300}{30 \times 280} = \underline{2.5 \text{ atmospheres}}$

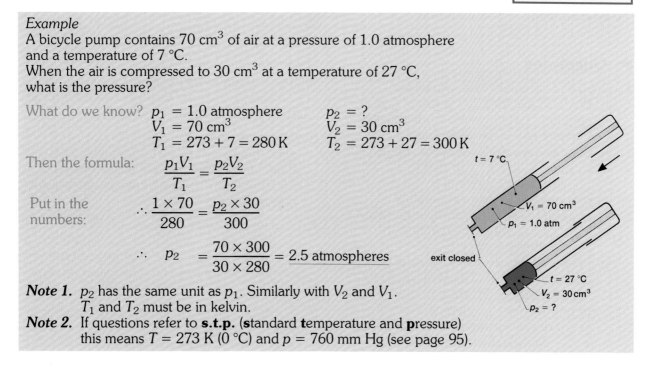

t = 7 °C
V_1 = 70 cm³
p_1 = 1.0 atm
exit closed
t = 27 °C
V_2 = 30 cm³
p_2 = ?

Note 1. p_2 has the same unit as p_1. Similarly with V_2 and V_1.
T_1 and T_2 must be in kelvin.
Note 2. If questions refer to **s.t.p.** (**s**tandard **t**emperature and **p**ressure) this means T = 273 K (0 °C) and p = 760 mm Hg (see page 95).

The gas laws and the kinetic theory

We saw on page 14 that air pressure is caused by the bombardment of billions of molecules.

Boyle's Law
If the volume of a fixed mass of gas is made smaller then the molecules will be closer together. If the volume is halved, then the number of molecules per cm³ will be doubled and so the pressure will be doubled (Boyle's Law).

Absolute zero
As the temperature falls, the molecules have less energy. Absolute zero is the temperature at which molecules have their lowest possible energy.

The Pressure Law
If the temperature rises, then the molecules have more energy and move faster.
If the volume of the container is kept constant then the molecules hit the walls harder and more often and so the pressure rises (the Pressure Law).

Charles' Law
If the pressure is to be kept constant even though the molecules begin to move faster, then the volume must increase. This lets the molecules move farther apart so that the pressure can stay the same (Charles' Law).

Summary

For a fixed mass of gas in each case:

Boyle's Law. At constant temperature: $p \propto \dfrac{1}{V}$ or $pV = \text{constant}$ or $p_1 V_1 = p_2 V_2$

Charles' Law. At constant pressure: $V \propto T$ or $\dfrac{V}{T} = \text{constant}$ or $\dfrac{V_1}{T_1} = \dfrac{V_2}{T_2}$

Pressure Law. At constant volume: $p \propto T$ or $\dfrac{p}{T} = \text{constant}$ or $\dfrac{p_1}{T_1} = \dfrac{p_2}{T_2}$

T must be in kelvin ($= 273 + $ temperature in °C)

Gas equation: $\dfrac{pV}{T} = \text{constant}$ or $\dfrac{p_1 V_1}{T_1} = \dfrac{p_2 V_2}{T_2}$

▶ Questions

1. Copy and complete these sentences:
   a) Boyle's Law: for a . . . . mass of gas, at constant . . . . , ... × . . . . is constant. Pressure is . . . . proportional to . . . . .
   b) Charles' Law: for a . . . . mass of gas, at constant . . . . , ... divided by . . . . is constant. Volume is . . . . proportional to . . . . temperature. A graph of volume against temperature is a . . . . line which intercepts the temperature axis at . . . . .
   c) Pressure Law: for a . . . . mass of gas at constant . . . . , ... divided by . . . . is constant. Pressure is . . . . proportional to . . . . temperature. A graph of pressure against temperature is a . . . . line which intercepts the temperature axis at . . . . .
   d) The temperature of absolute zero is . . °C or . . kelvin. At this temperature the molecules have their . . . . possible . . . . .
   e) S.t.p. means standard . . . . and . . . . (that is . . kelvin and . . mm Hg).

2. a) Convert these to kelvin: 27 °C, −3 °C, 150 °C, −90 °C.
   b) Convert these to °C: 373 K, 200 K, 1000 K.

3. A bubble of air of volume 1 cm³ is released by a deep-sea diver at a depth where the pressure is 4.0 atmospheres. Assuming its temperature remains constant ($T_1 = T_2$), what is its volume just before it reaches the surface where the pressure is 1.0 atmosphere?

4. To what temperature must 1 m³ of air at 27 °C be heated, at constant pressure, in order to increase its volume to 3 m³?

5. The pressure of air in a car tyre is 2.7 atmospheres when the temperature is −3 °C. If the temperature rises to 27 °C, what is the pressure then? (Assume the volume remains constant.)

6. A mass of gas has a volume of 200 cm³ at a temperature of −73 °C and a pressure of 1.0 atmosphere. What is its volume at a pressure of 3.0 atmospheres and a temperature of 27 °C?

7. A mass of gas has a volume of 380 cm³ at a pressure of 560 mm Hg and a temperature of 7 °C. What is its volume at s.t.p.?

8. The cylinder of a diesel engine (see page 71) contains 300 cm³ of air at 27 °C and 75 cm Hg. The air is rapidly compressed to 20 cm³ at a pressure of 3000 cm Hg.
   What is the temperature then?

9. A metal tank in a garage normally contained compressed air at 4 atmospheres and 7 °C. In a fire the tank exploded, although it was known to be safe up to a pressure of 14 atmospheres. If you were the detective of the insurance company, what value would you calculate for the minimum temperature of the fire near the tank?

Further questions on page 75.

measuring HEAT

We have seen that the moving molecules of an object have **internal energy**. When an object is heated, its internal energy increases. Like all other forms of energy, internal energy is measured in a unit called the **joule** (usually written J).

A joule is a small unit of energy – if a match burns up completely, it produces about 2000 J. This can also be written as 2 kJ where
   1 kJ = 1 kilojoule = 1000 joules.
For larger amounts of energy we use MJ where
   1 MJ = 1 megajoule = 1 million joules (1 000 000 J).

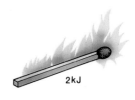

2kJ

Your body needs energy – energy to keep warm (at 37 °C) and energy to work and play. You get this energy from the fuel that you eat.

Can you work out the rhyme?   *Little Jack Horner,*
                              Sat in a corner,
                              Feeling so chilly and cool.
                              He said, "I should eat,
                              And so produce . . . . ,
                              The unit of which is a . . . . ."

Your body needs a total of about 10 MJ per day. Lying asleep, your body uses only about 4 kJ every minute, but for making your bed or dancing you use about 20 kJ every minute.

Too many joules, too few joules

The table below shows some energy values (called **calorific values**), measured in megajoules per kilogram of substance.

Butter	32 MJ/kg	Eggs	7 MJ/kg	Natural gas	55 MJ/kg
Sugar	16	Potatoes	4	Petrol	47
Meat	12	Fish	3	Coal	30
Bread	10	Fruit	2	Wood	15
Ice cream	9	Carrots	2	Dynamite	6

If you want to lose weight, which foods should you avoid?
Which food would you take on an expedition to the Arctic?

Did you hear about the cat burglar who stole the family joules?
He used them for (h)eating!

▶ Heating water

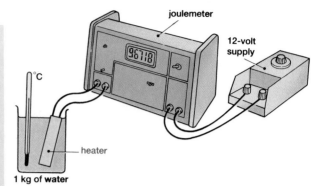

Experiment 7.1
Measure out 1 kg of cold water into a large beaker.
Take the initial temperature of the water accurately.
Put an electric heater in the water and connect
it to a 12-volt supply through a *joulemeter*, which
will measure how much energy we give the water.

Note the joulemeter reading and switch on.
When you have given the water 10 000 J, switch
off, stir gently and note the highest temperature
that is reached.
What is the rise in temperature of the water?

*(If you have not got a joulemeter, you can use a
50-watt heater switched on for 200 seconds.)*

Experiment 7.2
If you gave twice as much energy (20 000 J) to
the same 1 kg of water, what would be the
temperature rise?
Guess and then try it.

Is it exactly what you expected?
Would you get a better result if you wrapped cotton
wool round the beaker?

Do other substances need as much energy as water?

Experiment 7.3
Put your thermometer and heater into a 1 kg block
of aluminium. Give it the same amount of energy
(10 000 J) as in experiment 7.1, and find the
temperature rise.
Is it the same as when you had 1 kg of water?

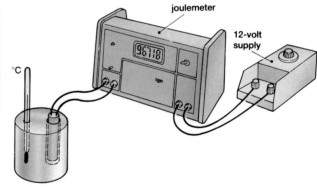

If the temperature rise is 5 times as much, it means
that aluminium needs only $\frac{1}{5}$th as much energy as
water (to give the same temperature rise to the
same mass).

Experiment 7.4
If you have time, repeat this experiment with other
solids and liquids and compare the results.

We find that different substances have different appetites or
capacities for energy.
Water is very greedy, it needs five times as much energy as the
same mass of aluminium to produce the same temperature rise.

▶ Specific heat capacity

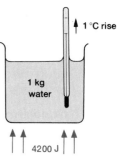

If you calculate how much energy 1 kg of water needs to become 1 °C hotter, you will find it needs about 4200 J. This number is called the *specific heat capacity* of water.

The specific heat capacity of a substance is the amount of energy (in joules) that is needed to raise the temperature of 1 kg of the substance by 1 °C.
Its unit is written J/kg °C or J/kg K.

Here are some values of this number – notice that the number for aluminium is about $\frac{1}{5}$th of the number for water.

Water	4200	J/kg °C	Ice	2100	J/kg °C
Meths	2500		Aluminium	880	
Paraffin	2200		Sand	800	
Mercury	140		Copper	380	

All these numbers are for 1 kg and 1 °C.
The energy needed to raise the temperature of 2 kg through 1 °C would be twice as much and the energy needed to raise 2 kg through 3 °C would be three times as much again – that is, six times as much.
So to calculate the energy needed, we multiply like this:

Energy needed = specific heat capacity × mass × change in temperature

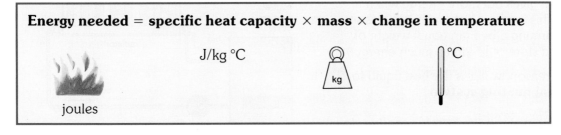

Example 1
How much energy is needed to heat 100 g of water from 10 °C to 30 °C?

What do we know?
Specific heat capacity of water = 4200 J/kg °C
mass of water = 100 g = 0.1 kg
temperature rise = 30 °C − 10 °C = 20 °C

Write down the formula
$$\text{Energy needed} = \text{specific heat capacity} \times \text{mass} \times \text{change in temperature}$$

Then put in the numbers
= 4200 × 0.1 × 20 joules
= 4200 × 2 joules
= 8400 J   (= energy from 4 matches)

In a similar way, using the same formula, you can work out **your** values for specific heat capacity from the results of your experiments.

If you met a girl called Julie, would you expect her to be full of energy?

Example 2

A 2 kW (2000 W) electric heater supplies energy to a 0.5 kg copper kettle containing 1 kg of water. Calculate the time taken to raise the temperature by 10 °C. (Use values from the table on page 43.)

2 kW = 2000 W = 2000 J/s = 2000 joules in each second (see also page 121).
∴ Energy supplied in t seconds = $(2000 \times t)$ joules

Assuming no energy is lost to the surroundings:

Energy lost by hot = **Energy gained by cold**
object (heater)   **objects** (water and kettle)

$$2000 \times t = \left(\text{mass} \times \frac{\text{specific heat}}{\text{capacity}} \times \frac{\text{rise in}}{\text{temp.}} \right)_{\text{of water}} + \left(\text{mass} \times \frac{\text{specific heat}}{\text{capacity}} \times \frac{\text{rise in}}{\text{temp.}} \right)_{\text{of copper}}$$

$$2000 \times t = (1 \times 4200 \times 10)_{\text{water}} + (0.5 \times 380 \times 10)_{\text{copper}}$$

$$2000 \times t = (42\,000) + (1900) = 43\,900$$

$$\therefore t = \underline{22 \text{ seconds}}$$

▶ Storing heat energy

You could see from the table that water has easily the highest specific heat capacity. It needs more energy to heat it up. It stores more energy when it is hot, and so it gives out more energy when it cools down. This is why a hot-water bottle is so effective in warming a bed (an equal weight of mercury would store only $\frac{1}{30}$th as much energy!).

For the same reason, water is the best liquid to use in a **central heating system**:

In a similar way, water is the best for *cooling down* the engines in cars and factories (see page 50).

Similarly, the specific heat capacity of water affects our weather. Islands, being surrounded by water, tend to stay at the same temperature while large areas of land tend to get very hot in summer and very cold in winter.

The water in your body helps to stop you cooling down or warming up too quickly.

Night storage heaters use concrete blocks to store heat. Although concrete has a lower specific heat capacity than water, it is more dense and so the same mass takes up less space in your house. The concrete blocks are heated by electric coils using cheaper night-time electricity. This stored energy is then released slowly during the day.

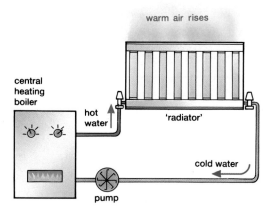

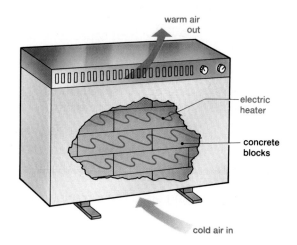

Summary

Thermal energy, like all other forms of energy, is measured in joules.
1 kJ = 1000 joules.
1 MJ = 1 000 000 joules.

$$\text{Energy needed to change temperature} = \text{specific heat capacity} \times \text{mass} \times \text{change in temperature}$$

▶ **Questions** (Use the tables on pages 41 and 43 if necessary.)

1. Copy out and complete:
   a) The unit of energy is the . . . . .
      1 kJ stands for one . . . . .
      1 MJ stands for one . . . . .
   b) The specific heat capacity of a substance is the amount of . . . . needed to raise one . . . . of the substance through . . °C.
      The unit of specific heat capacity is . . . . .
      The specific heat capacity of water in these units is . . .
   c) The formula for the energy needed to change the temperature of a substance is:
   d) Energy lost by . . . . object = energy . . . . by . . . . object.

2. *Jack Spratt could not eat fat,*
   His wife could not eat lean.
   While Jack ate fruit or carrots,
   His wife would lick ice-cream.
   Two kilograms they each put in
   And one grew fat and one grew thin.
   I ask you now to calculate
   How many joules of heat each ate.

3. Professor Messer fills his car. How many faults can you find with his idea (there are at least 4)?

4. Calculate the amounts of energy needed to change the temperature of
   a) 2 kg of water by 5 °C
   b) 500 g of water by 4 °C
   c) 100 g of aluminium from 20 °C to 30 °C
   d) 200 g of copper from 60 °C to 10 °C

5. A 2 kg block of iron is given 10 kJ of energy and its temperature rises by 10 °C.
   What is the specific heat capacity of iron?

6. At the sea-side, the sun shines down equally on the sand and the sea. Assuming they both absorb the same amount of energy, why is the sand hotter than the sea?

7. Professor Messer has an outdoor swimming pool, which contains 100 000 kg of water.
   a) He needs to heat the water from 15 °C to 20 °C. How much energy does this need?
   b) How much would this cost if the electricity board charge 3p for 1 MJ of energy?
   c) On a hot summer day the Sun shines for 10 hours and the water absorbs solar energy at a rate of 20 kW. How much energy is absorbed that day?
   d) By how much does the Sun heat up the pool?

Further questions on page 76.

chapter 8

Conduction, Convection and Radiation

There are **three** ways in which heat can travel.

> *Experiment 8.1* **Conduction**
> Get a piece of stiff copper wire about the same length as a match. Strike the match and hold the copper wire in the flame.
>
> What happens?
> Does the energy get to your hand quicker through wood or through copper?

We say that copper is a better **conductor** than wood. The heat has travelled from molecule to molecule through the copper.

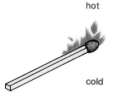

hot

cold

> *Experiment 8.2* **Convection**
> Hold your hand over and then under the flame of a match.
>
> What do you notice?

The hot air expands and then rises.
We say the heat is **convected** upwards.

Radiation

The thermal energy of the Sun is **radiated** to us in the same way that the light reaches us from the Sun.

Here is a 'model' to help us see the difference between the three ways. Three ways of getting a book to the back of the class:

1. *Conduction*: a book can be passed from person to person – just as heat is passed from molecule to molecule.

2. *Convection*: a person can walk to the back of the class carrying the book. This is the way hot air moves in convection, taking the energy with it.

3. *Radiation*: a book can be thrown to the back of the class rather like the way energy is radiated from a hot object.

CONDUCTION

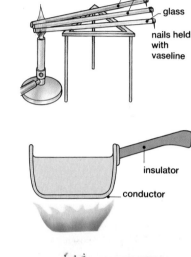

Experiment 8.3
Get some rods (all the same size) of different substances – for example, copper, iron and glass. Rest them on a tripod and fix a small nail near one end of each rod, using vaseline as 'glue'.

Heat the other ends of the rods equally with a Bunsen burner.

What happens? How many minutes does it take for the first and second nail to drop off?

All metals are good conductors of heat.
Copper is a very good conductor.
Glass is not a good conductor of heat – it is an **insulator**.

Pans for cooking are usually made with a copper or aluminium bottom and plastic handle. What would happen if the handle was made of copper?

The Prof. called his doctor by phone,
"I bought a new pan," was his groan,
"Twas badly constructed,
The handle conducted,
And my fingers are burned to the bone."

The heat energy is conducted from molecule to molecule. At the hot end the molecules are vibrating fast. This vibration is passed along to other molecules as they bump into each other.

A metal is a very good conductor because, as well as all these vibrating molecules, it also has many free-moving electrons to carry the energy. Insulators do not have these free electrons.

Experiment 8.4
Hold a large test tube of water at the bottom and aim a gentle Bunsen flame just below the water surface until the water boils at the top, without hurting your hand.

Is water a good conductor or a poor conductor?

Most liquids are poor conductors of heat.

▶ Insulators

Is air a conductor or an insulator?

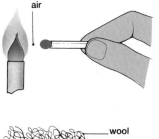

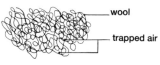

> *Experiment 8.5*
> Hold a live match about 1 cm away from a very hot Bunsen flame.
>
> Does the match get hot enough to burst into flame?

This shows that air is a very poor conductor – it is a very good insulator.

Many insulators contain tiny pockets of trapped air to stop heat being conducted away.

For example, wool feels warm because it traps a lot of air.

The air trapped in and between our clothes and blankets keeps us warm.

In the same way, the air trapped in fur and feather keeps animals warm. Birds fluff up their feathers in winter to trap more air.

A refrigerator has insulation material round it to keep it *cold*. The insulation reduces the amount of heat conducted to the inside from the warmer room.

Pipes and hot-water tanks should be lagged with insulation material to reduce the loss of energy.

▶ Keeping warm

Heating your house is expensive. The owners of this house pay £1000 a year for energy – look where the money goes!

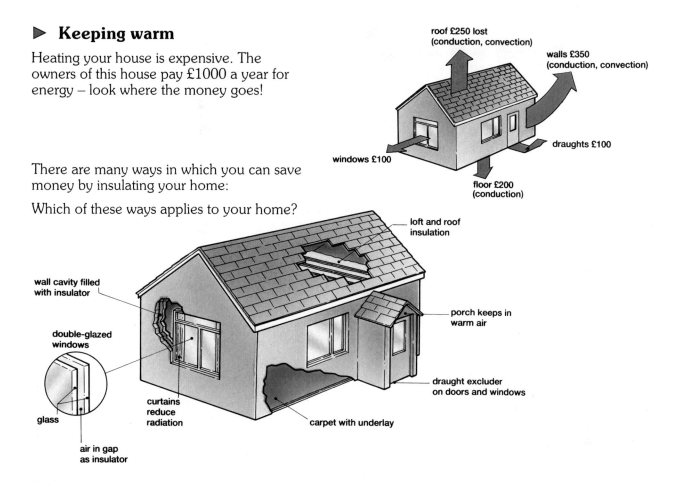

roof £250 lost
(conduction, convection)

walls £350
(conduction, convection)

draughts £100

windows £100

floor £200
(conduction)

There are many ways in which you can save money by insulating your home:

Which of these ways applies to your home?

loft and roof insulation

wall cavity filled with insulator

double-glazed windows

porch keeps in warm air

draught excluder on doors and windows

curtains reduce radiation

glass

carpet with underlay

air in gap as insulator

U-values

Architects can calculate the heat loss from a house by using *U-values* and a formula:

Example
An uninsulated roof measures 10 m by 10 m. How much heat is lost through the roof if the temperature inside the house is 20 °C and outside it is 5 °C?

What do we know? U-value of roof (see table) = 2.0
area of roof = 10 m × 10 m = 100 m^2
temperature difference = 20 − 5 = 15 °C

Then formula: Heat lost per second (W) = U-value × area × temperature difference

Then numbers: = 2.0 × 100 × 15

= 3000 joules per second

U-values	(in W/m^2 °C)
Roof, tiled, no insulation	2.0
Roof, tiled, insulated	0.4
Wall with air cavity	1.5
Wall with insulation in cavity	0.5
Window, single glazed	5.6
Window, double glazed	3.0

Now calculate how much heat is lost if the roof is insulated.

CONVECTION

Put a single crystal of potassium permanganate into a beaker of water, near to the side, as in the diagram.
Warm the water near the crystal with a small Bunsen flame.

The crystal colours the water so that you can see the movement of the **convection currents** in the water. The warm water rises in the same way that warm air rises in a flame.
The cooler water is heavier and falls to the bottom until it is heated.

Car engines are cooled by convection currents in the water pipes.
A pump is often used to help the water to circulate. This is 'forced convection'.

Water is a very good substance to carry the unwanted heat away from the engine to the 'radiator' (see page 44).
The 'radiator' is a **heat exchanger** where the hot water gives up its energy to the air.

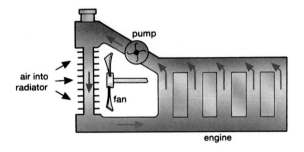

The diagram shows a simple domestic hot-water system as it might be in your home.

Convection currents take energy from the boiler up to the storage tank which is connected to the hot taps.

Where would you add lagging to insulate this system?

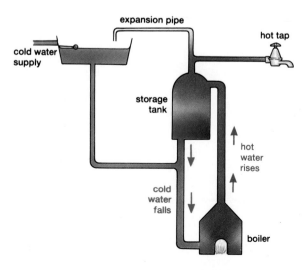

Experiment 8.7
Investigate convection currents in a model room like the one in the diagram, using smoke so that you can see how the air moves.

Hold the smouldering paper near the window, before and then after lighting the fire.

What do you notice?

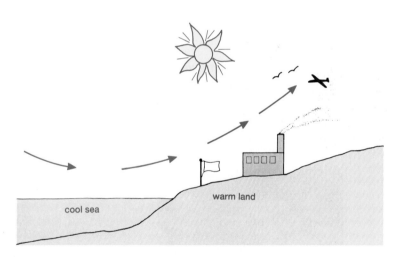

This shows that fires and chimneys help to ventilate rooms.

Experiment 8.8
Draw a spiral on paper as shown. Cut it out and hang it from cotton. Use it to investigate convection currents near fires and windows. It will rotate as convection currents move past it.

The Sun can cause very large convection currents which we call **winds**.

In the day-time the land warms up more than the sea. The warm air rises over the land and cool air falls over the sea. So we feel a sea breeze.

Rising convection currents over the land are used by glider pilots to keep their planes in the air.

Can you draw a diagram of what happens at night when the land cools quicker than the sea?

A rather small boy called Maguire,
Sat next to a roaring hot fire.
Life is full of surprises:
He found hot air rises,
And he floated up higher and higher!

Hot-air balloons rise in the air in the same way as convection currents.

RADIATION

Energy from the Sun reaches us after travelling through space at the speed of light. When this energy hits an object, some of it is taken in or **absorbed**. This makes the molecules vibrate more – and so the object is hotter.

Objects take in and give out energy (as radiation) all the time. Different objects give out different amounts of radiation, depending on their **temperature** and their **surface**.

Experiment 8.9
Get two cans of equal size, one that is polished and shiny on the outside and one that is black and dull.
Put a thermometer in each and pour in **equal** amounts of hot water. Let them cool, side by side. Stir the water and take their temperatures every minute.

Which can cools down more quickly?
Which can is losing energy more quickly?

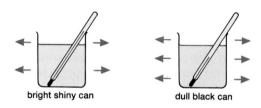

bright shiny can   dull black can

A dull black surface loses energy more quickly – it is a good radiator.
A bright shiny surface is a poor radiator.

Brightly polished kettles and teapots do not lose much energy by radiation – they keep warm.

Marathon runners need to keep warm at the end of the race. The shiny blanket reduces radiation (and convection and evaporation).

The cooling fins on the back of a refrigerator, in a car radiator and on a motor bike engine should be dull black so that they will radiate away more energy.

Central heating 'radiators' are not usually painted dull black because they lose most of their energy by convection and should be called 'convectors'.

If different surfaces give out different amounts of energy, do different surfaces take in different amounts of energy?

Experiment 8.10
Use the same two cans as in experiment 8.9, but this time pour in equal amounts of cold water. Place the two cans in full sunlight or place them equal distances from an electric fire. Stir and take the temperatures every minute.

Which can heats up more quickly?
Which can is absorbing energy more quickly?

Sun or fire

bright shiny can   dull black can

A dull black surface is a good absorber of radiation (as well as a good radiator). It takes in and gives out a lot of radiation.
A bright shiny surface is a poor absorber of radiation – it reflects the radiation away.

A fire-fighting suit is bright and shiny so that it does not take in a lot of energy and burn the fire-fighter.

In hot countries, people wear bright white clothes and paint their houses white to reduce absorption of energy from the Sun.

In the same way, petrol storage tanks (and some-times factory roofs) are sprayed with silver paint to reflect the Sun's rays.

▶ Infra-red rays

Energy to heat us up travels from the Sun at the speed of light, just like the light rays. The rays which cause the most heating are called **infra-red rays**. These are like light rays but have a **longer wavelength** than the light we use to see with (see page 213).

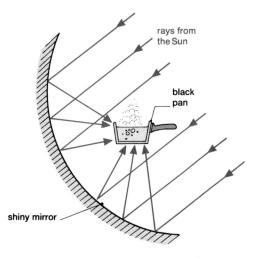

rays from the Sun

black pan

shiny mirror

Like visible light rays, infra-red rays can be reflected by mirrors. In hot countries, a large curved (concave) mirror can be used as a 'solar furnace' to collect the rays from the Sun and focus them on to a kettle or a pan for cooking.

(A TV 'dish' works in the same way – it collects radio waves from a satellite and passes the signals to your TV set; see page 184.)

In an electric fire, a mirror is used in the opposite way, to reflect rays out to the room.

Experiment 8.11
On a sunny day, use a concave mirror to heat a spoonful of water or to light a match.
Why does it help if you first blacken the head of the match with a pencil?

Greenhouses

The inside of a greenhouse is warmer than the outside because the rays from the very hot Sun have a short wavelength which can get through the glass. These rays are absorbed by the plants which get warmer and also radiate infra-red rays. However, because the plant is not very hot, the rays have a longer wavelength and cannot get through the glass. So energy is radiated in, but cannot radiate out again.

short wavelength gets in

long wavelength cannot get out

A '**greenhouse effect**' applies to planet Earth. Carbon dioxide and other gases emitted by factories and power stations act like the glass in the greenhouse. They may be making the whole Earth hotter and affecting our weather.

The bowl of my goldfish, (called Egbert by me),
Was left in the Sun for an hour or three.
It absorbed infra-red,
Till Egbert was dead,
So we all ate boiled Egbert for tea.

▶ The vacuum flask

A vacuum or 'Thermos' flask will keep tea hot (or keep ice-cream cold).
It does this by reducing or stopping conduction, convection and radiation.

It is a double-walled glass bottle. In the space between the two walls, both pieces of glass are coated with shiny bright 'silvering' and all the air is pumped out to form a vacuum.

A **vacuum** is used because it stops energy transfer, by stopping conduction and convection.
The **silvering** on one glass wall reduces radiation of energy and the silvering on the other glass wall reflects back any infra-red rays that may have been radiated.

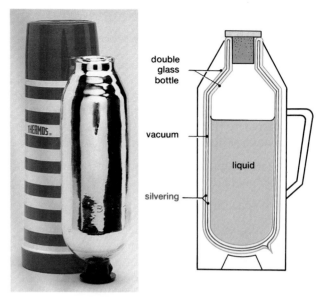

double glass bottle

vacuum

liquid

silvering

Summary

Conduction

The energy is passed from one vibrating molecule to the next.
All metals are good conductors.
Water is a poor conductor.
Air is a very poor conductor.

Convection

Hot liquids and gases expand and rise while the cooler liquid or gas falls (a convection current).

Radiation

Energy is transferred by infra-red rays which can travel through a vacuum, at the speed of light.
Dull black surfaces are good radiators and good absorbers.
Shiny bright surfaces are poor radiators and poor absorbers – they reflect the infra-red rays.

Professor Messer "RAYS-ES" the temperature!

▷ Physics at work: Heat radiation

Solar heating

Although the Sun's energy is free, it is not easy to make use of it. One way is to fit solar panels to the roof of a house, in order to make hot water. The hot water can be used for washing or for central heating radiators.

The photograph shows some houses with solar panels: Which way do you think is North?

Here is a simple design for a solar panel:

1. Why is the main surface dull black?
2. Why is it covered with glass? (See page 54.)
3. Why is the back insulated?
4. Should the pipes inside the box be made of copper or plastic?
5. Why is the pipe inside the box made as long as possible?
6. Why should the pipe CD be kept as short as possible?
7. Where would you add insulation?
8. Why does the water move through the pipes?
9. Why is the storage tank placed at a higher level than the solar panel?
10. Why would an electric immersion heater usually be added? (See page 262.)

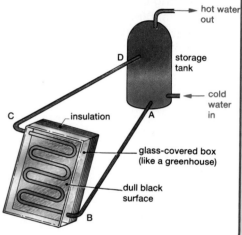

Thermography

Special photographs which are taken using infra-red rays are called ***thermographs***. Thermographs can be used by doctors because diseased parts of the skin are often hotter (and show up white or red on the thermograph).

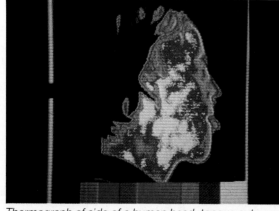

Thermograph of side of a human head, tongue out

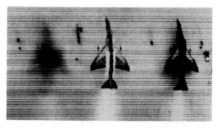

An infra-red photograph of an airfield. The plane at the right has warm fuel in its wings. The plane in the centre has its engines running. At the left, a 'heat shadow' shows where a plane has left the airfield.

▶ Physics at work: Keeping warm

When string vests are worn under clothes, they trap pockets of air and keep you warm. Why are string vests no good by themselves?

Why are the eskimos wearing their coats with the fur side inwards?

Why does it help to have fur at the neck, wrists and bottom?

New-born babies are sometimes wrapped in a blanket coated with shiny aluminium. This reflects the infra-red rays and keeps baby warm.

Why do mountain-rescue teams also carry some of these shiny blankets?

The latest padding for anoraks uses Physics to reduce conduction, convection *and* radiation:

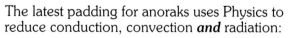

cloth

cloth

shiny plastic strips
trap air *and* reflect
back the infra-red rays

Physicists have found ways to show the energy we lose from our bodies.

Here the **Schlieren** method shows us the convection currents round a person's body:

Here is a photograph taken using infra-red rays:
This thief was walking away from the camera in total darkness!
You can see that more energy was lost from his warm bare hands and where his clothes were tight (at the shoulders and trousers).

Why does it help to have loose clothes?

Things to do:
1. Devise an experiment to compare the insulation of two anoraks. You can have a beaker of hot water placed in the open sleeve of an anorak. What else would you need? List all the things you would do to ensure this was a *fair* test.
2. A competition for the class: using only a yoghurt pot and anything that will fit *inside* it, who can keep an ice-cube longest? Can you beat 3 hours?

▶ Questions

1. Copy out and fill in the missing words:
   a) Thermal energy travels through the bottom of a pan by . . . . . The energy is passed from one vibrating molecule to the next. All metals are good . . . . . Plastics, water and air are poor . . . . (good . . . .).
   b) . . . . currents can form when liquids and gases are heated. The cold fluid . . . . and the hot fluid . . . . .
   c) Energy can travel through empty space by . . . . rays, which can be . . . . by mirrors like light rays. Dull black surfaces are . . . . radiators and . . . . absorbers. Shiny, bright surfaces are . . . . radiators and . . . . absorbers.
   d) A vacuum flask uses silvering to cut down heat transfer by . . . . and uses a vacuum to cut down heat transfer by . . . . and . . . . .

2. Explain the following:
   a) Copper or aluminium pans are better than iron ones.
   b) A fur coat is warmer if worn inside out.
   c) Sheets of newspaper can be used to keep ice-cream cold, and keep chips hot.
   d) In winter fat people keep warmer than thin people.
   e) A carpet feels warmer to bare feet than lino or concrete.
   f) Two thin blankets are usually warmer than one thick one.
   g) Eskimos build their igloos from snow not ice.

3. Explain seven ways by which energy loss from a house can be reduced.

   The Professor's **roof** loses £250 of energy per year. It can be insulated for £400 and would then lose £50 per year.
   The **walls** lose £350 per year, would cost £750 to insulate and would then lose £100 per year. Discuss what he should do.

4. a) Why do flames go upwards?
   b) Why do a chimney and a fire help to ventilate a room?
   c) Why would you crawl close to the floor in a smoke-filled room?
   d) Why is the element at the bottom in an electric kettle?
   e) Why is the freezer compartment at the top of a refrigerator?
   f) Why is there often a sea breeze during the day and a land breeze at night?
   g) How can a 500 kg glider stay up in the air?
   h) How have convection currents in the Earth moved the continents? (See p. 157)

5. Explain, with a diagram, how a domestic hot-water system works.

6. a) Why does a shiny teapot stay hotter than a dull brown teapot?
   b) Why are fire-fighting suits made of shiny material?
   c) What would probably happen if an astronaut's spacesuit was dull black?
   d) Why is it more likely to be frosty on a clear night than on a cloudy night?
   e) Why is the back of a refrigerator painted black?
   f) If pieces of black paper and white paper are laid on snow in sunshine, what is likely to happen?
   g) What might be the effect of sprinkling black soot over the ice in the arctic or the antarctic?

7. Explain why plants are warmer in a greenhouse than outside. How does the 'greenhouse effect' affect the Earth?

8. Explain, with a diagram, how a vacuum flask can a) keep tea hot and b) keep ice cold.

9. Transistors or 'microchips' that may get hot inside a computer often have a metal '**heat sink**' clipped to them.
   a) Why is it made of aluminium?
   b) Why has it got large 'fins'? c) Why is it black?

Further questions on page 77.

chapter 9

Changing State

SOLID *to* LIQUID

Before a solid can change state into a liquid, it must be given some energy.

In fact to melt 1 kg of ice at 0 °C into 1 kg of water at 0 °C, we must give it 340 000 joules of energy.

This energy is called 'hidden heat' or **latent heat** because it does not make it any warmer – it is still at 0 °C.

In an opposite way: to make 1 kg of ice at 0 °C, a refrigerator would have to take 340 000 J **away from** 1 kg of water at 0 °C.

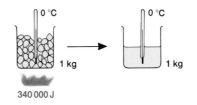

This number – 340 000 J/kg of water – is called the **specific latent heat of fusion of water**. (Fusion means melting.) Other substances have different values.

The specific latent heat of fusion of a substance is the amount of energy (in joules) needed to melt 1 kg of the solid to liquid without changing the temperature. Its unit is J/kg.

Molecules

The latent heat energy is given to the molecules in a solid, so that they can move more freely as in a liquid.
In ice there are strong forces between the molecules, so it needs a lot of energy to melt it.

Ice cubes cool down your drink because the ice needs a lot of latent heat in order to melt

▶ Calculating how much energy is needed

If 1 kg of ice at 0 °C needs 340 000 J to melt, 2 kg would need twice as much and 3 kg would need three times as much. We multiply like this:

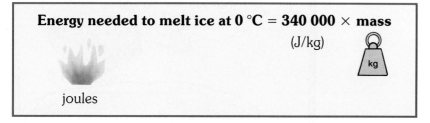

Energy needed to melt ice at 0 °C = 340 000 × mass

(J/kg)

joules

Example
How much energy is needed just to melt 100 g of ice at 0 °C?

What do we know?

Mass of ice = 100 g = 0.1 kg
specific latent heat = 340 000 J/kg

Formula:
Put in the numbers:

Energy needed = 340 000 × mass
= 340 000 × 0.1 joules
= 34 000 J   (= energy from about 20 matches)

Experiment 9.1
Put 100 g of ice in a beaker with a heater connected to a joulemeter.

Find how many joules are needed to just melt the ice without making it hotter.

When you have found the number of joules for 0.1 kg of ice, calculate the number for 1 kg of ice and see it agrees with the accepted value of 340 000 J/kg. If it is different, suggest reasons.

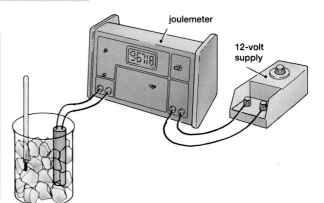

joulemeter

12-volt supply

0.1 kg ice

Experiment 9.2   To find the melting point of a substance
Put some of the substance (e.g. hexadecanol) in a test tube and melt it by putting the test tube in a beaker of boiling water.

Then take out the test tube and let it cool down, using a thermometer to take the temperature every $\frac{1}{2}$ minute.

Plot a graph of your results.

From your graph, work out the melting point of the substance.

Using the idea of latent heat, explain what is happening during the level part of the graph.

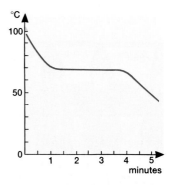

▶ Changing the melting point in two ways

1 Impurities

Experiment 9.3
Put some small pieces of ice in a beaker and sprinkle some salt on the ice.
Stir until the ice melts and then take its temperature.

What do you find?

ice, water and salt

This shows that when water is not pure, it melts at a temperature colder than 0 °C.

Impurities lower the melting point of water.

Why is salt put on icy roads by the Council?
Why do motorists add another liquid to the water in their car?
Why do motorists spray 'de-icer' liquid on windscreens?

Mrs Messer: How do you make anti-freeze?
Professor: Hide her coat!

2 Pressure

Experiment 9.4
Press two ice cubes together as hard as you can and then release the pressure.

What has happened?

Squeezing causes the ice to melt, but when you release the pressure, it re-freezes and 'glues' the ice cubes together.
We make snowballs in just the same way.
This shows that when the pressure is increased, ice melts at a temperature colder than 0 °C.

Increased pressure lowers the melting point of water.

Experiment 9.5
How to cut a block of ice in half without breaking it!

The high pressure of the thin copper wire lowers the melting point, so the ice melts and the wire moves down. When the water moves above the wire, it is not squeezed, so it re-freezes.

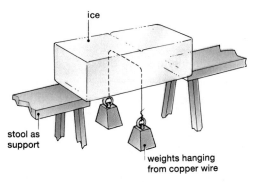

ice

stool as support

weights hanging from copper wire

This is called *regelation*.

In a similar way, icy glaciers flow very slowly down mountains.

Changing State

LIQUID to GAS

When water is heated by a Bunsen burner, it reaches 100 °C and then stays at that temperature as the water changes into steam.

The energy from the Bunsen is needed as latent heat – the energy is needed to overcome the forces between the molecules (and also to push back the surrounding air molecules).

In fact, to boil 1 kg of water at 100 °C into steam at the same temperature, we must give it 2 300 000 joules of energy.

In an opposite way, if 1 kg of steam at 100 °C condenses to hot water at 100 °C, it will give up 2 300 000 J of energy.

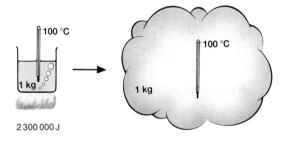

This number – 2 300 000 J/kg of water – is called the *specific latent heat of vaporisation of water*. Other substances have different values.

The specific latent heat of vaporisation of a substance is the amount of energy (in joules) needed to boil 1 kg of the liquid to gas without changing the temperature. Its unit is J/kg.

Experiment 9.6
You can do an experiment to find the number for water after reading question 8 at the end of this chapter.

To calculate the energy needed to boil a different mass of water, we use a formula like the last one:

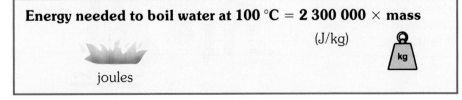

Energy needed to boil water at 100 °C = 2 300 000 × mass

(J/kg)

joules

How much energy would be needed to boil 2 kg of water at 100 °C?

▶ Changing the boiling point in two ways

1 Impurities

Experiment 9.7
Put some salt or other impurity into a beaker of water and heat it until it boils.
Measure the boiling point with a thermometer.

What do you find?

Impurities raise the boiling point of water.

Why do potatoes cook faster if you put salt in the water?

2 Pressure

Experiment 9.8   (Teacher demonstration)
Put some hot (but not boiling) water in a strong flask and use a pump to reduce the air pressure.

What happens?

What is the boiling point of the water now?

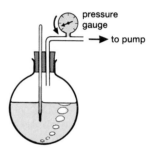

Reducing the pressure lowers the boiling point of water.

This is because the molecules find it easier to escape from the liquid when the pressure is less.

As you go higher in the air, the air pressure decreases. This means that it is difficult to make a good cup of tea on Mount Everest because the water boils at only 70 °C.

What would happen to the blood in an astronaut's body if he went into space without wearing his spacesuit?

The pressure cooker

In an opposite way, increasing the pressure raises the boiling point of water. This is used in the *pressure cooker*.

Because the pressure is about twice normal air pressure, the water boils at about 120 °C and the food cooks more quickly.

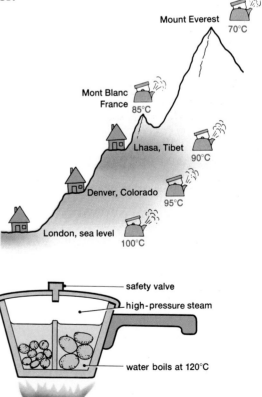

▶ Evaporation

A liquid can change state without boiling.
Even on a cool day, a rain puddle can dry up. We say the water is *evaporating*.

Experiment 9.9
Put some methylated spirit or petrol on the back of your hand.
What do you feel as it evaporates away?

How can you tell the wind direction by holding up a wet finger?
Why do you often feel cold after getting out of the swimming baths?
Why is it dangerous for you to be out on the hills in damp clothes?

Evaporation causes cooling

In each case the liquid takes latent heat from your body so that the liquid can change state by evaporating. The faster (hotter) molecules leave and the slower (colder) molecules stay behind until your body warms them.

Your body sweats more in hot weather. This is so that you keep cool as the sweat takes in latent heat from your body and evaporates.

When the water evaporates, it becomes a gas called *water vapour*.
Water vapour in the air can condense back to liquid water, to form clouds or fog.

See if you can decide the four factors that affect evaporation by thinking about drying clothes on wash-day:
1. Would you choose a dry day or a wet day?
2. Would you choose a cold day or a hot day?
3. Would you choose a still day or a windy day?
4. Would you fold the clothes or spread them out to show a large area?

Design four investigations to see if your four guesses (your 'hypotheses') are correct.
How would you make sure that each investigation is a *fair test* (see page 363)?

Freezing by evaporation

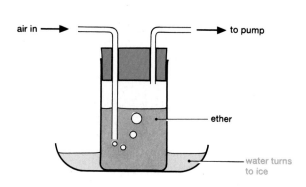

A similar idea is used in a kitchen refrigerator:

The refrigerator

An electric motor is used to pump a liquid called 'Freon' from a low-
pressure pipe to a high-pressure pipe, from where it squirts through
a narrow hole back into the low-pressure pipe and round again.

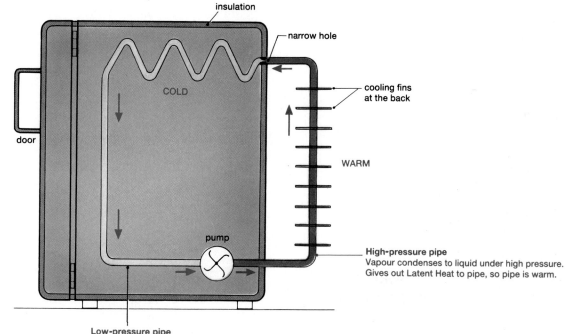

Why is the cold pipe put at the *top* of the box?
Why are the cooling fins made of *metal*, with
a *large* area? What *colour* should they be?

The refrigerator is a **heat pump**. The same idea
can be used to heat houses (with the hot pipes
inside the house and the cold pipes outside).

How to get water from the ground – even in the desert

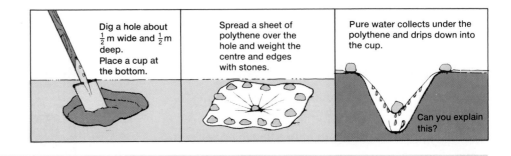

Dig a hole about $\frac{1}{2}$ m wide and $\frac{1}{2}$ m deep.
Place a cup at the bottom.

Spread a sheet of polythene over the hole and weight the centre and edges with stones.

Pure water collects under the polythene and drips down into the cup.

Can you explain this?

Summary

Substances must be given latent heat before they can melt or boil. This energy is needed to overcome forces between the molecules.

The specific latent heat of fusion (melting) of a substance is the amount of energy (in joules) needed to melt 1 kg of the solid to the liquid, without changing the temperature.
Its unit is J/kg.

The specific latent heat of vaporisation (boiling) of a substance is the amount of energy (in joules) needed to boil 1 kg of the liquid to the gas, without changing the temperature.
Its unit is J/kg.

Evaporation depends on the temperature, the dryness of the air, the wind and the surface area.

Impurities or a change of pressure affect the melting point and the boiling point of water as the diagram shows:

adding impurities
or
increasing pressure

100 °C

0 °C

$$\text{Energy need to change state} = \text{specific latent heat} \times \text{mass}$$

joules

J/kg

kg

When liquids evaporate, they cool down. This is used in the refrigerator.

Professor Messer's boiling mad!
He feels quite sure that he's been had.
The mercury just won't go high,
Please work it out and tell me why:

I'LL JUST CHECK THIS THERMOMETER IN BOILING WATER

DRAT! IT READS ONLY 90°C – IT MUST BE A FAULTY ONE....

....I'LL HAVE TO THROW IT AWAY AND JUST GO AND BUY A NEW ONE

► Questions

1. Copy out and complete:
   a) A solid must be given .... heat before it can melt into a liquid and a liquid must be given .... heat before it can boil into a gas. The energy is used to overcome the forces between the .... so that they can move more freely.
   b) The specific latent heat of fusion of a substance is the amount of .... (measured in ....) needed to melt .. kg of a solid into a .... without changing the ..... Its unit is .... / .....
   c) The formula is:
   $$\frac{\text{Energy needed}}{\text{to change state}} = \text{specific} \dots \text{heat} \times \dots$$
   d) Impurities .... the melting point of water. Increased pressure .... the melting point of water.
   e) Impurities .... the boiling point of water. Increased pressure .... the boiling point of water.
   f) Evaporation causes cooling because only the .... (hotter) molecules escape, leaving the slower (....) molecules in the liquid.

2. Explain the following, using the idea of molecules where possible.
   a) Lakes are more likely to freeze than the sea.
   b) Salt and sand are sprinkled on roads in winter.
   c) Motorists put 'anti-freeze' in the cooling system in winter.
   d) Snowballs cannot be made in very cold weather.
   e) Icebergs take a long time to melt.

3. When water freezes, it **expands** (unlike most other substances). How does this explain:
   a) water pipes often bursting in cold weather,
   b) not finding out about them immediately?

4. What factors affect the rate of evaporation of a liquid?

5. Explain the following, using the idea of molecules where possible:
   a) A steam burn is worse than a hot-water burn.
   b) A pressure cooker cooks food faster.
   c) A car cooling system is usually pressurised.
   d) An astronaut needs a pressurised space-suit all round his body.
   e) It is difficult to make tea on Mount Everest.
   f) Wet clothes feel cold.
   g) Tea is cooled more rapidly by blowing on it.
   h) Athletes should put on a tracksuit soon after finishing a race.
   i) A swimming pool is colder on a windy day.

6. Explain, with a labelled diagram, how a refrigerator works.

7. If the specific latent heat of ice is 340 000 J/kg and the specific latent heat of steam is 2 300 000 J/kg, calculate
   a) the energy needed to melt 2 kg of ice at 0 °C
   b) the energy needed to melt 500 g of ice at 0 °C
   c) the energy needed to boil 3 kg of water at 100 °C
   d) the energy needed to boil 100 g of water at 100 °C.

8. An electric kettle (marked 3 kW) produces 3000 joules of heat every second. It is filled with water, weighed and switched on. After coming to the boil, it is left on for a further 80 seconds and then switched off. It is found to be 100 g lighter. Calculate the specific latent heat of steam. Suggest reasons why your answer does not agree with the accepted value.

9. If the specific latent heat of ice is 340 000 J/kg, and the specific latent heat of steam is 2 300 000 J/kg and the specific heat capacity of water is 4200 J/kg K, calculate the heat needed to change 2 kg of *ice* at 0 °C to *steam* at 100 °C.

Further questions on page 79.

chapter 10

HEAT ENGINES

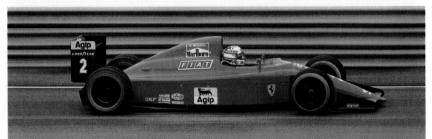

All these engines convert fuel energy (chemical energy) into movement energy (kinetic energy).

They do this by burning the fuel at a high temperature so that the molecules have a lot of movement energy and this energy is used to push the engine forwards.

These engines are not very efficient because a lot of energy is wasted (as heat energy).

Note – these three experiments could be dangerous.
Do* not *attempt them yourself unless your teacher tells you to.

Demonstration 10.1
Get an empty tin can with a press-on lid. Push on the lid but not too tightly.
Place the tin over a Bunsen burner and stand well back.   What happens?

hot air

As the temperature rises, the molecules move faster and faster and beat against the lid until they force it off violently.

Demonstration 10.2
Repeat the experiment with a few drops of water inside the tin.

Does the lid come off sooner?

This is because the molecules of steam are hitting the lid as well as the molecules of hot air.

The high pressure of steam was used in steam engines but they were only about 8% efficient – that is, for every 100 joules of energy in the fuel (coal), only 8 joules were used to push the engine and the other 92 J were wasted.

Demonstration 10.3
Make a small hole in the lid of the tin and a larger hole in the bottom so that it just fits over an unlit Bunsen burner.
Use the Bunsen to fill the tin with gas and then immediately put the tin on a tripod, light the gas at the small hole in the lid and stand well back.

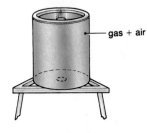
gas + air

The flame burns lower and lower, using up gas, until with the correct mixture of air and gas there is a violent explosion.
This is called ***internal combustion***.

In the ***internal combustion engine*** of a motor-car or motor-bike, the explosion is made even more violent by:
1. using petrol vapour instead of gas
2. squeezing the petrol vapour to a high pressure before exploding it.

▶ The 4-stroke petrol engine – suck, squeeze, bang and blow

1 Suck

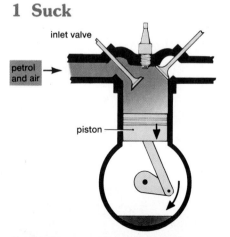

Induction stroke
The inlet valve is open and the piston is moving down. A mixture of petrol vapour and air is sucked in from the carburettor.

2 Squeeze

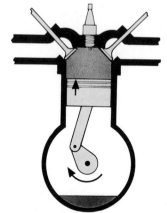

Compression stroke
Both valves are closed and the piston is moving up to *squeeze* the mixture of petrol and air to about $\frac{1}{8}$th of its original volume. It gets *hotter*.

3 Bang

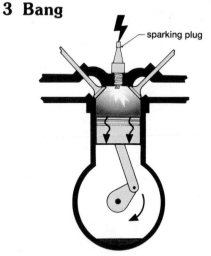

Power stroke
An electric spark from the sparking plug ignites the mixture which burns rapidly, and the hot gases force the piston down.

4 Blow

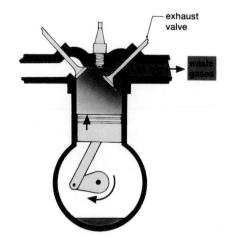

Exhaust stroke
The exhaust valve is open and the piston is moving up, to push out the waste gases. The cycle then begins again.

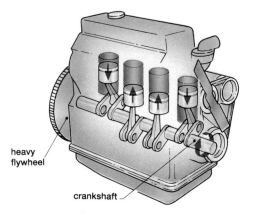

heavy flywheel

crankshaft

Motor-cars usually have four pistons connected by a crankshaft to a flywheel. The flywheel smooths out the jerks and keeps the engine turning over between explosions.

Motor-cars are only about 25% efficient – three-quarters of the chemical energy of petrol is wasted! Petrol comes from oil, which is getting scarce (see page 11).

▶ Diesel engines

Diesels are built almost the same as petrol engines, but there are five differences:

1. They use diesel oil instead of petrol.

2. There is no carburettor – only air is sucked in. Then the fuel is squirted in through a valve when the piston is at the top of its compression stroke.

3. There is no sparking plug.

4. The engine is designed to squeeze the air more than in a petrol engine. It squeezes the air to about one-sixteenth of its volume and this makes it so hot that the fuel catches fire and explodes as soon as it is squirted in.

5. The diesel engine is about 40% efficient.

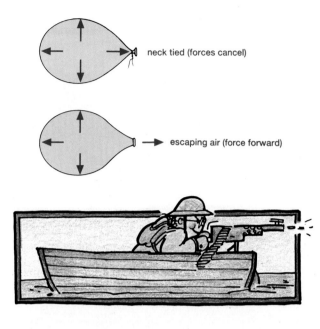

neck tied (forces cancel)

escaping air (force forward)

▶ Jets and rockets

Experiment 10.4
Blow up a balloon and then release it.

What happens?

The balloon moves forward because the escaping air is rushing backwards.

In the same way, jet engines and rockets move forward because of the hot gases rushing out of the back.

Which way would Professor Messer's machine-gun boat move?

These are all examples of Newton's Third Law (see page 100).

▶ Jet engine – suck, squeeze, burn and blow

Jet engines are used on aeroplanes.

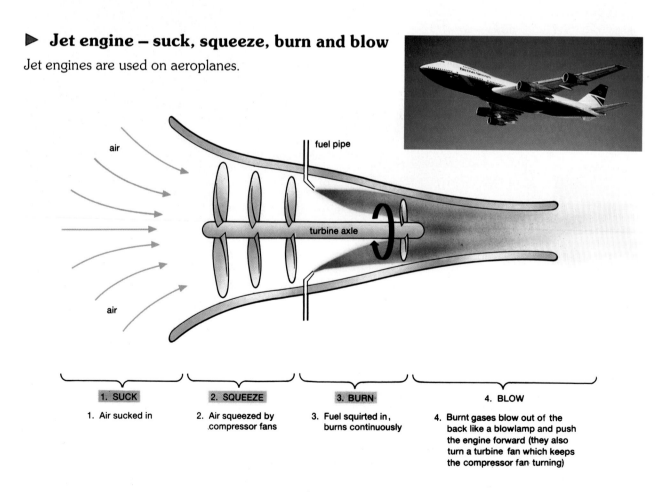

air

fuel pipe

turbine axle

air

1. SUCK	2. SQUEEZE	3. BURN	4. BLOW
1. Air sucked in	2. Air squeezed by compressor fans	3. Fuel squirted in, burns continuously	4. Burnt gases blow out of the back like a blowlamp and push the engine forward (they also turn a turbine fan which keeps the compressor fan turning)

▶ Rocket engine

A rocket engine uses the same idea but carries its own compressed air or oxygen with it, so that it can go outside the Earth's atmosphere. (See also page 101.)

astronauts and cargo

fuel (liquid hydrogen)

liquid oxygen

▶ Questions

1. a) In a heat engine, ... energy is converted to . . . . energy, because at high temperature, the . . . . move more violently and exert a larger . . . . .
   b) The 4-stroke cycle in a petrol engine or a diesel engine is 1. Suck (induction), 2. . . . . (. . . .), 3. . . . . (. . . .), 4. . . . . (. . . .). A petrol engine is only about . . % efficient but a diesel engine is about . . % efficient because a diesel engine squeezes the air . . . . than a petrol engine.
   c) A jet engine burns fuel continuously. The four stages are suck, ... , ... , ...
   d) A rocket engine carries its own . . . . and so can work . . . . the Earth's atmosphere.

2. Copy out and complete the rhymes, using the words in the correct order.

   Down the road, fast or slow,
   (Suck, ... , ... , ...)
   Car or lorry, van or truck,
   (Squeeze, ... , ... , ...)
   Don't forget to oil and grease,
   (Bang, ... , ... , ...)
   Petrol burns and pistons clang,
   (Blow, ... , ... , ...)
   Efficiency is awful low!
   (Suck, ... , ... , ...).

3. Why do motor cars have a) a carburettor b) sparking plugs c) a crankshaft d) gears e) a clutch f) four or six small cylinders rather than one large cylinder?

4. Professor Messer wants to buy a car – it could be petrol or diesel. He expects to travel 15 000 km in a year.
   Using the data shown:
   a) How much fuel would each one use in a year?
   b) What would be the fuel cost per year for each one?
   c) How long would it take to recover the extra cost of a diesel car?
   d) Which one would you buy?

	Cost of car	Cost of fuel	Car travels:
Petrol car	£10 000	40p per litre	10 km on 1 litre
Diesel car	£10 500	35p per litre	15 km on 1 litre

73

Further questions on heat

▶ Energy resources

1. There are five main sources of energy used in the world. These are: coal, nuclear fuel, natural gas, oil, water power.

a) Write down the **three** energy sources in the above list which are known as fossil fuels. [3 marks]

b) The pie-chart shows how the total energy used in the world in 1983 was made up from these sources.

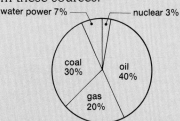

water power 7% — nuclear 3%

coal 30% oil 40% gas 20%

i) Which energy source made the biggest contribution? [1]

ii) Only a small fraction of the energy comes from water power. A similar pie-chart for the UK in 1983 would show an even smaller fraction. Why? [1]

c) The table below shows the estimated world reserves of coal, oil and natural gas, and the quantity which was used in 1983.

	Estimated reserves	Quantity used in 1983
Coal	500 000	1250
Oil	100 000	3000
Natural gas	90 000	1500

(The figures are in 'million tonnes of oil equivalent')

i) If natural gas continues to be used at the 1983 rate, how many years will the reserves last? [2]

ii) If coal continues to be used at the 1983 rate, how many years will reserves last? [2]

iii) It is always difficult to make accurate predictions of how long reserves will last. Why? [1]

d) The pie-chart for the year 2020 is likely to be different from the one shown for 1983. Suggest **two** likely differences and give your reason for each. [4] (SEG)

Further questions on energy on page 152.

▶ Molecules

2. The density of a solid is much greater than that of a gas because the molecules of a solid are:

A closely packed together
B in random motion
C vibrating rapidly
D arranged in a regular pattern
E at rest. (NI)

3. An industrial gas storage tank has a fixed volume. When the temperature of the gas is raised from 20 °C to 60 °C, the molecules of the gas will:

A hit the sides of the tank harder
B move more slowly
C become further apart
D have less kinetic energy
E become larger. (MEG)

4. Some perfume is spilled in a corner of a room in which the doors and windows are closed to stop draughts. The scent slowly spreads by *diffusion* to all parts of the room.

a) By writing about molecules, explain how *diffusion* occurs. [2]

b) Why does the scent spread slowly even though the molecules move fast? [3]
(MEG)

5. a) The properties of materials can be explained by using a particle model of matter.

i) What are the 3 states (phases) of matter?

ii) Describe, briefly, how the particles (molecules) are arranged in the 3 states (phases) of matter. [6]

b) Brownian motion is one piece of evidence in support of this model of matter.

i) Draw a *simple* diagram of apparatus used to observe Brownian motion.

ii) Describe and explain what is observed with the apparatus. [6] (NI)

6. An oil drop of volume 10^{-9} m^3 spreads out on water to form a film of area 0.2 m^2.

Estimate the length of an oil molecule from this information. What assumption have you made in your calculation?

Describe what could be done to make the film of oil clearly visible in this experiment.

What precautions should be taken to achieve a reliable result?

▶ Expansion

7. The temperature of some water is controlled by a device called a thermostat:

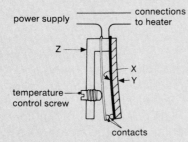

X and Y form a bi-metallic strip.
Z is an electrical insulator.
a) Name a suitable material for Z.
b) Explain why the bi-metallic strip bends when it becomes hot.
c) Explain what effect this has on the current through the heater.
d) If the temperature control screw is turned so that it moves to the left, does the temperature of the water increase, decrease or stay the same? (NEA)

8. You are trying to measure the boiling point of water using a thermometer as shown:

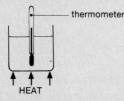

a) i) What does a thermometer measure?
    ii) What liquid is in this thermometer?
    iii) What liquid would be used in a thermometer to measure a boiling point of about −50 °C? [3]
b) What is the boiling point of water
    i) in degrees Celsius, ii) in kelvin? [2]
c) A nurse may use a clinical thermometer to measure a patient's temperature.
    i) Give **two** reasons why this is more suitable than the one used in part a).
    ii) For a healthy person the instrument will read 37 °C. Change 37 °C to kelvin. [4]
d) Write down the name of an instrument that could be used to measure up to 1500 °C as in a furnace. [1] (NEA)

▶ The gas laws

9. The diagram shows apparatus in which a fixed mass of air was compressed in a calibrated syringe, which was approximately half full of air at atmospheric pressure and 17 °C.

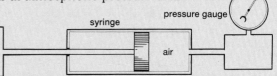

Corresponding values of volume and pressure of the trapped air are shown in the table.

Pressure (kPa)	50	60	75	90	105	120
Volume (m³)	0.000 48	0.000 40	0.000 32	0.000 27	0.000 23	0.000 20
$\frac{1}{\text{volume}}$ (m⁻³)		2500		3704		5000

a) Copy and complete the table by calculating values for $\frac{1}{\text{volume}}$. [3]
b) On graph paper, plot a graph of pressure on the **y**-axis against $\frac{1}{\text{volume}}$ on the **x**-axis. [8]
c) What relationship between pressure and volume of the trapped air can be deduced from your graph? Explain your answer. [3]
d) If the temperature of the air is increased to 27 °C, what volume will be occupied by the air at a pressure of 100 kPa? [6] (NEA)

10. a) Describe how you would show experimentally the relationship between the volume and pressure of a fixed mass of gas at constant temperature. Your answer should include i) a diagram; [3]
    ii) a statement of the readings made; [2]
    iii) one precaution taken to ensure a reliable result. [1]
b) The following readings for the pressure and temperature of a fixed mass of gas were obtained in a different experiment.

Pressure (kPa)	3.80	4.08	4.36	4.64	4.92	5.20
Temperature (°C)	0	20	40	60	80	100

    i) Plot a graph of pressure (**y**-axis) against temperature (**x**-axis). [4]
    ii) From the graph find the temperature when the pressure is 4.50 kPa. [1]
    iii) Find the pressure when the temperature is 150 K. [2]
    iv) Comment on the reliability of values obtained in parts ii) and iii). [2] (LEAG)

▶ Specific heat capacity

11. 8000 J of energy are supplied to a mass of 2 kg of a metal. If the temperature rise is 10 °C then the specific heat capacity of the metal, in J/kg °C, is

  A 400   **B** 800   **C** 1600
  D 4000   **E** 16 000   (NI)

12.

thermometer

immersion heater

block of aluminium, mass 1 kg

  a) In an experiment, a student used a block of aluminium of mass 1 kg. The block was heated with an immersion heater supplying 900 J of energy every minute. The student found that in 5 minutes, the temperature of the block rose by 4.8 K. Calculate the specific heat capacity of aluminium using the formula

$$\text{specific heat capacity} = \frac{\text{energy supplied}}{\text{mass} \times \text{temperature rise}} \quad [3]$$

  b) The student then went on to take a series of readings of time and temperature rise.

Time (minutes)	0	5	10	15	20	25	30	35
Temperature rise (K)	0	4.8	9.9	14.7	19.8	24.6	27.6	30.1

  On a sheet of graph paper,
    i) label the horizontal axis, [1]
    ii) plot the points carefully, [6]
    iii) draw the best straight line through the points between temperature rise 0 K and 25 K. [1]
    iv) From your graph, find what time was needed to give a temperature rise of 18.0 K. How much energy does the heater supply in this time? [2]
    v) Use the information in b) iv) and the equation in part a) to calculate a second value for the specific heat capacity of aluminium. [2]
    vi) Suggest one reason why the graph is **not** straight for higher temperature rises. [1] (SEG)

13. A 0.5 kg pizza is put in a 500 W microwave oven for 40 seconds.
  a) How much energy is given to it?
  b) If the temperature rises by 10 °C, what is the specific heat capacity of the pizza?

14. a) A tank contains 40 kg of water at 15 °C. The heater is switched on until all the water is at 65 °C.
    i) What is the temperature rise? [1]
    ii) The specific heat capacity of water is 4200 J/(kg K). Work out how many joules of energy are needed to heat the water using the formula [4]

  heat energy = mass × specific heat capacity × temp. rise.

  b) One electricity board unit is 3.6 MJ and costs 8p.   i) 3.6 MJ = . . J [1]
    ii) How many electricity board units are needed to heat the 40 kg of water? [1]
    iii) How much does it cost to heat the 40 kg of water? [1] (SEG)

15. Here is a specification for an electric iron:
  Power of heating element = 800 W
  Mass of alloy plate to be heated = 0.4 kg
  Melting point of alloy = 620 °C
  Specific heat capacity of alloy = 480 J/(kg K)
  a) How much heat energy is supplied per second to the iron? [1]
  b) How much heat energy is needed to take the base plate from 20 °C to its melting point? [3]
  c) If the current were supplied continuously, how many seconds would it take for the melting point to be reached? (Assume that all the energy supplied is used to raise the temperature of the plate.) [1] (SEG)

16. An electric shower unit is rated at 7 kW. Cold water enters at 12 °C and emerges as hot water at the rate of 4.0 kg per minute.
  a) If the specific heat capacity of water is 4200 J/kg K, calculate the temperature of the hot water.
  b) By making a sensible guess at the quantities involved, compare the cost of taking a shower and having a bath.
  c) Describe **two** features of such an electric shower necessary for the safety of a user. [2,2,2] (WJEC)

See p. 275 for more questions on electrical heating.

▶ Conduction, convection, radiation

17. In which one of the following examples is convection likely to be the main method of heat transfer?

   A Food being cooked in a microwave oven
   B Energy from the Sun reaching the Earth
   C A room being heated by a radiator
   D Heating a wire with a soldering iron
   E Food being cooked under an electric grill
                            (LEAG)

18. The table below gives information about the rate of heat flow through different surfaces of a room in very cold weather.

Surface	Rate of heat flow in kJ/h
Single glazed window	1200
Wall area between room and outside	3460
Wall area between inside rooms	860
Door	200
Ceiling	480
Floor	1820

  a) What is the meaning of 'k' in kJ?
  b) Which surface gives the lowest rate of energy loss?
  c) Suggest why less energy is lost through the wall area between inside rooms than is lost through wall area to the outside.
  d) Find the total heat loss per hour from the room.
  e) If one bar of an electric fire supplies 2700 kJ/h, how many bars would be needed?   (SEG)

19. This question is about solar panels, devices that are sometimes seen on the roofs of houses and are used to provide hot water. An example is shown on page 56.

  a) What is the purpose of the following:
    i) the insulation behind the absorber panel?   [1]
    ii) having the absorber panel painted black?   [1]
    iii) having a glass cover on the top of the panel?   [1]

  b) i) Name suitable materials for making the absorber panel and water-ways. (Do not use brand names.)   [2]
    ii) Give your reasons for the choice of such materials.   [2]
  c) The pipe connecting the water outlet from the panel to the hot-water storage tank should be kept short. Why is this desirable?   [1]

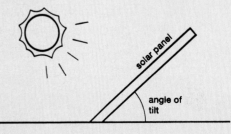

  d) The angle of tilt of a solar panel greatly affects the amount of energy it receives at different times of the year. The diagram shows what is meant by the angle of tilt. The table of data below shows the effect of different angles of tilt for the summer months.

Maximum daily input of energy in megajoules to a 1 m² panel:

Month	Angle of tilt of panel to the horizontal					
	20°	30°	40°	50°	60°	70°
Apr.	23.8	24.9	24.8	24.1	22.7	20.5
May	28.4	28.8	27.4	25.2	23.0	19.8
Jun.	29.2	29.2	27.4	25.2	22.3	19.1
Jul.	28.8	29.2	27.4	25.6	23.0	20.2
Aug.	25.6	25.9	26.3	24.8	22.7	20.5
Sept.	20.5	21.6	22.3	22.7	21.6	20.5

Use the table above to answer the following:
    i) What angle of tilt would be ideal for a solar panel in April?   [1]
    ii) Is it better to have the solar panel tilted at an angle of 40° or at an angle of 50° for all the months shown in the table? Give the reasons for your answer.   [3]
    iii) What is the maximum amount of energy that a 4 m² panel could receive during a day in July?   [2]
                           (LEAG)

▶ Conduction, convection, radiation (continued)

20. Electronic components can be damaged by overheating. One method of overcoming this problem is to mount the component on a **heat sink**. This is often in the form of a piece of aluminium of mass much greater than that of the component.

The purpose of the *heat sink* is to 'soak up' surplus heat produced by the electric current in the component and transfer this heat to the surroundings.

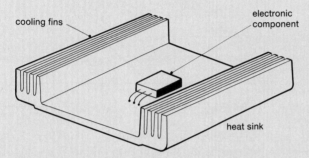

cooling fins — electronic component — heat sink

a) i) List the processes by which the heat sink **could** transfer heat to the surroundings.

ii) Suggest reasons for:
making the heat sink from aluminium; painting the heat sink black; having cooling fins on the heat sink.   [6]

b) Read the passage and then answer the questions that follow.

Manufacturers describe the way a certain heat sink behaves by stating its *thermal resistance*. The heat sink in the diagram has a *thermal resistance* of 4 K/W. This means that for every watt of power dissipated (given out/converted) by the component, the temperature of the heat sink will rise by 4 K. The temperature of the heat sink will then remain steady at this new temperature.

i) When the component on the heat sink dissipates 5 W calculate the temperature rise in the heat sink.   [2]

ii) The component produces heat continuously. Explain why the temperature of the heat sink becomes constant after a short time.   [2]

iii) The mass of the heat sink is 0.1 kg. The specific heat capacity of aluminium is 900 J/kg K. How much heat is required to produce the temperature rise in i)?   [2]

iv) Approximately how long does it take for the temperature to become constant after switching on? (You may assume that the component dissipates 5 W throughout.)   [2]

(LEAG)

21. This question is about **heat exchangers**. The name 'heat exchanger' suggests a complicated piece of machinery but, in fact, you have almost certainly seen one at home or at school. For example, a simple central heating 'radiator' is a heat exchanger.

a) From what does the 'radiator' take thermal energy?   [1]

b) To what does it give thermal energy?   [1]

c) 'Radiator' is not really the right name for this device. By what thermal energy transfer method does a 'radiator' **mainly** heat a room?   [1]

d) Why are heat exchangers usually made of metal?   [1]

e) What would be a suitable colour to paint the 'radiator' if it is to behave like a true radiator?   [1]
Explain your choice of colour.   [1]

f) The insides of 'radiators' sometimes get coated with 'scale', which is a deposit from hard water (see diagram). What effect do you think that this would have on their efficiency?   [1]
Explain your answer.   [1]

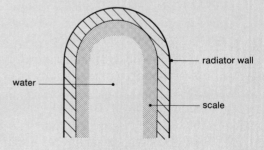

water — radiator wall — scale

(NEA)

▶ Conduction, convection, radiation (continued)

22. A heated house loses heat in several ways.

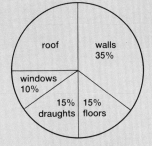

a) What accounts for the largest heat loss?
b) What percentage is lost through the roof?
c) Why does the roof lose more heat than the floors?
d) Suggest what could be done to reduce heat loss through the floors.
e) What factors would result in heat losses from a house being greater? Give your reasons. (MEG)

23. a) A tiled roof with no insulation has a **U**-value of 2.2 W/m^2 °C. Explain what is meant by the phrase underlined.

b) A room in a house has a wall area 15 m^2 and a window area of 4 m^2. The room temperature is 20 °C and the outside is 8 °C. Use the **U**-values below to calculate the saving in energy per second when the house owner decides to insulate the cavity walls *and* install double glazing.

Material	**U**-value (W/m^2 °C)
Double brick wall with air cavity	1.7
Double brick wall with foam filled cavity	0.6
Single glazed window	5.6
Double glazed window	2.9

c) Night storage heaters heat up during the night, using cheap, off-peak electricity. During the day the stored heat is given out to heat a room.
   i) One such heater uses a concrete block of volume 0.1 m^3. If the density of concrete is 1100 kg/m^3, calculate the mass of the concrete block.
   ii) The specific heat capacity of concrete is 1050 J/kg °C. If this heater cools from 80 °C to 40 °C, how much heat energy does it give out? (NI)

▶ Change of state

24. The following readings were taken in an experiment in which hot naphthalene was allowed to cool down to room temperature.

Reading	A	B	C	D	E	F	G	H	I	J	K
Time (minutes)	0	1	2	3	4	5	6	7	8	9	10
Temperature of naphthalene (°C)	105	95	85	81	81	81	81	80	78	76	74

Plot a graph of temperature against time.
What is happening between readings A and D?
What is happening between readings D and G?
What is happening between readings H and K?
What property of naphthalene can be deduced from the experimental results?

25. a) Use the molecular theory to explain why evaporation causes cooling.
b) State **two** ways by which the rate of evaporation of a liquid may be increased.
c) State **two** ways in which boiling is different from evaporation. [2,2,2] (WJEC)

26. A firm is going to sell a new aftershave called 'Macho'. They say it makes you feel fresher than other aftershaves. Men use aftershave because of its smell and to make them feel fresher after shaving. They feel fresh because aftershave cools their skin as it evaporates. You are going to carry out a laboratory test to see if 'Macho' does cool objects down more than 'Brand X'. You are given: cotton wool; clamp stand; 2 thermometers; dropper (dropping pipette); samples of each aftershave; stopclock; any other equipment you think that you will need.
a) i) Draw a labelled diagram of how you would set up your apparatus. [4]
   ii) How would you carry out the test? [3]
   iii) Write down **two** ways in which you would make sure that your results are a fair comparison. [2]
b) i) You sweat if you get hot. How does sweating help you to cool down? [2]
   ii) Hair on the body stands up when the air is cold. How does this help to keep heat in the body? [3]
      (NEA)

Parachutists have two forces acting on them.
One is the force upwards due to the air resistance.
What is the other force?

The metal weights have two forces acting on them.
One is their weight (the pull of gravity downwards).
What is the other force?

The rope has two forces acting on it.
One is the rock pulling upwards.
What is the other force?

The ice-skater has two forces on her.
One is the man pushing upwards.
What is the other force?

chapter 11

Pushes and pulls are *forces*. Whenever we are pushing or pulling, lifting or bending, twisting or tearing, stretching or squeezing, we are exerting a force.

Experiment 11.1
Exert some push and pull forces on different objects including a ball, a spring, a piece of plasticine and an elastic band. What do you notice? Make a list of the things that forces can do to the objects.

Forces can
a) change the speed of an object
b) change the direction of movement of an object
c) change the size or shape of an object.

Weight is a very common force. Weight is the force of gravity due to the pull of the Earth. This is the force that is pulling you downwards now. It acts towards the centre of the Earth.

An apple fell on Newton's head,
Moved <u>downwards</u> from the tree,
And that was when Sir Isaac said:
"The force is gravity."

The weight of an object varies slightly at different places on the Earth.

If an object is taken to the Moon, it weighs only about one-sixth as much because the Moon is smaller than the Earth.

In outer space, well away from any planets or stars, objects become weightless.

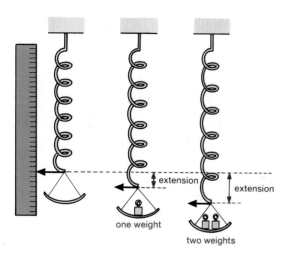

one weight

two weights

extension

extension

▷ Stretching a spring

Experiment 11.2
Find out how forces change the length of a spring.
Hang a spring from a retort stand and fix a pointer at the bottom of the spring so that it points to the scale of a ruler which is clamped vertically.

You will need several *equal* weights. Hang one weight on the spring and measure the *extension* of the spring.

Then add another weight so that there are two pulling down on the spring. Again measure the extension (from the original position).
Increase the load by adding more weights and measure the extension each time.

What do you notice?
If you double the load, does it double the extension?
If you put on three times the load, do you get three times the extension?

This means that:

> **The extension is directly proportional to the load.**

This is called **Hooke's Law**. This law also applies to the stretching of metal wires and bars.

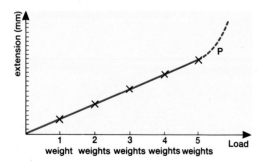

extension (mm)

1 weight 2 weights 3 weights 4 weights 5 weights Load

P

From your results, plot a graph of extension against load.

What shape is the graph?

A straight line through the origin of the graph confirms that the extension is *directly proportional* to the stretching force (see page 374).

What happens with very heavy loads?
Hooke's Law only applies to the straight part of the graph (up to the *limit of proportionality*).

The point P is called the *elastic limit*. If a spring is taken beyond this limit, it will not return to its old shape. It is permanently *deformed*.

Hooke's Law also applies to your bed-springs, to car springs, and to the steel girders used in bridges and buildings.
Graphs of extension : load are important to construction engineers.

▷ Measuring forces

A *spring balance* uses a spring to measure the weight of an object or to measure the strength of any pulling force. The marks on the scale are equally spaced because of Hooke's Law.

Forces are measured in units called **newtons** (often written N), named after Sir Isaac Newton.

An apple weighs about 1 newton (1 N).
The weight of this book is about 10 newtons (10 N).

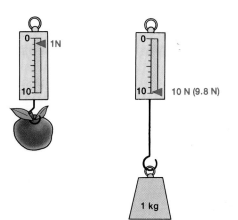

Experiment 11.3
Use a spring balance to find the weight of a 1 kg mass.

A mass of 1 kg (here on Earth) weighs almost 10 newtons. (To be more exact, 1 kg on Earth weighs 9.8 newtons.)
A mass of 2 kg weighs 20 N, and so on:

How much would a 1 kg mass weigh on the Moon, if the pull of gravity there is about one-sixth of the pull of gravity here?

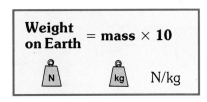

Weight on Earth = mass × 10

Experiment 11.4
Use a weighing machine to find the weight of your body in newtons.

Your weight changes slightly from place to place on the Earth – at the North Pole you would weigh about 3 newtons heavier (like having 3 extra apples in your pocket).

How much would you weigh on the Moon?

How many press-ups could you do on the Moon?

Experiment 11.5
Use a force-meter (a very strong spring balance) to measure the strength of your arm muscles when you are pushing and when you are pulling.

Here is a weighty question from Professor Messer: does 1 newton of lead weigh more than 1 newton of feathers?

▷ Mass and inertia

If this book (which has a weight of about 10 newtons here on Earth) is taken into outer space, well away from the Earth or any other object, it will become weightless. But it would still be the same book with the same *mass*.

The mass of an object is the amount of matter in it. It is measured in *kilograms* (kg). The mass of this book is about 1 kg.

When an object is stationary, it needs a force to make it move. The bigger the mass, the bigger the force needed to start it moving. We say that masses have *inertia*, a reluctance to start moving.

Experiment 11.6
Hang two tins from long pieces of string. Fill one with wet sand and leave the other empty. Try pushing the cans.

Which is the harder to push?
Which has the larger inertia?
Which has the larger mass?

Since this experiment depends on the *mass* of the object (not the weight), you would feel just the same effect on the Moon or out in space – if you pushed a large mass you would feel a large amount of inertia.

Experiment 11.7
Place a piece of smooth card on top of a milk bottle and place a coin on top. Flick the card away with your finger.

What happens?
Why does the coin stay when the card moves?

Explain the physics of this cartoon.

In a similar way, moving objects need a force to *stop* them moving. Their inertia tends to keep them moving.

Your inertia can kill you!

Experiment 11.8
Pull aside the cans used in experiment 11.6 and when they swing back, stop them with your hand.

Which is the harder to stop?

Passengers in a car have a lot of inertia and so they need seat-belts. If the car stops suddenly, the people will tend to keep on moving (through the windscreen), unless the seat-belts exert large forces to stop them.

Cars are designed to crumple safely in crashes

What happens when the car turns a corner? Your body, because of its inertia, will tend to travel straight on. You can feel your body sway as the car turns the corner, but fortunately your seat exerts a force on you and this pulls you round the corner with the car.

If you are standing on a bus, what happens to you when the bus:
a) starts moving?
b) stops moving?
c) turns a corner to the left?

Sir Isaac Newton stated all this in **Newton's First Law of Motion:**

If a mass has _no_ resultant force on it, then
- **if it is at rest, it stays at rest**
- **if it is moving, it keeps moving at a constant speed in a straight line.**

Professor Messer puts his skates on. Explain the physics of this cartoon.

▷ Circular motion

Experiment 11.9
Fasten a 'conker' or a cork to a piece of string, and whirl it round your head in a circle.
Can you feel the force in the string?

Newton's First Law says that the conker will continue to move **in a straight line**, unless a force is applied. The force in this case is due to your hand pulling in the string and the conker. This force is acting **towards the centre** of the circle and is called the **centripetal force**. You can also feel this force when you are on a roundabout.

The Moon is held in orbit round the Earth because the centripetal force is provided by the pull of gravity. The Earth and other planets orbit round the Sun for the same reason.

What happens if the force is not strong enough? If the string breaks, the conker moves **in a straight line**:

This idea is used in a spin-drier. The clothes spin very fast in a metal drum and are held in orbit by the drum. The water can leave through holes in the drum and travels **in a straight line**.

A car on a bend is held by the friction of the tyres (unless it skids). The centripetal force needs to be greater if the car travels faster, if it gains a greater mass, or if the curve is tighter.

The man is pulling on the string to provide the centripetal force

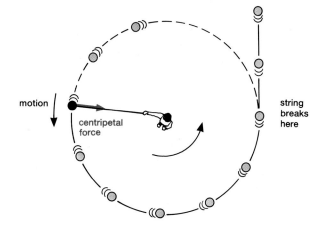

Summary

Forces can
a) change the speed of an object
b) change the direction of movement of an object
c) change the size or shape of an object.

Hooke's Law: The extension of a spring is proportional to the force pulling it (up to the limit of proportionality).

Weight is the force due to the pull of gravity. It can be measured in newtons (N) by a spring balance and varies from place to place.

Mass is the amount of matter in an object. It can be measured in kilograms (kg) and does not change from one place to another.
A mass of 1 kg here on Earth weighs about 10 N (9.8 N).

A mass has a reluctance to change its motion, called inertia.

Newton's First Law:
Every mass stays at rest or moves at constant speed in a straight line <u>unless</u> a resultant force acts on it.

▷ Questions

1. Copy out and fill in the missing words:
   a) When a spring is pulled, the . . . . is proportional to the . . . . (called . . . . Law).
   b) Weight is the force of . . . . on an object due to the pull of the . . . . .
   c) Weights and other forces are measured in . . . . .
   d) Here on Earth, the pull of gravity on a mass of 1 kg is . . newtons.
   e) Weight is measured by a . . . . balance and . . . . from place to place.
   f) Newton's First Law says: every mass stays at . . . . or moves at constant . . . . in a . . . . line unless a resultant . . . . acts on it.

2. An object has a mass of 4 kg. What is its weight (in newtons) here on Earth?

3. An astronaut has a mass of 60 kg.
   a) What is her weight here on Earth?
   b) What is her weight on the Moon?
   c) What is her mass on the Moon?

4. A spring is 20 cm long when a load of 10 N is hanging from it, and 30 cm long when a load of 20 N is hanging from it.
   Draw diagrams and work out the length of the spring when
   a) there is no load on it
   b) there is a load of 5 N on it.

5. In a spring experiment, the results were:

Load (N)	0	1	2	3	4	5	6	7
Length (mm)	50	58	70	74	82	90	102	125
Extension (mm)								

   a) What is the length of the spring when unstretched?
   b) Copy and complete the table.
   c) Plot a graph of extension : load.
   d) One of the results is wrong. Which is it? What do you think it should be?
   e) Mark the elastic limit on your graph.
   f) What load would give an extension of 30 mm?
   g) What would be the spring length for a load of 4.5 N?

6.

Should Professor Messer pull the tablecloth quickly or slowly? Explain the physics of this trick, using the correct scientific words.

7. An engineer needs to know how far a long steel beam will sag under a load. The table shows some results:

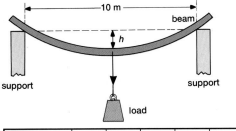

Load (N)	1000	2000	3200	4400	5200	6500
Sag, h (cm)	2.0	4.0	6.6	8.8	10.4	13.4

   a) Plot a graph of the sag, h, against load.
   b) One of the measurements of h is wrong. Which is it? What do you think it should be?
   c) What would be the sag for a load of 4500 N?
   d) What load would give a sag of 52 mm?
   e) Would a longer beam sag more or less? Sketch its graph on the same axes.

8. Could the convict escape this way?

Could he lift the ball more easily in a prison on the Moon?
Could he jump higher in a prison on the Moon?

Further questions on page 150.

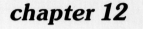

chapter 12

Which is heavier – iron or wood? Many people say iron – and yet an iron nail is lighter than a wooden tree!

What people mean is that iron and wood have different **densities**. To measure density, we need to measure the mass of a definite volume of the substance. In fact:

$$\textbf{Density} = \frac{\textbf{mass}}{\textbf{volume}}$$

If the mass is measured in **kg** (kilograms) and the volume in **m^3** (cubic metres), the density is measured in **kg/m^3** (kilograms per cubic metre).

Sometimes the mass is measured in **g** (grams) and the volume in **cm^3** (cubic centimetres) so the density is measured in **g/cm^3** (grams per cubic centimetre).

1 m^3 of iron
mass = 8000 kg
∴ density = 8000 kg/m^3

iron

1 m^3 of oak wood
mass = 700 kg
∴ density = 700 kg/m^3

wood

Here are the densities of several substances:

Substance			Density	
Solid	Liquid	Gas	kg/m^3	g/cm^3
Gold			19 000	19
	Mercury		14 000	14
Lead			11 000	11
Iron			8 000	8
	Water		1 000	1
Ice			920	0.92
	Petrol		800	0.80
		Air	1.3	0.0013

What do you notice about the numbers in the table?

Polystyrene has a very low density

Example
An engineer needs to know the mass of a steel girder which is 20 m long, 0.1 m wide and 0.1 m high. (Density of steel = 8000 kg/m^3)

Calculate the volume first:
Volume of girder = length × width × height
= 20 m × 0.1 m × 0.1 m
= 0.2 m^3

Formula:
$$\textbf{Density} = \frac{\textbf{mass}}{\textbf{volume}}$$

Then put in the numbers:
$$8000 = \frac{\text{mass}}{0.2\,\text{m}^3}$$

∴ mass = 8000 × 0.2 = <u>1600 kg</u>

▷ Measuring density

Experiment 12.1
Find the density of a **solid** in three stages:

1. Find the mass of the solid (for example, a stone) by using a top-pan balance.

2. Find the volume of the solid by using a measuring cylinder. Remember to put your eye at the right level, and to read the **bottom** of the meniscus (see page 19).

3. Use the formula: density $= \dfrac{\text{mass}}{\text{volume}}$

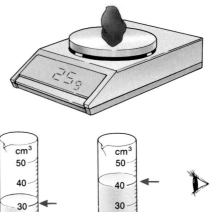
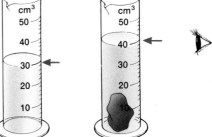

What is the density of the stone shown here?

Geologists need to measure the density of rocks. It helps them to identify the kind of rock, and where it comes from (see page 156).

Milk inspectors and beer inspectors measure the density of the liquid to see if it has been watered down (see also page 93).

Experiment 12.2
Find the density of **water** in five stages:
1. Find the mass of a dry empty beaker using a beam balance or a top-pan balance.
2. Use a measuring cylinder to pour exactly 100 cm³ of water into the beaker.
3. Now find the mass of the beaker with the water in it.
4. Work out the mass of the water from the results of steps 1 and 3.

5. Use the formula:

density $= \dfrac{\text{mass}}{\text{volume}}$ to calculate the density of water.

What unit is your result measured in?
How does your result compare with the value shown opposite?
What would your result be in units of kg/m^3?

Experiment 12.3
Repeat experiment 12.2 to find the density of another liquid.

Can you find Professor Messer's mistake? What is the correct answer?

Summary

1. $\text{Density} = \dfrac{\text{mass}}{\text{volume}}$

Changing round the formula (see page 374):

2. $\text{Volume} = \dfrac{\text{mass}}{\text{density}}$

3. $\text{Mass} = \text{volume} \times \text{density}$

Density of water $= 1000 \text{ kg/m}^3 = 1 \text{ g/cm}^3$.

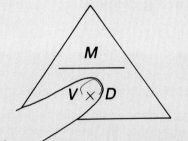

To find the one you want, cover up that letter in the triangle and the remaining letters show you the formula.

▷ Questions

1. An object has a mass of 100 grams and a volume of 20 cubic centimetres. What is its density?

2. An object has a mass of 40 000 kilograms and a volume of 5 cubic metres. What is its density?

3. An object has a volume of 3 cubic metres and a density of 6000 kilograms per cubic metre. What is its mass?

4. Copy and complete this table.

Object	Density (kg/m³)	Mass (kg)	Volume (m³)
A		4000	2
B	8000		4
C	2000	1000	
D		2000	4

a) Which object has the greatest mass?
b) Which has the smallest volume?
c) Which objects could be made of the same substance?
d) Which object would float on water?

5. A water tank measures 2 m × 4 m × 5 m. What mass of water will it contain?

6. An object has a mass of 20 000 kilograms and a density of 4000 kilograms per cubic metre. What is its volume?

7. A stone of mass 30 grams is placed in a measuring cylinder containing some water. The reading of the water level increases from 50 cm³ to 60 cm³. What is the density of the stone?

8. Professor Messer wants to load some bricks into his van. There are 1000 bricks, and when stacked neatly they measure 2 m by 1 m by 1 m.
a) What is the volume of the stack?
b) What is the volume of one brick?
c) If the density of brick is 2500 kg/m³, what is the mass of the stack?
d) If his van's maximum load is 1000 kg, how many bricks can he load?

9. The density of air is 1.3 kg/m³. What mass of air is contained in a room measuring 2.5 m × 4 m × 10 m? How does this compare with your mass?

10. Describe, in detail, with diagrams, how you would find the density of a piece of coal.

11. Professor Messer has been offered a 'gold medallion' by a woman in the market.
a) How (in detail) could he test it?
b) He found its mass was 550 g and its volume was 50 cm³. What do you think? (See the table on page 88.)
c) What is the connection with Archimedes in ancient Greece?

Further questions on page 151.

chapter 13

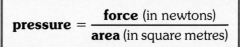

You can push a drawing pin into a piece of wood – but you cannot push your finger into the wood even if you exert a larger force. Why? What is the difference between a sharp knife and a blunt knife?

The difference in each case is a difference of **area** – the point of the drawing pin and the edge of the sharp knife have a small area.

A force acting over a small area gives a large **pressure**. Pressure is **force per unit area**, or

$$\text{pressure} = \frac{\textbf{force} \text{ (in newtons)}}{\textbf{area} \text{ (in square metres)}}$$

Its unit is newtons per square metre (N/m^2). This unit is also called the pascal (Pa), named after Blaise Pascal, who investigated air pressure.

Why do eskimos wear snow-shoes?

Example 1
An elephant weighing 40 000 N stands on one foot of area 1000 cm^2 ($= \frac{1}{10}$ m^2). What pressure is exerted on the ground?

Formula first: $\text{Pressure} = \dfrac{\text{force}}{\text{area}}$

Then put in the numbers:
$$= \frac{40\,000\,\text{N}}{\frac{1}{10}\,\text{m}^2}$$
$$= 40\,000 \times 10 \quad \text{N/m}^2$$
$$= 400\,000\,\text{N/m}^2$$

Example 2
What is the pressure exerted by a girl weighing 400 N standing on one 'stiletto' heel of area 1 cm^2 ($= \frac{1}{10\,000}$ m^2)?

Formula first: $\text{Pressure} = \dfrac{\text{force}}{\text{area}}$

Then put in the numbers:
$$= \frac{400\,\text{N}}{\frac{1}{10\,000}\,\text{m}^2}$$
$$= 400 \times 10\,000 \quad \text{N/m}^2$$
$$= 4\,000\,000\,\text{N/m}^2 \text{ (ten times bigger!)}$$

So the elephant exerts a larger **force** (because it is heavier) but the girl's heel exerts a larger **pressure** (because of its smaller area). Her heel would sink farther into the ground.

Why do camels have large flat feet?

Sir jumps quickly to his feet
He's got the point (– he's got a scar!)
The pressure, acting on his seat,
Is force per unit areaaaaaagh!

▷ Pressure in liquids

Squeezing the top of the bag causes the water to
squirt out. This means that:
Pressure is transmitted throughout the liquid.

Also the water squirts out in **all** directions. This
means that:
Pressure acts in all directions.
In the picture above you can see that the red
arrows are acting in **all** directions on the divers –
up and down, and left and right.

Where does the water squirt out fastest?
Where is the pressure greater – at the top hole or
at the bottom hole?
Pressure increases with depth.

As well as depth, the pressure also depends on:
– the density of the liquid,
– the pull of gravity, **g** = 10 N/kg (see page 83).

In fact you can calculate the pressure by:

Pressure =	10	× depth × density
(N/m²)	(N/kg)	(m) (kg/m³)

▷ Floating and sinking

When you walk into the water at a swimming pool or at the seaside, do you feel lighter or heavier?

If you walk in deep enough (up to your chin), your feet come off the ground and you start to float. This is because the water is exerting on you a force called **upthrust**.

Experiment 13.3
Hang a piece of metal on a spring balance and measure its weight.

Then lower the metal into a beaker of water. How much does it appear to weigh now?

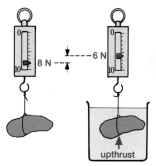

The difference between the two readings is equal to the upthrust.
How much is the upthrust in the diagram?

Upthrust and pressure

All liquids exert an upthrust like this because the pressure inside the liquid increases as you go deeper (see opposite). This means that the pressure on the bottom of an object is greater than on the top, and so there is a resultant force upwards.

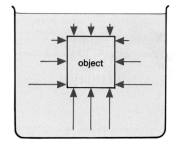

Upthrust and density

If the object has exactly the same density as the liquid, then it will stay still, neither sinking nor floating upwards, just as the liquid nearby stays still.

But if the object is denser than the liquid, then the same volume will weigh more and it will sink, because the upthrust will not be enough to support it. If the object is less dense, the upthrust will make it float up to the surface.

Look at the table on page 88 and decide which substances will sink in water. What will float on mercury? Will ice float on petrol?

The hydrometer

This is used to measure the density of milk and other liquids. It floats to different depths in different liquids, depending on their densities.

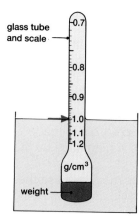

▶ Hydraulic machines

The pressure in a liquid can be used to work machinery.

The diagram shows a narrow syringe A, connected by a tube to a wider syringe B. They are filled with a liquid.

What would happen if you pushed on piston A?

The pressure is transmitted from one piston to the other, so piston B moves out a little bit. The pressure is the **same** at both ends: at piston A it is caused by a **small** force acting over a small area. At piston B it exerts a **larger** force acting over the larger area. The force has been magnified.

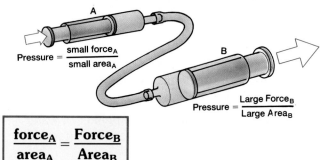

$$\frac{force_A}{area_A} = \frac{Force_B}{Area_B}$$

Hydraulic disc brakes

Experiment 13.1 showed that pressure is transmitted throughout a liquid. This idea is used in a car, to apply equal forces to all four wheels, and to magnify the force on the brake.

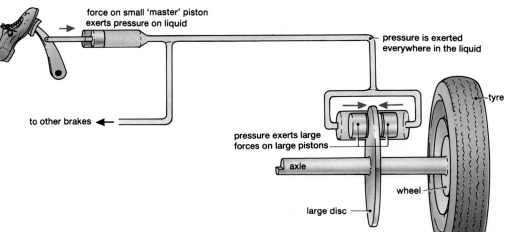

The driver's foot pushes the piston to exert pressure on the liquid. This pressure is transmitted to pistons on each side of a large disc on the axle. The pressure makes the pistons squeeze the disc (like the brakes on a bicycle) to slow down the car. Exactly the same pressure is applied to the other brakes on the car.

If the pistons at the disc have **twice** the area of the master piston, they will each exert **twice** the force that the driver applies with her foot. The force is magnified by the increased area of the pistons.

Other hydraulic machines use the same principle – for example, a hydraulic car jack, and the moving arms on this mechanical digger:

▷ Atmospheric pressure

We are living at the bottom of a 'sea' of air called the atmosphere, which exerts a pressure on us (just as the sea squeezes a diver).

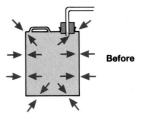
Before

Experiment 13.4
Remove the air from the inside of a can by connecting it to a vacuum pump. (Alternatively, the air can be removed by boiling some water in the can so that the steam drives out the air before a bung is used to seal the can.)

What happens?

Before the pump is switched on, molecules are hitting the outside and the inside of the can with equal pressure.

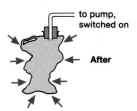
to pump, switched on

After

After the pump is switched on, there are almost no molecules inside the can and the pressure of the molecules outside the can crushes it!

Molecules of air are hitting us all the time to cause this pressure (see page 14). Fortunately the pressure inside our bodies is equal to this, so we are not crushed. In fact, astronauts must keep this air pressure round them by wearing spacesuits.

Experiment 13.5
Put together two **Magdeburg hemispheres** and use a pump to take all the air out and so form a **vacuum**.

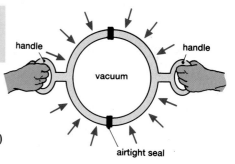
handle   handle
vacuum
airtight seal

Hold the handles and pull. Can you pull them apart?
Why not?
What happens if you let the air back in?

The pressure exerted on us by the atmosphere is about 100 000 newtons per square metre (100 kN/m^2)!
On a dining table of area 1 square metre, the air pushes down with a force of 100 000 newtons (the weight of more than 100 people). Fortunately, the air under the table exerts an equal force upwards.

The mercury barometer

A mercury barometer can be made by filling a long glass tube with mercury, and then turning it upside down in a bowl of mercury.

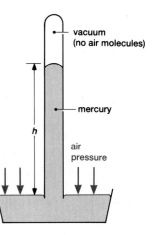
vacuum (no air molecules)

mercury

h

air pressure

The column of mercury is held up by the air pressure. As the air pressure varies from day to day (depending on the weather), the height of the mercury varies.

The distance to measure is shown in the diagram. A height of 760 mm is called **Standard Atmospheric Pressure** (written 760 mm Hg).

If you suck on a straw, the drink moves up for the same reason – atmospheric pressure pushes the liquid up into your mouth.

▷ Pressure Gauges

1 Bourdon gauge

A Bourdon gauge is often used to show the pressure in gas cylinders and boilers (see p. 35). It works in the same way as the paper tubes you may have blown out straight at parties.

Higher pressure makes the tube straighten out slightly, and this movement is used to turn a pointer.

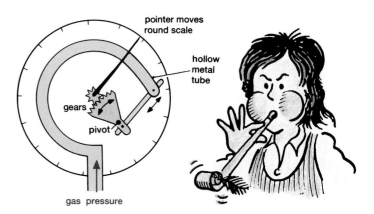

2 Manometer

A manometer is a U-tube containing some liquid, usually water.

Experiment 13.6
Use a manometer to measure the pressure of the gas supply.

The pressure of the gas changes the levels of the water in the manometer. The pressure can be found (in 'centimetres of water') by measuring the height marked **h**.

Experiment 13.7
How hard can you blow?
Under your teacher's supervision, use a large manometer to measure the pressure you can exert with your lungs.

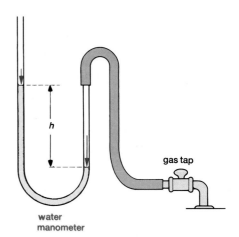

3 Aneroid barometer

This kind of barometer uses a flexible metal can which has had air taken out of it. A strong spring stops the atmosphere from completely crushing the can.

If the air pressure increases, the top of the can is squeezed down slightly. If the air pressure decreases, the spring pulls up the top of the can. This small movement is magnified by a long pointer.

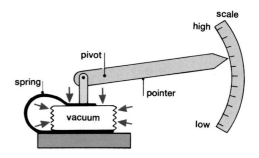

If a barometer is taken up a mountain, it shows a reduced pressure – because it is going nearer to the top of our 'sea' of air. Mountaineers and aircraft pilots can measure their height by using an aneroid barometer as an *altimeter*.

Summary

Pressure (in N/m^2) = $\dfrac{\text{force (in N)}}{\text{area (in m}^2)}$

Pressure is transmitted throughout a fluid and acts in all directions.

Pressure in a fluid increases with depth.

Pressure = 10 × depth × density

Atmospheric pressure is about 100 000 N/m^2 (or about 760 mm Hg on a mercury barometer) but depends on the weather and the altitude.

▷ Questions

1. a) Pressure = ⋯⋯ . Its units are . . . . per square . . . . (or . . . .).
   b) In fluids (liquids and gases), pressure acts in . . . . directions, and pressure . . . . as the depth increases.
   c) In order to sink in a fluid, the density of an object must be . . . . than the . . . . of the fluid.
   d) Air pressure can be measured by a mercury . . . . . The . . . . height of the mercury is usually about . . mm.

2. A box weighs 100 N and its base has an area of 2 m^2. What pressure does it exert on the ground?

3. If atmospheric pressure is 100 000 N/m^2, what force is exerted on a wall of area 10 m^2?

4. Explain the following:
   a) Stiletto heels are more likely to mark floors.
   b) Eskimos wear snowshoes.
   c) It is useful for camels to have large flat feet.
   d) Tractors have large tyres, bulldozers have caterpillar tracks and heavy lorries may need eight rear wheels.
   e) It hurts to hold a heavy parcel by the string.
   f) It is more comfortable to sit on a bed than on a fence.
   g) A ladder would be useful if you had to rescue someone from an icy pond.
   h) An Indian fakir can lie on a bed of nails if there is a large number of nails.

5. Explain the following:
   a) You can fill a bucket from a downstairs tap quicker than from an upstairs tap.
   b) Deep sea divers have to wear very strong diving suits.
   c) A dam is thicker at its base than at the top.
   d) A hole in a ship near the bottom is more dangerous than one nearer the surface.
   e) A giraffe must have a stronger heart than a human.
   f) A barometer will show a greater reading when taken down a coal mine.
   g) Aeroplanes are often 'pressurised'.
   h) Astronauts wear spacesuits.

6. Referring to the table of densities on page 88, which of those substances will   a) sink in water   b) float on water   c) sink in mercury   d) float on mercury?
   A solid has a mass of 2000 kg and a volume of 4 m^3. Will it sink or float in water?

7. In a hydraulic brake, a force of 500 N is applied to a piston of area 5 cm^2.
   a) What is the pressure transmitted throughout the liquid?
   b) If the other piston has an area of 20 cm^2, what is the force exerted on it?

8. Explain, with diagrams, how each of the following uses atmospheric pressure:
   a) a drinking straw   b) a syringe
   c) a rubber sucker   d) a vacuum cleaner
   e) an altimeter   f) a rubber plunger used for clearing blocked sinks.

Further questions on page 151.

more about

FORCES

▷ **Friction** Friction is a very common force. Whenever one object slides over another object, friction tries to stop the movement.

Friction always **opposes** the movement of an object.

Friction is often a nuisance, because it converts kinetic energy into heat and wastes it.

Experiment 14.1
Place the palms of your hands together and rub hard. What do you notice? This waste of heat energy reduces the efficiency of engines (chapter 10) and other machines (chapter 17).

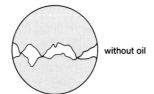

without oil

views through a microscope

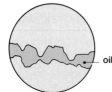

oil

Reducing friction

1. The slide in the park is polished smooth so that you can slide down easily.

    However even highly polished objects look rough through a microscope, so there is always some friction (as the lumps on one object catch and stick to the lumps on the other object).

2. A second way of reducing friction is to **lubricate** the surfaces with oil. The oil separates the surfaces so that they do not catch each other.
    Ice is usually slippery because it is usually covered with water, which lubricates like oil.

    If workers puff
    That rough
    Is tough,
    Reduce the toil
    By adding
    Oil.

3. Another way of reducing friction is to separate the two surfaces by air. This is how a hovercraft works.

4. A fourth way of reducing friction is to have the object **rolling** instead of sliding. This is what happens with ball bearings.

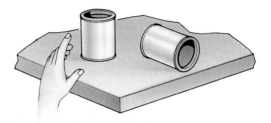

Experiment 14.2
Slide and then roll a can along the bench, using equal pushes. Which way does it go farther?

5. A fifth way: cars, planes and rockets are **streamlined** to reduce friction with the air. Even so, rockets can get very hot when they enter the Earth's atmosphere.

Dolphins are streamlined to reduce friction:

Advantages of friction

Although it is often a nuisance, friction is also very useful. Friction allows you to pick up this book with your hands.
Our lives depend on the friction at the brakes and tyres of cars and bicycles.
Air friction slows down the parachute of a falling man so that he can land safely.

You are able to walk only because of friction with the floor (try walking on wet ice!).
Knots in string and the threads in your clothes are held together by friction.
Nails and screws are held in wood by friction.
What things do you think would happen if there was suddenly no friction at all in this room?

▷ Other forces

Frictional forces only act when two objects are in contact. Other forces, like the pull of gravity, can act at a distance, when the objects are not touching each other.

Experiment 14.3
Place a bar magnet on a smooth table and then bring another magnet near it.
What happens?
Turn one of them round. What happens now?

Magnetic forces can act at a distance.

Experiment 14.4
Hang up two dry balloons side by side and touching each other. Rub each balloon on a dry cloth so that they become charged with static electricity.
What happens?

Electric forces can act at a distance.

You will learn more about magnetic and electric forces later (see pages 239 and 276).

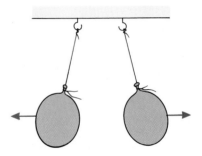

▷ Newton's Third Law

The two teams are having a tug-of-war contest. If the rope is not moving, what can you say about the forces exerted by the two teams?

Experiment 14.5
Use two spring balances to have a tug of war.

What do you notice about the readings on the balances?

Newton noticed that forces were **always** in pairs and that the two forces were **always** equal in size but opposite in direction. He called the two forces *action* and *reaction*. **Newton's Third Law of Motion** is:

> **The action force and the reaction force are equal and opposite.**

In the tug-of-war picture, if the rope is not moving,

the force		the force
of team A	=	of team B
on team B		on team A

You can see that the way we get the sentence describing the second force is by **changing round the words from the first sentence**. We use the same method in the following examples.

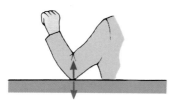

a) Rest your elbow on the table. What is the pair of forces here?

The force of your elbow on the table (downwards)	=	The force of the table on your elbow (upwards)

b) What are the forces when you start to walk or run?

The force of your feet on the Earth (moves the Earth slightly backwards)	=	The force of the Earth on your feet (moves you forwards)

Yes! The Earth moves backwards slightly as you move forwards.

100

c) Now consider a block of wood resting on a table. There are four forces here, in **two** pairs:

First pair (red)

The force of the block on the table (downwards)	=	The force of the table on the block (upwards)

These forces act where the block touches the table.

Second pair (blue)

The force of gravity of the Earth pulling down on the block (this weight in effect acts at the centre of the block – see p. 108)	=	The force of gravity of the block pulling up on the Earth (this force in effect acts at the centre of the Earth)

When the block is on the table, these four forces are equal. But if the block is allowed to fall to the floor, only the second pair of forces exists and as the block moves downwards, the Earth moves slightly upwards!

Experiment 14.6
Get a plastic water rocket and follow the instructions to prepare it. What happens when you release it and let the compressed air force out the water?

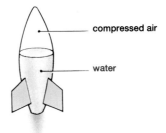

Copy and complete this equation:

The force of the rocket on the water (downwards)	=	The force of the . . . . on the . . . . (. . . .wards)

This is why rockets and jet engines move forward (see page 72 and page 148).

Experiment 14.7
Blow up a balloon and release it. What happens?

Why? (See page 72.) Write down an equation for the pair of forces.

When a gun is fired, why does it recoil?
Write down an equation for the pair of forces (see the cartoon on page 72).
Why does the bullet move farther than the gun?

▷ Adding forces

If you want to move this book to another place you will have to apply a force. **Two** things about the force are important: the **size** of the force and the **direction** of the force.
We show the direction of a force by an arrow and show the size of a force by the length of the arrow drawn to a chosen scale – for example, 1 cm for 1 newton.

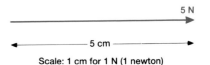

Scale: 1 cm for 1 N (1 newton)

Because both **size** and **direction** are important, force is called a **vector** quantity. Some other vector quantities are: displacement, velocity, and acceleration (see page 132).

Other quantities are called **scalars**. Scalars have only a size, no direction. For example: temperature, money, mass, and volume.

Adding scalars

For scalar quantities, 2 + 2 = 4, always.

Adding vectors

For vector quantities, like forces, 2 and 2 *sometimes* equals 4!   For example:

a) Forces of 2 N and 2 N acting in the same direction, add up to give a resultant force of 4 N:

b) Forces of 2 N and 2 N acting in opposite directions, cancel out to give no resultant force at all:

c) What would be the resultant force of 5 N opposed by 2 N? The 2 N force cancels out part of the 5 N force. How much is left?

d) When two forces are not in the same straight line, the resultant force can be found by drawing a *parallelogram of forces*:

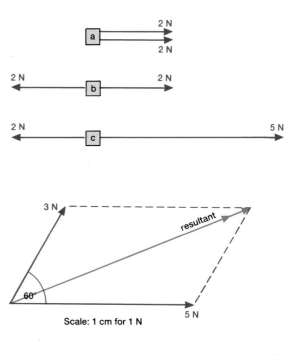

Scale: 1 cm for 1 N

The diagonal line shows the direction and size of the resultant force.

How big is the resultant force in the diagram? How can you describe its direction accurately?

Two velocities can be added in the same way.

Summary

Friction always opposes movement.
Vectors have size and direction.
Scalars have size only.

Forces are always in pairs.
Newton's Third Law: the action and reaction forces are equal and opposite.

▷ Questions

1. Explain the following:
   a) Walking on wet ice is difficult.
   b) It is difficult to strike a match on a smooth surface.
   c) It is more difficult to pull a boat on the beach than in the sea.
   d) Wet floors and wet roads are dangerous.
   e) Cars are likely to skid on loose gravel.
   f) Aborigines light fires by rubbing pieces of wood together.
   g) Spaceships get hot when they return to Earth.
   h) Sliding down a rope can burn you.
   i) Speedboats have sharper bows than barges.
   j) Racing cyclists wear smooth tight clothes.

2. One night Professor Messer dreamed that all friction suddenly disappeared. Write a story about what might happen to the Professor. Why did his dream turn into a nightmare?

3. The table shows the air resistance on a bicycle at different speeds:

Speed (m/s)	0	1	2	3	4	5
Friction (N)	0	10	30	70	130	200

   a) Draw a smooth graph of this data.
   b) How fast can the cyclist travel if she can exert a forward push of 180 N?
   c) When the cyclist crouches lower, she travels faster. Explain why.

4. Explain this cartoon – how many pairs of forces can you find?

5. For each of the following forces, describe the reaction, giving its direction and stating what it acts on.
   a) the push (north) of a boot on a football
   b) the push (west) of a crashing car on a wall
   c) the push (backwards) of a swimmer on the water
   d) the pull of gravity on a falling apple.

6. Explain the following:
   a) A gun recoils when it is fired.
   b) Firemen have to brace themselves when aiming a fire hose.
   c) You, by yourself, can move our planet.
   d) An astronaut is drifting away from his spaceship. How can he return, using only an aerosol spray?

7. Explain the difference between a vector and a scalar. Divide the following into vectors and scalars: force, temperature, velocity, displacement, volume, money, acceleration, mass, and weight.
   Find the sum of
   a) a mass of 6 kg and a mass of 8 kg
   b) a force of 6 N acting in opposition to a force of 8 N.

8. Tom, Dick and Harry are all pulling on a hoop. Tom pulls north with a force of 90 N and Dick pulls east with a force of 120 N.
   a) What is their resultant force?
   b) If the hoop does not move, what force is Harry exerting?

▷ Physics at work: Building Bridges

Experiment 1   Bending a beam
Put a ruler as a bridge between two books. Then press down gently on the middle so that the 'beam' bends.
What must be happening to the molecules of the ruler?

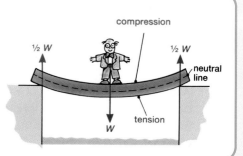

In the lower part of the ruler the molecules are pulled farther apart (as the ruler bends). This part of the ruler is in **tension**.

In the upper part of the ruler the molecules are pushed closer together. This part of the ruler is in **compression**.

The middle section of a bent beam (the neutral line) is neither stretched nor compressed. This means that the material there is having little effect – so it can be removed, to make the beam lighter.

I-shaped steel girders are often used because they are strong at the top and the bottom to resist the tension and compression forces, but they are light and economical (steel is expensive).

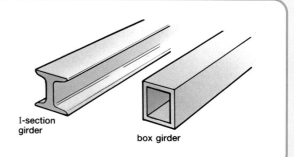

I-section girder

box girder

Concrete is often used as a building material. It is cheaper than steel and does not rust, but it has a disadvantage. Can you see from the figures in the table what that disadvantage is?

Although concrete is fairly strong when in compression, it is weak when it is in tension!

Material	Tensile strength (stretching) (N/mm²)	Compressive strength (crushing) (N/mm²)
Steel	800	500
Human bone	100	150
Concrete	4	20

This means that it is likely to crack and break on the underside of a bridge:

The answer is to **re-inforce** the concrete with steel bars set inside near the bottom edge.
As the bridge bends, the steel bars are in tension and take the strain.

In some bridges, the steel rods are stretched tightly before the concrete sets. When the concrete has set hard, the rods are released and keep the concrete compressed so it is strong. This is **pre-stressed** re-inforced concrete.

Steel and concrete expand by the same amount. What would happen if they didn't?

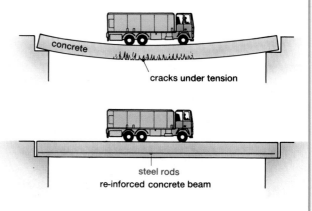

concrete

cracks under tension

steel rods
re-inforced concrete beam

104

▷ Physics at work: Designing Bridges

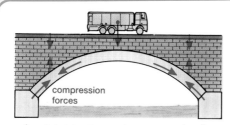

An **arch** bridge has only compression forces. It can be safely built from concrete (or stone or brick) because they are strong in compression.

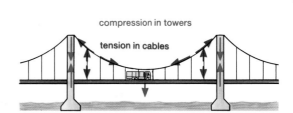

A **suspension** bridge. Here the towers (in compression) can be made from concrete, but the cables (in tension) are made from steel.

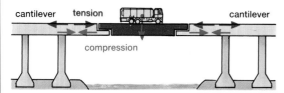

A **cantilever** bridge is an extension of a simple beam bridge (see opposite page). The cantilevers allow a wider gap to be bridged.

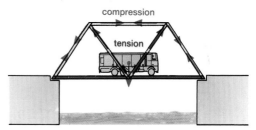

A **girder** bridge is a strengthened beam bridge. It uses steel triangles to make it rigid. Longer bridges need more triangles (see also page 22).

Survey your local bridges: what are they made of; which parts are in tension, or in compression?

Experiment 2   Testing girders
For girders you can use drinking straws, with pins to fasten them (or you can use balsa wood, or wooden 'spills', Lego, or Meccano). Use girders to build these shapes:

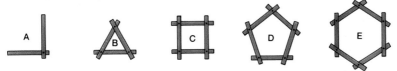

B is much more rigid than the others. Why? How can you add extra girders to make them more rigid?

Experiment 3   Building a girder bridge: competition 1
Task: use your girders to build a bridge to cross a 10 cm gap.
You can use up to 10 straws but imagine each one costs £1000. You can test each bridge (until it collapses) by hanging a paper cup from it and gently adding weights or sand to it.
How can you ensure your tests are *fair*? Who can build the cheapest bridge to support a 500 g mass? Who can build the strongest bridge for £10 000?

Experiment 4   Building a girder bridge: competition 2
Task: build a girder bridge to span a 60 cm gap, with a 5 cm wide roadway.
For girders use 3 mm × 3 mm balsa wood strips, with thin card and glue.
Weigh your bridge and then carefully hang weights at the centre point to test it to destruction.

Then calculate the **best model ratio** $= \dfrac{\text{maximum load (N)}}{\text{weight of bridge (N)}}$ Who can achieve the highest ratio?

chapter 15

TURNING FRCES

If you want to undo a very tight nut on your bicycle, would you choose a short spanner or a long spanner?

Why is the handle on a door placed a long way away from the pivot? Can you close a door easily by pushing near the pivot? Who would win if a strong man pushed *at* the pivot and a small boy pushed the opposite way at the handle? Try it.

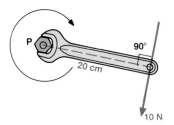

Experiment 15.1
Hold a rod horizontally and hang a weight on the rod near your hand.
Move the weight to different distances from your hand and try to keep the rod horizontal.

What happens?
Now try it with a heavier weight.

You can feel that the turning effect on your hand depends on the *size* of the force (the weight) and the *distance* from your hand. The turning effect of a force is called a **moment** or a **torque**.
It is calculated by:

Moment of a force = force $\times$ **perpendicular distance (from the force to the pivot)**
(newtons)   (metres)

The distance used is always the *shortest* (perpendicular) distance.
Moments are measured in newton-metres (often written N m).

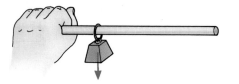

Example
What is the turning effect of the force in the diagram, about the nut at point P?

Perpendicular distance from the force to P = 20 cm = 0.20 m

Formula first: Moment of the force = force $\times$ perpendicular distance from the force to the pivot

Then numbers: = 10 N $\times$ 0.20 m
= 2 N m   turning *clockwise*.

A car mechanic might need to apply a torque of 40 N m to a nut.

You know that a see-saw can be balanced even when the two people have different weights, by sitting them at different distances from the pivot.

When this see-saw is balanced and is not moving (**in equilibrium**), the moment of the girl (a clockwise moment) must **equal** the moment of the man (an anti-clockwise moment).
This is called the **principle of moments**:

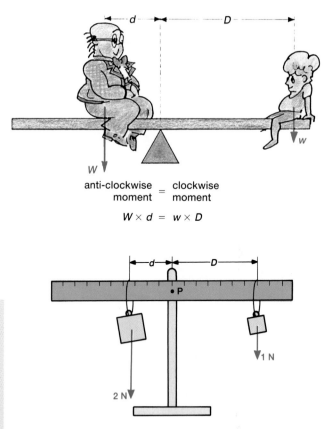

anti-clockwise moment $=$ clockwise moment

$$W \times d = w \times D$$

> **In equilibrium,**
> | total anti-clockwise moment | = | total clockwise moment |

You can check this by an experiment:

> ### Experiment 15.2
> Get a metre rule with a hole drilled at its centre; push a large pin through and clamp it to a retort stand. If the rule is not balanced, weight the higher end with plasticine until it balances.
> You need weights of 1 newton (a 100 g mass will do) and 2 newtons (a 200 g mass will do) to hang from loops of string.
>
> Adjust the weights until the rule balances. Measure the distances **d** and **D** and note them in the table. Repeat with these and other weights at different distances. Repeat with two weights on one side (see the example below).

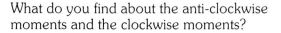

What do you find about the anti-clockwise moments and the clockwise moments?

Anti-clockwise			Clockwise		
Weight W (N)	Distance d (cm)	Moment W × d	Weight w (N)	Distance D (cm)	Moment w × D
2	20	40	1	40	40
2	15		1		

> ### Example
> If the ruler in the diagram is balanced, what is the weight **W**?
>
> Equation first:
> In equilibrium,
> | total anti-clockwise moments | = | total clockwise moments |
>
> Then put in numbers (see diagram):
> $$W \times 25 = (4 \times 15) + (1 \times 40)$$
> $$W \times 25 = 60 + 40$$
> $$W \times 25 = 100$$
> $$W = \underline{4 \text{ newtons}}$$

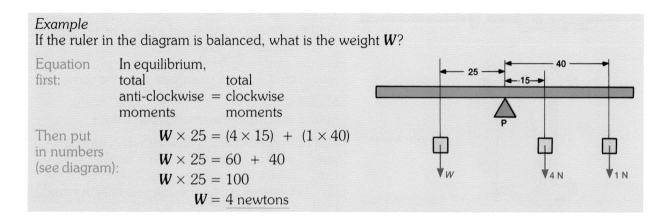

▷ Centre of mass (centre of gravity)

What part of a ladder would you rest on your shoulder in order to carry it most comfortably?

Although the force of gravity pulls down on every little molecule, the weight of the ladder seems to act at its centre. This point on the ladder, where it balances, is called the **centre of gravity**, or **centre of mass**.

The centre of gravity is the point through which the whole weight of the object seems to act.

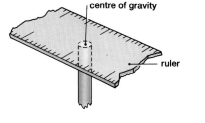

centre of gravity

ruler

Experiment 15.3
Find the centre of gravity of a ruler by moving it until it balances on the end of a pencil.

Where is the centre of gravity of a ruler?
Where is the centre of gravity of a plank?

Use a similar experiment to find the centre of gravity of a book, a retort stand or a hammer.

Experiment 15.4
Tie a small weight to a piece of string to form a *plumbline*. Let it hang down from your hand.

Why does it always hang vertically?
What can a plumbline be used for?

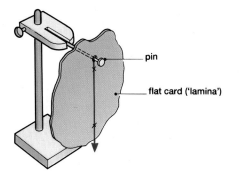

pin

flat card ('lamina')

Experiment 15.5
Find the centre of gravity of a piece of flat card (a *lamina*) using a plumbline.
Let the card hang *freely* from a large pin held in a retort stand. Hang a plumbline from the same pin.

Mark the position of the plumbline by two crosses on the card. Join the crosses with a ruler.

Just as the plumbline hangs with its centre of gravity vertically below the pivot, so also will the card. This means that the centre of gravity of the card is *somewhere* on the line you have marked. How can you find where it is on this line?

To find just where the centre of gravity is on this line, re-hang the card with the pin through **another** hole and again mark the vertical line. The only point that is on **both** lines is where they cross, so this point must be the centre of gravity.

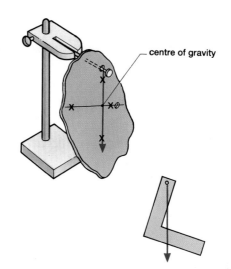
centre of gravity

How can you check that this point really is the centre of gravity?

Experiment 15.6
Repeat experiment 15.5 with an L-shaped piece of card. Where does the centre of gravity seem to be?

▷ Stability

This plumbline is said to be in **stable equilibrium** because if you push it to one side, it returns to its original position. It does this because when you push it to one side its centre of gravity rises and gravity tries to pull it back to its lowest position.

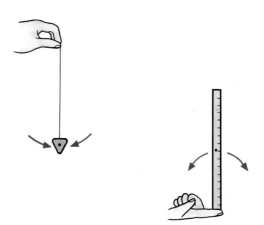

If you carefully balance a ruler vertically on your finger it is in **unstable equilibrium**, because if it moves slightly, its centre of gravity falls and keeps on falling down.

A billiard ball on a perfectly level table is in **neutral equilibrium**, because if it is moved, its centre of gravity does not rise or fall.

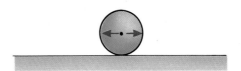

Is a tightrope walker in stable or unstable equilibrium?

A bottle is shown in equilibrium in three positions:

Name the kind of equilibrium for each position.

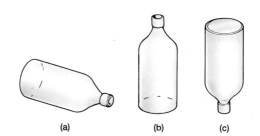

(a)   (b)   (c)

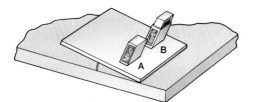

▷ Stable and unstable objects

Most of the objects we use each day are in stable equilibrium.
How can objects be made more stable?

Why is a rowing-boat less stable if you stand up?

Experiment 15.7
Place a box or matchbox upright on a rough piece of wood (A). Slowly tilt the wood until the box topples over. Note the angle of the wood.

Raise the centre of gravity by weighting the top of the box with plasticine or sliding up the matches (B). Tilt the wood again. Is the box more stable or less stable?

Place the box in other positions and see when it topples.

To make the box more stable, should it have
a) a high or a low centre of gravity?
b) a narrow or a wide base?

Which is more stable, a racing car or a double-decker bus? How could the bus be made more stable? Why are passengers not allowed to stand upstairs?

How are these objects made stable?

Bunsen burner   retort stand   wine glass

Discuss the physics of these cartoons.

Where do you think her centre of gravity is?

A dangerous stunt! What kind of stability is this?

A bus being tested for stability.
Where do you think its centre of mass is?

Doing the 'Fosbury Flop'. She is keeping her
centre of gravity as low as possible – why?

Summary

$$\begin{array}{cccc} \text{Moment of} & & & \text{perpendicular distance} \\ \text{a force} & = & \text{force} & \times & \text{from the force to the pivot} \\ \text{(newton-metre)} & & \text{(newtons)} & & \text{(metres)} \end{array}$$

Principle of moments:

In equilibrium,

total anti-clockwise = total clockwise
moments   moments

The **centre of gravity** is the point through which the whole weight of the object seems to act. An object can be in stable, unstable or neutral equilibrium depending on the position of its centre of gravity. To be more stable an object needs a low centre of mass and a wide base.

Discuss the physics of these cartoons.

I'LL PUT ALL THESE HEAVY THINGS IN THE TOP DRAWER

HOW CAN SHE DO THAT UP THERE BLINDFOLDED?

I THINK THE DOOR LOOKS NICER WITH THE HANDLE AT THIS SIDE!

▷ Questions

1. Copy and complete these sentences.
   a) The turning effect (or . . . .) of a force is equal to the force multiplied by the . . . . distance from the . . . . to the . . . . . Its unit is . . . . .
   b) The principle of moments states that, in . . . ., the total . . . . are equal to the total . . . .
   c) The centre of gravity (centre of . . . .) is the point through which the whole . . . . of the object seems to act.
   d) A stable object should have a . . . . centre of . . . . and a . . . . base.
   e) If the centre of . . . . of an object
      i) is raised by tilting, it is in . . . . equilibrium
      ii) is lowered by tilting, it is in . . . . equilibrium
      iii) remains at the same height, it is in . . . . equilibrium.

2. Explain the following:
   a) A mechanic would choose a long spanner to undo a tight nut.
   b) A door handle is placed well away from the hinge.
   c) It is difficult to steer a bicycle by gripping the centre of the handlebars.

3. A mechanic applies a force of 200 N at the end of a spanner of length 20 cm. What moment is applied to the nut?

4. The diagrams show rulers balanced at their centres of gravity. What is the weight of:
   a) X?   b) Y?

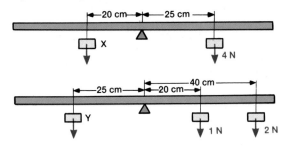

5. A crane driver has a chart in his cab which tells him how far he can safely extend the jib of his crane:

Size of load (tonnes)	Maximum length of jib (m)	Moment (load × length)
10	12	
20	6	
30	4	
40	3	

   a) Copy and complete the table.
   b) What do you notice? Why is this?
   c) What length of jib could he have with a load of 24 tonnes?

6. A, B, C and D are pieces of hardboard. Copy the diagrams and mark with a cross the position of the C of G of each.

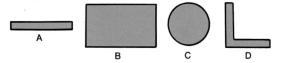

7. Sketch diagrams and mark the approximate positions of the centres of mass of
   a) a hammer   b) a boomerang
   c) a tea cup   d) a human being.

8. Explain the following:
   a) A Bunsen burner has a wide heavy base.
   b) Racing cars are low, with wheels wide apart.
   c) A boxer stands with his legs well apart.
   d) A wine glass containing wine is less stable.
   e) Lorries are less stable if carrying a load.

There was an old teacher called Grace,
Who often fell flat on her face.
The reason you see,
Was her C of G,
Its place was too high for her base.

Further questions on page 152.

chapter 16

WORK, ENERGY & POWER

If the man in the picture is moving the van, we say he is doing **work**.
He is doing work only if there is **movement** against an **opposing** force.

The opposing force is often friction (as in the picture), or gravity.
The man can only do this work if he has some **energy**.

If the man pushes with a stronger force or moves it a longer distance he will do more work. In fact,

Work done	=	force	×	distance moved
		(newtons)		(metres)

The distance used in this formula must be the distance moved *in the direction of the force*.

If the force is in newtons and the distance is in metres then the work is measured in *joules*, often written J (see page 41).

1 joule is the work done when a force of 1 newton moves through 1 metre (in the direction of the force).

Example

A man lifts a brick of mass 5 kg from the floor to a shelf 2 metres high. How much work is done?

The opposing force in this case is the *weight* of the brick. Here on Earth a mass of 1 kg weighs 10 newtons (see page 83).
∴ a mass of 5 kg weighs 50 newtons.

Formula first: **Work done = force × distance moved**
| | (J) | (N) | (m) |

Then put in the numbers:
= 50 N × 2 m
= 100 joules

Where does he get the energy to do this work?
Could he do this work if he did not eat?

If a man pushes a van against a friction force of 300 newtons for a distance of 10 metres how much work does he do?

Large amounts of work may be measured in kilojoules or even megajoules (see page 41).

▷ Forms of energy

We have seen already (on page 8) that there are several forms of energy.
We have studied **thermal** energy and found that it is really the movement energy of the molecules.

All moving objects have movement energy, called **kinetic** energy. A heavy lorry travelling fast has a lot of kinetic energy (and if you get in the way it will use some of this energy to damage you).

Energy which is stored is called **potential** energy.

When an object is lifted to a higher place, it is given more **gravitational** *potential energy*. This is stored energy which it can give out if it falls down.

A second kind of potential energy is **elastic** *potential energy* which is stored in the elastic of a catapult or in a stretched bow.

Chemical energy is energy stored in food and other fuels. Your body gets its energy from food you eat.

Sound energy and *light* energy are common types of energy.

Electrical energy is common and useful. Like other forms of energy, it can be very dangerous. *Magnetic* energy is always connected with electrical energy.

Nuclear energy is stored in the centre (or nucleus) of an atom. The Sun, like a hydrogen bomb, runs on nuclear energy.

Changing energy from one form to another

We have seen already (on page 9) that energy can be changed from one kind to another. When this happens, the *amount* of energy stays the *same.* We say it is '*conserved*'. This is because energy cannot be made or destroyed.

This fact is called:

> **the Principle of Conservation of Energy:**
> energy can be changed from one form to another,
> *but it cannot be created or destroyed.*

▷ Energy and work

Energy is the ability to do work.
The amount of work that is done tells us how much energy has been converted from one form to another.

> **Work done = Energy changed**

Working against gravity

In the example on page 113 we calculated that the work done by the man in lifting the brick to the shelf was 100 joules. This means that 100 joules of his chemical (food) energy was converted to 100 joules of gravitational potential energy.

Falling under gravity

The diver in the diagram uses 6000 joules of his chemical energy to climb to the top where he has 6000 J of gravitational potential energy (PE) but no kinetic energy (KE).
As he falls down, his potential energy is converted into an *equal* amount of kinetic energy.
The total amount of PE + KE is always 6000 J.

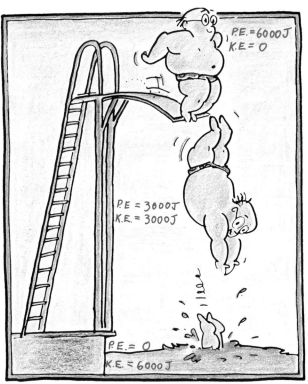

The pendulum

When a pendulum swings to and fro, its energy is constantly changing from potential energy to kinetic energy and back again.

This energy is gradually converted to heat by friction with the air. This heat is 'low-grade' energy and we cannot make use of it.
All energy eventually becomes low grade.

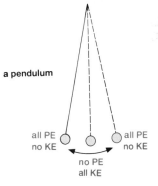

a pendulum

What are the energy changes in these examples?

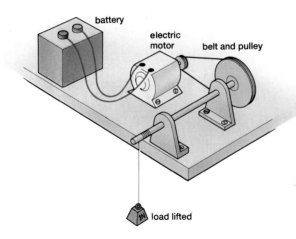

battery
electric motor
belt and pulley
load lifted

▷ Energy experiments

Experiment 16.1
Arrange an electric motor, as in the diagram, so that it can turn an axle from which a load is hanging.

What happens when you connect the motor to a battery?

What are the energy changes here?

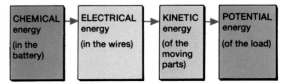

| CHEMICAL energy (in the battery) | → | ELECTRICAL energy (in the wires) | → | KINETIC energy (of the moving parts) | → | POTENTIAL energy (of the load) |

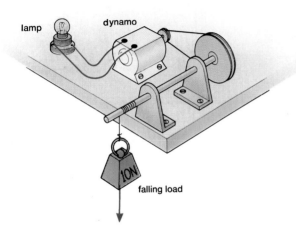

lamp
dynamo
falling load
10N

Experiment 16.2
Disconnect the battery and connect a lamp in its place. Start with a heavy load raised high and let it fall to drive the 'motor' which will now act as a **dynamo** and produce electricity to light the lamp.

What are the energy changes here?

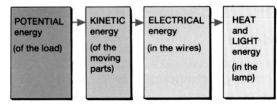

| POTENTIAL energy (of the load) | → | KINETIC energy (of the moving parts) | → | ELECTRICAL energy (in the wires) | → | HEAT and LIGHT energy (in the lamp) |

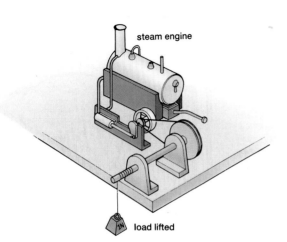

steam engine
load lifted

Experiment 16.3
Use a model steam engine to lift a load.

What are the energy changes here?

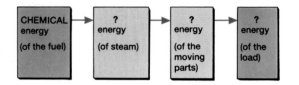

| CHEMICAL energy (of the fuel) | → | ? energy (of steam) | → | ? energy (of the moving parts) | → | ? energy (of the load) |

Measure the weight of the load and the height it is lifted.
Then calculate the work done on the load.

Experiment 16.4
Build a model power station using a steam engine and a dynamo. Connect them to a lamp to represent all the lights in your home.

What are the energy changes here? (See also p. 351.)

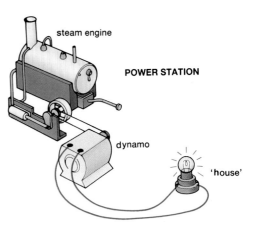

steam engine

POWER STATION

dynamo

'house'

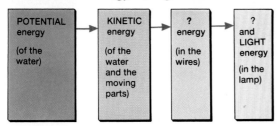

| **?** energy (of the fuel) | **?** energy (of the steam) | **?** energy (of the moving parts) | **?** energy (in the wires) | **HEAT** and **LIGHT** energy (in the lamp) |

Experiment 16.5
Build a model hydroelectric power station using the energy of falling water to turn a turbine and drive a dynamo.

What are the energy changes here?

| **POTENTIAL** energy (of the water) | **KINETIC** energy (of the water and the moving parts) | **?** energy (in the wires) | **?** and **LIGHT** energy (in the lamp) |

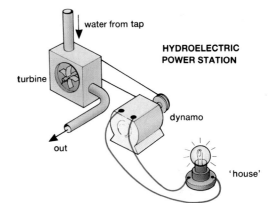

water from tap

HYDROELECTRIC POWER STATION

turbine

out

dynamo

'house'

▷ Sources of energy

Where does our energy come from? The answer is that it almost all comes from the **Sun**!
Study the different parts of this diagram and you will see how.

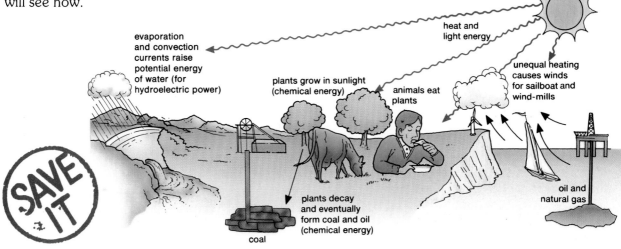

evaporation and convection currents raise potential energy of water (for hydroelectric power)

heat and light energy

plants grow in sunlight (chemical energy)

animals eat plants

unequal heating causes winds for sailboat and wind-mills

plants decay and eventually form coal and oil (chemical energy)

coal

oil and natural gas

SAVE IT

▷ Calculating Potential Energy

Gravitational Potential Energy

The weight-lifter is lifting a mass of 200 kg, up to a height of 2 metres.
We have already seen (on page 113) how to calculate the potential energy of his weights:

$$\text{Potential energy} = \text{work done}$$
$$= \text{weight} \times \text{height lifted}$$

But here on Earth (see page 83):
weight (in N) = **mass** (in kg) × **10**   so:

PE = mass × 10 × height moved

or

Gravitational PE = mass × g × height
(joules)   (kg)   (N/kg)   (m)

because $g = 10$, here on Earth, but g has different values on other planets (see page 161).

Elastic Potential Energy (strain energy)

This is the kind of energy stored in a bow or in a catapult or in a spring in a clock.

Energy stored = work done to stretch the bow, so:

Strain energy = average force × distance
(joules)   (newtons)   (metres)

Example
Robin Hood exerts an average force of 100 N in pulling back his bow by 0.5 m. He fires the arrow (mass = 0.2 kg) vertically upwards. How much energy is stored in his bow, and how high does the arrow go?

$$\text{Strain energy} = \text{average force} \times \text{distance}$$
$$= 100\,\text{N} \times 0.5\,\text{m}$$
$$= \underline{50\ \text{joules}}$$

Since this energy is conserved (see page 114), then this must be the kinetic energy as the arrow leaves the bow, and it must also equal the gravitational PE at its highest point:

$$\text{mass} \times 10 \times \text{height} = 50\ \text{joules}$$
$$0.2 \times 10 \times \text{height} = 50$$
$$\text{height} = \underline{25\ \text{metres}}$$

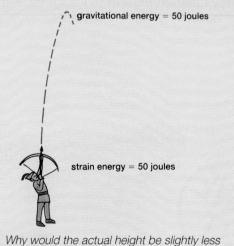

gravitational energy = 50 joules

strain energy = 50 joules

Why would the actual height be slightly less than the one we have calculated?

118

▷ Calculating Kinetic Energy

A running elephant has more kinetic energy than a running man, because it has more mass.

A racing car has more kinetic energy than a family car because it has a higher speed.

In fact, the formula for the kinetic energy is:

25 000 joules of kinetic energy

$$\text{Kinetic Energy} = \frac{1}{2} \times \text{mass} \times \text{speed squared}$$
$$\text{(joules)} \qquad\qquad \text{(kg)} \qquad \text{(m/s)}^2$$

Example 1
An elephant of mass 2000 kg travelling at 5 m/s has
KE $= \frac{1}{2} \times 2000 \times 5 \times 5 = \underline{25\,000 \text{ joules}}$.

Example 2
Galileo drops a stone from the leaning tower of Pisa, which is 45 metres high.
At what speed does the stone hit the ground?

The energy is **conserved** (see page 114) so, assuming no air resistance:

gravitational energy = **kinetic energy**
at the top   **at the bottom**

$$\text{mass} \times 10 \times \text{height} = \frac{1}{2} \times \text{mass} \times \text{speed}^2$$
$$10 \times 45 = \frac{1}{2} \times \text{speed}^2$$
$$\text{speed}^2 = 10 \times 45 \times 2 = 900$$
$$\text{speed on impact} = \underline{30 \text{ m/s}}$$

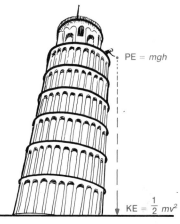

PE = mgh

KE = $\frac{1}{2}mv^2$

When the stone hits the ground, the KE is converted to heat and sound

Road safety

The formula for KE has the speed **squared**. So, if a car goes at three times the speed, it has **nine** times the energy and so the braking distance is **nine** times as long!

Too much kinetic energy

Study this table:

Speed of car (m/s)	Driver's thinking distance, assuming an alert driver and a good reaction time (m)	Car's braking distance, assuming good brakes and tyres and a dry road (m)	Total stopping distance in metres	in car-lengths
10	6	6	12	3
20	12	24	36	9
30	18	54	72	18
40	24	96	120	30

Notice how the braking distance increases rapidly if the car goes faster.

▷ **Power**   If two cars of the *same* weight climb up the *same* hill, then they do the *same* amount of work. (See page 113.) But if car A climbs the hill in a shorter time than the other car, we say it has a greater *power*. Power is the *rate of working* (the rate of converting energy).

$$\textbf{Power} = \frac{\textbf{work done} \text{ (in joules)}}{\textbf{time taken} \text{ (in seconds)}} = \frac{\textbf{energy converted}}{\textbf{time taken}}$$

The work done is measured in joules, the time taken is measured in seconds and so power is measured in *joules per second* or *watts* (W).
1 watt (1 W) = 1 joule per second

The power of very powerful engines may be measured in kilowatts (kW) or even megawatts (MW).
1 kW = 1 kilowatt   = 1000 watts
1 MW = 1 megawatt = 1 000 000 watts.
Engines may also be measured in horsepower (h.p.) (1 h.p. = 750 W)

Example
A crane lifts a load weighing 3000 N through a height of 5 m in 10 seconds.
What is the power of the crane?

Calculate it in two parts:

Formula first: **Work done = force × distance moved**
(see page 113)
Then put in the numbers:
= 3000 N × 5 m
= 15 000 joules

Formula first:
$$\textbf{Power} = \frac{\textbf{work done}}{\textbf{time taken}}$$
Then put in the numbers:
$$= \frac{15\,000 \text{ joules}}{10 \text{ seconds}}$$
= 1500 W = 1.5 kW

A family car might develop a power of 40 kW, a racing car perhaps 400 kW and a Moon rocket 100 000 MW.

Personal power

How much power can you develop with your legs?

Experiment 16.6
You need a flight of stairs to run up, so that you
are doing work against the force of gravity (your
weight).
First measure your weight in newtons (see page 83).
Then, since you are going to lift your body against
the **vertical** force of gravity, you must measure
the **vertical** height of the stairs (in metres).
Then ask someone to use a stop-clock to time how
long it takes you to run from the bottom to the
top of the stairs.

Now use the method shown on the opposite page
to calculate
1. the amount of work you did
2. the power developed by your legs.

Who develops the most power in your class?
What are the energy changes in this experiment?

Wattsisname?
Q. *What's Watt?*
A. *What watt? 'Watt' or 'watt', or what?*
Q. *Yes, what Watt is the watt named after?*
A. *James Watt. He knew what was watt!*

▷ Electric power

Electric power is also measured in watts. Electric lamps are marked
with the electrical power that they convert.
Look at the writing on a lamp. '60 W' means that the lamp converts
60 joules per second (from electrical energy to heat and light energy).

60 W

Example
An electric kettle is rated at 2 kW.
How many joules of energy are converted in 10 seconds?

Power = 2 kW = 2000 W = 2000 joules per second
∴ In 10 seconds, energy converted to heat = 2000×10 joules
                                          = 20 000 joules (= 20 kJ)

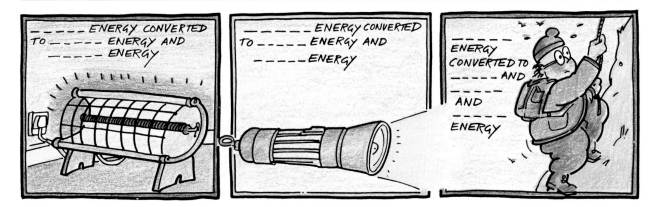

Summary

Work done = force × distance moved = energy changed
(joules)   (newtons)   (metres)

Principle of conservation of energy:
energy can be changed from one form to another, but it cannot be created or destroyed.

$$\text{Power} = \frac{\text{work done (joules)}}{\text{time taken (seconds)}} = \frac{\text{energy converted}}{\text{time taken}}$$

1 watt is a rate of working of 1 joule per second.

Gravitational Potential Energy = mgh joules.   Kinetic Energy = $\frac{1}{2}mv^2$ joules.

▷ Questions

1. Copy and complete these sentences.
   a) The work done (measured in . . . .) is equal to the . . . . (in newtons) multiplied by the . . . . moved (in metres).
   b) 1 joule is the work done when a force of one . . . . moves through a distance of one . . . . (in the direction of the . . . .).
   c) The principle of . . . . of energy says that energy can be . . . . from one form to another, but it cannot be . . . . or . . . . .
   d) Work done = ... changed.
   e) The nine forms of energy are:
   f) Power is the rate of doing . . . . .

   $$\text{Power (in . . . .)} = \frac{\text{work done (in . . . .)}}{\text{. . . . (in seconds)}}$$

   g) 1 watt is a rate of working of one . . . . per . . . . .

2. A man lifts a parcel weighing 5 newtons from the ground on to a shelf 2 metres high. How much work does he do on the parcel?

3. Name the energy changes in each one in the list below. For example, in an electric lamp:

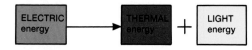

   a) a torch battery   b) a falling stone
   c) a Bunsen burner   d) a dynamo
   e) a microphone   f) a pendulum
   g) bicycle brakes   h) a human body
   i) an electric bell   j) a car crash.

4. A girl weighing 500 N climbs 40 m vertically when walking up the stairs in an office block. How much work does she do against gravity? What are the energy changes here?

5. Name the forms of energy stored in:
   a) a slice of bread   b) stretched elastic
   c) a torch battery   d) a turning flywheel
   e) petrol   f) water in a dam.

6. Name the forms of energy at each stage of:
   a) a car engine (see page 70)
   b) a coal-fired power station
   c) a nuclear power station (see page 351)
   d) a hydroelectric power station.

7. Draw an 'energy chain diagram' to show how the energy in a cheese sandwich comes from the Sun to:
   a) the bread   b) the cheese.

8. Professor Messer dreamed one night that the world had suddenly changed so that energy could not be converted from one form to another. Why did his dream turn into a nightmare?

9. Look at the table on page 119. A car is travelling at 20 m/s (45 m.p.h. or 72 km/h).
   a) What is the thinking distance at this speed?
   b) How does it change if the driver is tired?
   c) What is the braking distance?
   d) How does this change if the road is wet, icy or made of loose gravel?
   e) What is the total stopping distance?

10. What are the energy changes in these examples?

11. Use the table on page 119.
    a) Plot a graph of the thinking distance data against speed. What do you notice?
    b) Plot the braking distance data on the same graph. What do you notice?
    c) Plot the total stopping distance also.
    d) The UK motorway speed limit is 31 m/s (70 m.p.h. or 112 km/h). What is the stopping distance at this speed?

12. A girl does 1000 joules of work in 5 seconds. What power does she develop?

13. A man lifts a weight of 300 N through a vertical height of 2 m in 6 seconds. What power does he develop?

14. A man weighing 1000 N runs up some stairs, rising a vertical height of 5 m in 10 seconds. What power does he develop? What are the energy changes here?

15. A crane lifts a load of 3000 N through a vertical height of 10 m in 4 seconds. What is its rate of working in a) watts b) kilowatts?

16. What are the energy changes in these pictures?

17. An electric lamp is marked 100 W. How many joules of electrical energy are transformed into heat and light
    a) during each second
    b) during a period of 100 seconds

18. A polevaulter has a mass of 50 kg.
    a) What is her weight in newtons? (g = 10 N/kg)
    b) If she vaults to 4 metres high, what is her gravitational potential energy?
    c) How much kinetic energy does she have just before reaching the ground?

19. A car of mass 1000 kg is travelling at 30 m/s.
    a) What is its kinetic energy?
    b) It slows to 10 m/s. What is the KE now?
    c) What is the change in kinetic energy?
    d) If it takes 80 metres to slow down by this amount, what is the average braking force?

20. A girl throws a ball upwards at a velocity of 10 m/s. How high does it go? (g = 10 N/kg)

21. Find out all you can about the work of
    a) James Joule   b) James Watt.

Further questions on page 152.

chapter 17

Machines change energy from one form to another.
We know that the total amount of energy put into a machine must equal the total amount of energy output. This is the principle of conservation of energy (see page 114).
However, only *some* of the output energy is useful to us. The rest is wasted energy. This affects the *efficiency* of the machine.

A car is not very efficient. For every 100 joules of chemical energy (petrol) that is put into a car, only 25 joules appear as useful movement energy. The other 75 J are wasted as heat. It is low-grade energy and we cannot use it. The efficiency is calculated by:

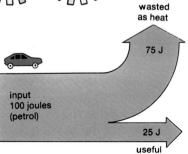

$$\text{Efficiency} = \frac{\text{useful energy output}}{\text{total energy input}} \quad \textbf{or} \quad \text{Efficiency} = \frac{\text{power output}}{\text{power input}}$$

Example 1
For this car, the efficiency $= \dfrac{25}{100} = 0.25$ or 25%

Because of friction in a machine there is always some wasted energy. This means the efficiency is **always less than 100%**.

The energy in the fuel is concentrated and useful. Whenever energy changes from one form to another, some of it becomes less concentrated and so less useful. In any machine, energy tends to spread out more and more, into less useful forms.

Efficiency will soon decrease
If you forget the oil and grease

Example 2
An athlete in a race exerts a force of 80 newtons for a distance of 100 metres, while she uses up 40 000 joules of food energy. What is her efficiency?

Formula first: Useful energy = work done
= force × distance moved (see p. 113)
Then numbers: = 80 N × 100 m = 8000 joules

Formula: Efficiency $= \dfrac{\text{useful energy output}}{\text{total energy input}}$

Then numbers: $= \dfrac{8000}{40\,000} = 0.20$ or 20%

The rest of the energy (32 000 J) appears as useless heat – this is why athletes sweat a lot!

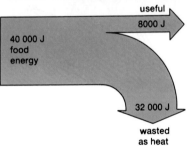

▷ Levers

A lever is a very common machine. It helps us to do work more easily.
Here is a crowbar being used to lever up a large stone. It is like the see-saw on page 107.
The man is 3 m from the pivot, but the stone is only 1 m from the pivot.

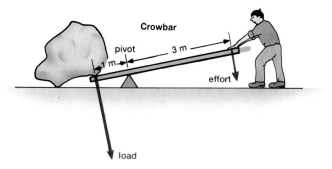

He is exerting an *effort* force of only 100 N, against a larger *load* force of 300 N (the weight of the stone).
The lever is acting as a *force magnifier*.

The principle of moments (page 107) shows us why:
$$300 \, \text{N} \times 1 \, \text{m} \; = \; 100 \, \text{N} \times 3 \, \text{m}$$

To lift the stone by 0.1 m, the man must move $3 \times 0.1 \; = \; 0.3$ m
The work done on the stone $= \; 300 \, \text{N} \times 0.1 \, \text{m} \; = \; 30$ joules
The work done by the man $\; = \; 100 \, \text{N} \times 0.3 \, \text{m} \; = \; 30$ joules

This assumes that there is no friction, so the efficiency is 100%.
In practice, if 10% energy is lost by friction, the efficiency $= \; 90\%$.

Here are some examples of levers:

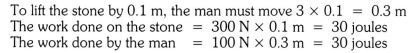

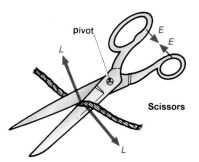

This is a force magnifier. The effort is farther away from the pivot, just like in the crowbar at the top of the page.

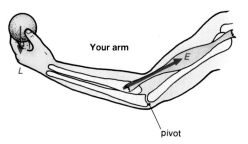

This is a force magnifier. The effort force needed is about half the load because it is twice as far from the pivot.

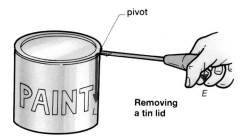

This is a force magnifier, because the effort is farther from the pivot. What if you try to cut string with the very tip of the scissors?

This is not a force magnifier – it is a distance magnifier. A small contraction of your arm muscle gives a big movement of your hand.

▷ The inclined plane (or ramp)

A ramp, a slope and a hill are examples of inclined planes. They are force magnifiers.

In the diagram the man is moving a heavy load into the van using a small effort force.
As in all these machines, the smaller force has to go a longer distance – in this case the effort has to move the full length of the slope while the load moves a shorter distance vertically.

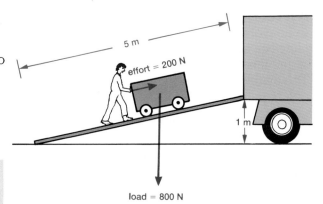

Example
Find the efficiency of the ramp in the diagram:

Useful energy out = work done against gravity
    = weight × height lifted
    (see page 113)
    = 800 N × 1 m
    = 800 joules

Total energy in = work done by man
    = force × distance moved
    = 200 N × 5 m
    = 1000 joules

$$\text{Efficiency} = \frac{\text{useful energy output}}{\text{total energy input}}$$

$$= \frac{800 \text{ joules}}{1000 \text{ joules}} = \underline{0.80} \quad \text{or} \quad \underline{80\%}$$

The other 200 joules (20%) are lost as heat (due to friction).

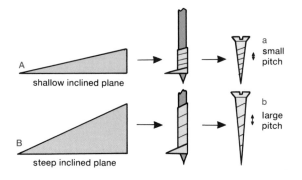

How did the Egyptians use ramps to build the pyramids?

▷ The screw

The thread of a screw is really an inclined plane wrapped round the screw.

Cut out two pieces of paper so that they look like ramps, one shallow and one steep.
Wrap each piece of paper round a pencil and you will see that each one forms a screw.

Which ramp (A or B) would need less effort?
Which screw (a or b) would need less effort?

A screw is used to exert a large force in a nut and bolt, a workshop vice, and a car-jack.

▷ Gears

Gears can be used as force magnifiers or as distance magnifiers. It depends on the size of the wheels.

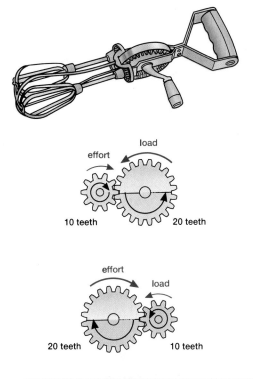

In the diagram, if the smaller (effort) wheel has 10 teeth and the larger (load) wheel has 20 teeth:
When the small wheel moves through one complete turn, the large wheel moves *anti*-clockwise through only *half* a turn.

The larger wheel always moves more slowly, but with a larger force. This is a force magnifier.

If the wheels are the other way round, the smaller (load) wheel moves faster (but with a smaller force). This is a distance magnifier and a speed magnifier.

In a hand whisk and in a bicycle, gears are used as speed magnifiers:

In a car, the driver can change the gear ratio for force (climbing hills) or for speed (on a flat motorway).

Look at the cogs at the top of page 124. If the 'm' gear turns clockwise, which way does the 's' gear turn?

▷ Wheel and axle

This force magnifier is really a kind of continuous lever.
A small effort on the wheel gives a large turning force on the axle.
The diagrams show some examples:

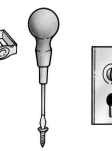

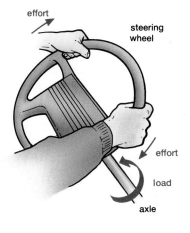

127

▷ Pulleys

Pulleys are very useful for lifting loads vertically.

Even a single pulley is useful because it changes the direction of the force. It is easier to pull downwards (with your weight helping you) than it is to pull upwards.

Two pulleys

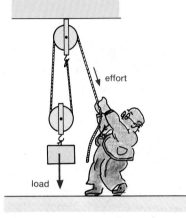

It is easier to lift a load when the rope goes round **two** pulleys. You can see that one end of the rope is fixed and the man is pulling on the other end to lift up the bottom pulley and the load.

Because there are **two** ropes pulling up the load, the effort needed is only about **half** as much.

How far would you have to move the effort to lift the load by one metre?
Because the bottom pulley is held up by two ropes and because they **both** have to be shortened by 1 metre, the effort has to move 2 metres.

As in all these machines, the smaller force has to move a longer distance.

Three pulleys

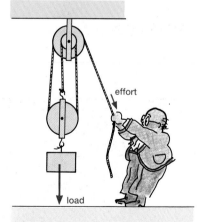

The diagram shows how **three** pulleys can be connected by a rope to make the lifting even easier. Now there are two pulleys in the top block and the rope is fixed to the bottom pulley.
(In practice the two top pulleys are of equal size, but here one is drawn smaller so that you can see where the rope goes.)

The efficiency of a pulley system is **always** less than 100%, for two reasons:
– friction changes some of the energy to heat,
– energy is needed to lift up the bottom pulleys and the ropes.

Pulleys are used in cranes to lift heavy loads. They are used in garages to lift the engine out of a car.

Experiment 17.1 Finding the efficiency of a machine
Set up a pulley system with four pulleys and hang
a known weight on the bottom pulley block. Use
a spring balance to measure the effort needed to
lift the load. The example below shows how to
calculate the efficiency.

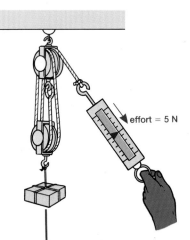

effort = 5 N

load = 10 N

Example
Calculate the efficiency of this pulley system:

Suppose the load is lifted by 1 metre. Each of
the four sections of rope has to be shortened
by 1 m, so the effort force has to move 4 m.

Useful energy output = work done on the load
                    = weight × height lifted
                    = 10 N × 1 m = 10 joules

Total energy input = work done by the effort
                = force × distance moved
                = 5 N × 4 m = 20 joules

$$\text{Efficiency} = \frac{\text{useful energy output}}{\text{total energy input}}$$

$$= \frac{10 \text{ joules}}{20 \text{ joules}} = \underline{0.50} \quad \text{or} \quad \underline{50\%}$$

Summary

$$\text{Efficiency} = \frac{\text{useful energy output}}{\text{total energy input}} = \frac{\text{power output}}{\text{power input}}$$

Machines can be force magnifiers or distance magnifiers.
Hydraulic (liquid) machines are described on page 94.

Because of friction, the efficiency is
always less than 100%.

The smaller force always moves the
longer distance.

Barrows for sale – but would you buy? Please work it out and tell me why.

▷ Questions

1. Copy out and complete:

   a) $\text{Efficiency} = \dfrac{\ldots \ldots \text{energy} \ldots \ldots}{\ldots \ldots \text{energy} \ldots \ldots}$

   $= \dfrac{\text{power} \ldots \ldots}{\ldots \ldots \text{input}}$

   b) The energy change is calculated by:
      work done = ... × . . . . moved

   c) Because of friction, some . . . . is always
      wasted. This means that the . . . . is
      always . . . . than 100%.

2. Here are data for four heat engines. Copy
   and complete the table:

Engine	Fuel	Input	Useful output	% efficiency
Petrol engine		80 kJ	20 kJ	
Diesel engine		20 kJ	7 kJ	
Steam engine		100 MJ	10 MJ	
Cyclist		500 W	100 W	

   Which is the most efficient? Which is the
   most environmentally-friendly?

3. A 100 W electric lamp uses 100 joules of
   electrical energy each second. It produces
   only 2 W of light. What is its efficiency?
   What happens to the other 98 W?

4. A man uses a crowbar 1.5 metres long to lift
   a rock weighing 600 newtons. If the fulcrum
   is 0.50 metre from the end of the bar touching the rock, how much effort must the man
   apply? (Draw a diagram first.)

5. Make a list of all the examples of the lever
   and the wheel-and-axle in your home.

6. A trolley is being pulled up a ramp:

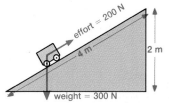

   Calculate:
   a) the work done on the load (see page 113)
   b) the work done by the effort
   c) the efficiency of the machine.

7. In the diagram, gear A is rotating at
   30 revolutions per second.

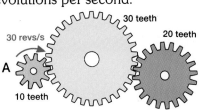

   Sketch these gears and mark on each one:
   a) the direction in which it is turning
   b) the number of revolutions per second.

8. How many examples of simple machines can
   you find in the photograph? In each case
   name the type of machine and say whether
   it is a force magnifier or a distance magnifier.
   Why does a cyclist often zig-zag up a hill?

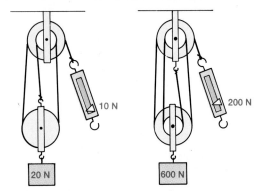

9. For each of these pulley systems, calculate:
   a) the work done in lifting the load 1 metre
   b) the work done by the effort in this case
   c) the percentage (%) efficiency.

Further questions on page 152.

▷ Mechanics crossword

(Do not write on this book – copy the grid with tracing paper first.)
For some of these questions you may need to read the next chapter.

Clues across:

1. The pull of the Earth on one ____ = 9.8 newtons.
8. Change in velocity divided by time taken.
10. A fresh weight for Sir Isaac?
13. One joule is the energy needed to move 1 N through ____ metre.
14. Velocity equals distance travelled divided by ____ .
15. Source of wisdom!
17. Like action or reaction?
18. Velocity and force are ____ .
21. Your most valuable equipment in experiments.
23. Physics is ____ special!
25. Power is work done divided by ____ .
26. A ____ converts electric energy to kinetic energy.
27. Not the first unit of time?
28. A wound-up spring has ____ energy.
31. It might make you move.
32. Shortened unit of electric current.
35. What is the unit!
36. You have been, all term.
37. Newton's First Law: a stationary object, with no resultant force acting on it, remains at ____ .
38. 3.141 592 653 589 793 238 462 643 383 279 502 884 197 169 399 375 105 8 . . . .
39. Speed? Not quite.
40. If a velocity–time graph is a straight line, the acceleration is ____ .

Clues down:

1. and 9. A heavy lorry travelling at high speed has a lot of this.
2. Newton's Third Law: apply a force in, the reaction is ____ .
3. In a little while you have the turning effect of a force!
4. Newton's Third Law: apply a force to the west; the reaction is to the ____ .
5. Force is measured in ____ .
6. A man walking at 2 m/s for 4 seconds covers a distance of ____ metres.
7. It keeps you going when the car crashes.
9. See 1 down.
11. The pull of gravity.
12. A joule is a unit of power – yes or no?
16. When there is a resultant force on a body, it ____ .
19. Push or pull to ____ a force.
20. When you divide by 'time', you are calculating a ____ .
21. An experimental inaccuracy.
22. A dynamo converts kinetic energy to ____ energy.
24. Cold place with little friction.
28. Rate of working.
29. Take in fuel.
30. An electric fire converts electric energy to heat and ____ energy.
33. A city where you might be inclined to experiment!
34. An engine converts ____ energy to kinetic energy.
36. If the mass is 6 kg and the volume is 3 m^3, then the density is ____ kg/m^3.

chapter 18

VELOCITY and ACCELERATION

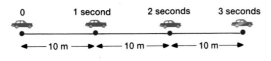

A constant speed of 10 m/s

If a car travels down the road at a **constant speed** of 10 metres per second (10 m/s), it means that it travels 10 metres in every second.

The speed can be found by

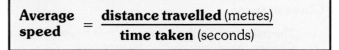

$$\text{Average speed} = \frac{\text{distance travelled (metres)}}{\text{time taken (seconds)}}$$

Velocity is almost the same thing as speed but it is a **vector** quantity (see page 102). Velocity has a **size** (called speed) **and** a **direction**.

If a car is travelling at a constant speed **in a straight line**, then it has a constant velocity. If it turned a corner then its velocity would change, even though its speed remained constant.
A car travelling in the opposite direction has a negative velocity.

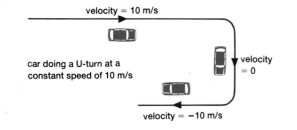

car doing a U-turn at a constant speed of 10 m/s

▷ Acceleration

If a car-driver presses the accelerator, the car goes faster – it accelerates.
The acceleration can be found by

$$\text{Acceleration} = \frac{\text{change in velocity (m/s)}}{\text{time taken for the change (s)}}$$

If the change in velocity is measured in m/s and the time is measured in seconds, then the acceleration is measured in m/s² (metres per second squared).

An acceleration of 2 m/s² means that the velocity increases by 2 m/s every second.

If a car is slowing down then it has a **deceleration** or a **retardation** or a **negative acceleration**.

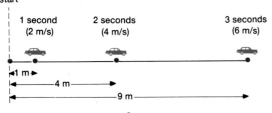

An acceleration of 2 m/s²

▷ Ticker-timer experiments

A ticker-timer is simply a vibrator that puts little black dots on to a paper tape at the rate of 50 dots in each second.

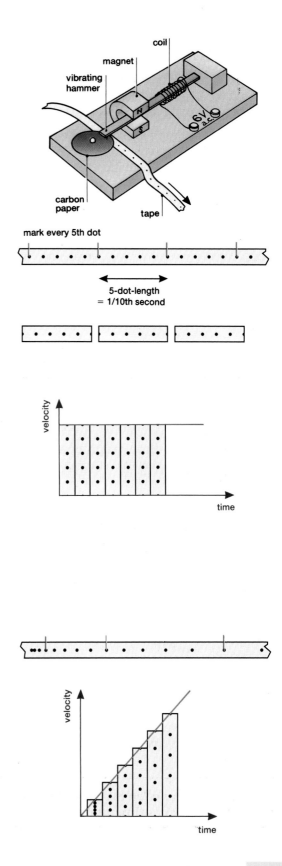

Experiment 18.1
Pull a length of tape through the ticker-timer at a constant velocity.

What do you notice about the spacing of the dots?

a) Count along the tape, marking off every fifth dot. Because there are 50 dots in each second, each 5-dot length is produced in $\frac{1}{10}$th second.
b) Use scissors, at each mark, to cut the tape into several 5-dot lengths (carefully keeping them in the right order).
c) Stick the 5-dot lengths side by side in the right order to get a strip chart.
     Since each strip is the distance travelled in $\frac{1}{10}$th second, you have made a strip chart of *velocity* against *time*.

The constant height of the strips means that the tape was moving at a constant velocity.

Example
Find the speed of the tape in the diagram:

Measure the tape shown. Do you find that it has travelled 1.9 cm in each 5-dot length?

$$\text{Speed} = \frac{\text{distance travelled (5 dots)}}{\text{time taken for 5 dots}}$$
$$= \frac{1.9 \text{ cm}}{0.1 \text{ s}} = \underline{19 \text{ cm/s}}$$

Experiment 18.2
Repeat experiment 18.1 but make the tape accelerate as you pull it.

What do you notice about the spacing of the dots? Repeat steps a), b) and c) to build a velocity–time strip chart for this accelerating tape.

You can see that the velocity is increasing (because the tape was accelerating).

What would be the velocity–time graph for an object which was slowing down (decelerating)?

▷ Velocity–time graphs

The diagram shows a **velocity–time graph for a car** travelling at constant velocity.
It is not accelerating and not decelerating.

What would the graph look like if the car accelerated gently?
Because its velocity would increase, the graph would slope upwards:

In this graph, the car starts off from rest (velocity is zero) and accelerates uniformly (steadily).
A straight graph means uniform, constant acceleration.

What would the graph look like if the car accelerated more rapidly?
Since its velocity would increase more rapidly, the graph would be *steeper*. In fact,

> the *acceleration* is shown by the *slope* (or gradient) of the velocity–time graph.

Page 364 explains how to find the slope of a graph.

Here is a velocity–time graph of a car starting off at one set of traffic lights and stopping at the next set of lights.
Can you see that the car accelerates from A to C, travels at constant velocity between C and D and then decelerates (brakes) rapidly to stop at E?

At which point (A, B or C) is it accelerating most?

Here is a velocity–time graph of a motor-bike.
Between which points is the motor-bike accelerating most rapidly?
Which part of the graph might show the motor-bike changing gear? When is the bike travelling fastest?

When does the rider start to brake? Which part of the graph shows the bike hitting a solid brick wall?

If this was a ticker-tape strip chart you could find the total *distance travelled* by measuring the total amount of tape under the graph. In fact,

> the *distance travelled* is shown by the *area* under the velocity–time graph.

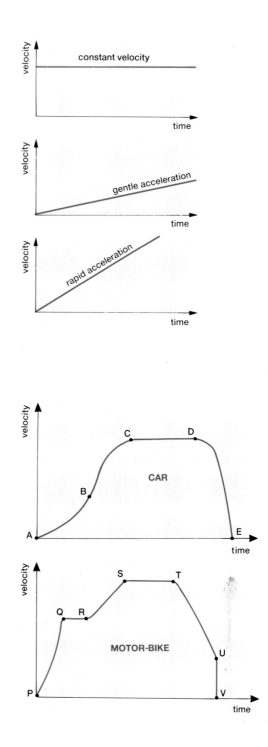

Here is a velocity–time graph of a 'stockcar' starting a race, crashing into another car, and then reversing.

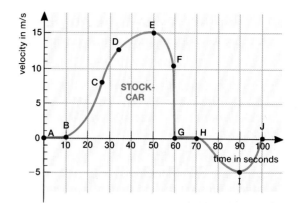

How long do you think the driver had to wait before the starter waved his flag to start the race?

Is his acceleration greatest at B, C, D or E?
At which point is he travelling fastest?
What is his velocity then? How much time has passed?

When does he start to brake?
At what time does he come to a stop after hitting the other car? How long has he been travelling then?

For how long does he stay at rest before reversing?
At which point (G, H, I or J) does he start to reverse his car?
When the car is reversing, it has a **negative** velocity (remember velocity is a **vector** quantity, see p. 102).

What is his maximum velocity in reverse?
For how many seconds does he reverse the car?
When does he finally stop the car?

How do the areas tell you whether the distance travelled is farther in forward gear or reverse?

Here is a velocity–time graph of a lift climbing from the ground floor to the top of a building.

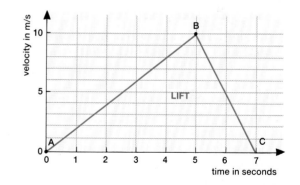

At which point (A, B or C) is it on the ground floor?
At which point (A, B or C) is it travelling fastest?
What is its maximum velocity in m/s?
How long does it take to reach this velocity?

It then decelerates to stop at C, the top of the building. How long did the journey take?

Example 1
We can find its acceleration from the slope:

$$\textbf{Acceleration} = \frac{\textbf{change in velocity}}{\textbf{time taken for the change}}$$

$$= \frac{10\,\text{m/s}}{5\,\text{s}}$$

$$= \underline{2\,\text{m/s}^2}$$

Can you see why the deceleration is 5 m/s²?

Example 2
To find the distance travelled between A and B:

Distance travelled = **area under graph**

$$= \text{area of triangle under AB}$$
$$= \tfrac{1}{2} \times \text{base} \times \text{height}$$
$$= \tfrac{1}{2} \times 5 \times 10$$
$$= \underline{25\,\text{metres}}$$

Do you agree the distance from B to C is 10 m?

▷ Displacement–time graphs

We have seen that velocity is almost the same thing as speed, but it is a vector quantity (see page 102).

In a similar way, *displacement* is almost the same thing as distance, but it is a *vector* quantity. Displacement has a *size* (called distance) *and* a *direction*.

If you move to a different part of the room you might have a displacement of 2 metres *in an easterly direction*. The direction is important.

'The direction is important'

Ticker-timer measurements are sometimes used to draw **displacement–time graphs** (although they are not as useful as velocity–time graphs).
The diagram shows a displacement–time graph for a car which is *at rest* (stationary, not moving):

If the car moves away at constant velocity then the displacement will increase:

If the car moved away at a higher velocity then the line would be steeper. In fact,

> the *velocity* is shown by the *slope* (or gradient) of the displacement–time graph.

If the car accelerates, then the velocity increases, and so the slope of the graph increases:

If the car decelerated, then the velocity would decrease, and so the slope of the graph would decrease. The graph would curve the other way.

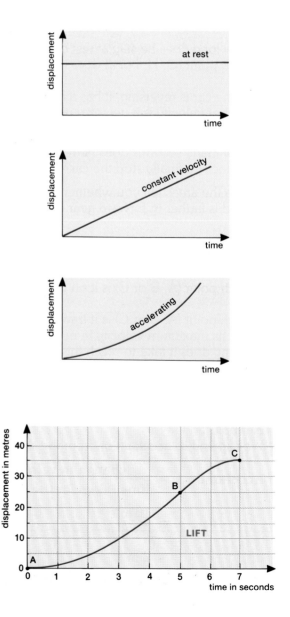

Here is a displacement–time graph for the same lift that is shown on the previous page. It has the same points (A, B, C) marked:

The lift accelerates from A to B, and is travelling fastest at B.

Where is it decelerating?

What is the total distance travelled?
How long does it take the lift to travel 25 m?

▷ Equations of motion

These are 4 equations which can be used whenever an object travels with **constant, uniform acceleration** in a straight line.

We write these equations using 5 symbols: **s, u, v, a, t.**
Suppose an object is travelling at a velocity **u** and then moves with a uniform acceleration **a** for a time **t.**
Its velocity is then **v** and it has travelled a distance **s.**

s = distance travelled (metres)
u = initial velocity (m/s)
v = final velocity (m/s)
a = acceleration (m/s^2)
t = time taken (seconds)

Then from page 132:   **Acceleration** = $\dfrac{\text{change in velocity}}{\text{time taken}}$

or, in symbols: $a = \dfrac{v - u}{t}$

$$\therefore \quad \boxed{v = u + at} \quad \dots (1)$$

Also from page 132:   **Average speed** = $\dfrac{\text{distance travelled}}{\text{time taken}}$

or, in symbols: $\dfrac{u + v}{2} = \dfrac{s}{t}$

$$\therefore \quad \boxed{s = \dfrac{(u + v)}{2}\,t} \quad \dots (2)$$

Using equation (1) to replace v in equation (2): $s = \left[\dfrac{u + (u + at)}{2} \right] t$

$$\therefore \quad \boxed{s = ut + \tfrac{1}{2}at^2} \quad \dots (3)$$

Using equation (1) to replace t in equation (2): $s = \dfrac{(u + v)}{2}\,\dfrac{(v - u)}{a}$

$$\therefore \quad \boxed{v^2 = u^2 + 2as} \quad \dots (4)$$

You can remember the 5 symbols by 'suvat'. If you know any three of 'suvat' the other two can be found.
In questions, 'initially at rest' means $u = 0$. A negative number for the acceleration means the object is slowing down (decelerating).

Example
A cheetah starts from rest, and accelerates at 2 m/s^2 for 10 seconds. Calculate:
a) the final velocity and b) the distance travelled.

First write 'suvat' and show what you know:
$s = ?$
$u = 0$
$v = ?$
$a = 2$ m/s^2
$t = 10$ s

a) Use equation (1): $v = u + at$
   put in numbers: $= 0 + 2 \times 10$
   $= 20$ m/s

b) Use equation (2): $s = \dfrac{(u + v)}{2}\,t$
   put in numbers:
   $= \dfrac{(0 + 20)}{2} \times 10$
   $= 100$ m

A cheetah can accelerate up to a speed of 27 m/s

▷ Acceleration due to gravity

The force of gravity pulls down on all objects here on Earth (see page 81). If objects are allowed to fall, they **accelerate** downwards.

If there is no air resistance or friction then all objects accelerate downwards at the **same** rate. You may have heard of a famous experiment that Galileo is supposed to have done from the leaning tower of Pisa: if a heavy stone and a light stone are dropped together, they accelerate at the same rate and land at the same time.

Accurate measurements show that:
the **acceleration due to gravity** = **9.8 m/s^2.**

For simple calculations we usually use 10 m/s^2.

This means that for an object falling with no air resistance, the velocity after 1 second is 10 m/s, and after 2 seconds the velocity is 20 m/s, and so on.
If a man fell for 5 seconds with no air resistance, what would his speed be?

In practice, there is usually air resistance.
If the girl in the picture falls a long way without a parachute, then because of air friction she reaches a final or **terminal** velocity of about 50 m/s – the speed of a fast racing car.

There was a young man who had heard,
That a person could fly like a bird.
To prove it a lie
He jumped from the sky,
– His grave gives the date it occurred!

A parachute is designed to make the air resistance as large as possible.
With a parachute the terminal velocity is only 8 m/s.

Raindrops, snowflakes and sycamore seeds – they all fall at their own terminal velocity.

At the terminal velocity, the forces on the object are **balanced**:

    force of gravity (weight) = force of air resistance
        (downwards)   (upwards)

This is an example of Newton's First Law (page 85): because there is no resultant force on the object, it continues to move at constant speed in a straight line.

*Experiment 18.3   Measuring the acceleration of free fall, **g***
An electric stop-clock (accurate to $\frac{1}{1000}$ second) is used to
measure the time taken for a small steel ball to fall through a
known distance, **s**.

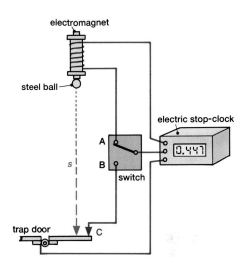

electromagnet

steel ball

electric stop-clock

A

s

B

switch

trap door

C

When the switch is in position A, the electromagnet holds
up the ball:

When the switch is moved quickly to B, the electromagnet
releases the ball and the clock starts timing.
When the ball hits the trap door, the circuit is broken at
contact C, and the clock stops.

To calculate **g**, use: $s = ut + \frac{1}{2}at^2$ (see page 137).
But $u = 0$ and $a = g$, acceleration due to gravity

$\therefore\ s = \frac{1}{2}gt^2$ or $g = \dfrac{2s}{t^2}$

How could this experiment be improved?

Example
A ball is thrown vertically upwards at 20 m/s.
Ignoring air resistance and taking $g = 10$ m/s^2, calculate
a) how high it goes   b) the time taken to reach this height
c) the time taken to return to its starting point.

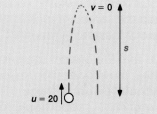

v = 0

s

u = 20

When travelling upwards it is **de**celerating, so $a = -10$ m/s^2.
At the moment when it reaches its highest point, $v = 0$.

Write $s = ?$ a) From p. 137: $v^2 = u^2 + 2as$ b) From p. 137: $v = u + at$
'suvat' $u = 20$ m/s $0 = 20^2 + 2(-10)s$ $0 = 20 + (-10)t$
first: $v = 0$ $20s = 20^2$ $\therefore\ t = \underline{2\ \text{seconds}}$
      $a = -10$ m/s^2 $\therefore\ s = \underline{20\ \text{m}}$
      $t = ?$ c) It also takes 2 seconds to fall down
                                                        again. $\therefore$ total $= \underline{4\ \text{seconds}}$

Vertical *and* horizontal motion

Experiment 18.4
Set up a ruler and two coins as shown.
Press on the ruler and tap the end so that A
falls vertically while B is projected sideways.

Listen to the coins hitting the ground.
Do they hit at the same time?

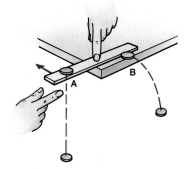

A

B

The vertical accelerations of the coins are
exactly the same – even though one is moving
horizontally as well! (See also page 144.)
In calculations it is better to deal with the
vertical and horizontal movements separately.

> The horizontal and vertical motions of a body
> are independent and can be treated separately.

▷ Force, Mass and Acceleration

When the forces on an object are **un**balanced, there is a resultant force. This resultant force causes the object to change its velocity. It accelerates or it decelerates.

The bigger the force on the javelin, the more it accelerates. The heavier the javelin, the less it accelerates.

Experiment 18.5
Investigating acceleration
You can use a ticker-timer (see page 133) to investigate the acceleration of a trolley:

First, you can keep the mass of the trolley constant, and vary the force using elastic bands as shown: (You must keep the elastic bands stretched the same amount throughout these experiments.)

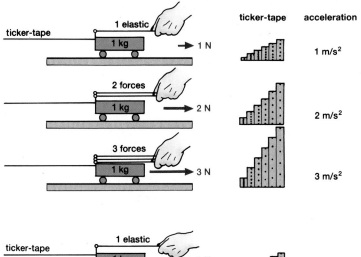

You see that the greater the force, the greater the acceleration. In fact:

acceleration ∝ force

where ∝ means 'proportional to'.

Second, you can keep the force constant (just one elastic band) and vary the mass of the trolley. You see that the greater the mass, the *less* the acceleration. In fact:

$$\text{acceleration} \propto \frac{1}{\text{mass}}$$

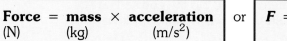

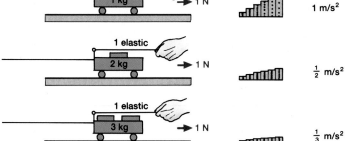

Combining these two results:

$$\text{acceleration} \propto \frac{\text{force}}{\text{mass}}$$

or, **force ∝ mass × acceleration.**

Provided we measure the force in newtons, the formula becomes:

$$\begin{array}{ccc} \textbf{Force} & = & \textbf{mass} \times \textbf{acceleration} \\ \text{(N)} & & \text{(kg)} \qquad \text{(m/s}^2) \end{array} \quad \text{or} \quad \boxed{\textbf{\textit{F} = \textit{ma}}}$$

This is **Newton's Second Law of Motion**.
Remember: • *F* is the resultant (or unbalanced) force on the mass *m*,
       • the force *F* must be measured in newtons.

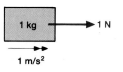

One newton (1 N) is defined as the force which gives to a mass of 1 kg, an acceleration of 1 m/s².

Example

When a force of 6 N is applied to a block of mass 2 kg, it moves along a table at constant velocity.

a) What is the force of friction?

When the force is increased to 10 N, what is:

b) the resultant force?

c) the acceleration?

d) the velocity, if it accelerates from rest for 10 seconds?

a) When the block is moving at constant velocity (no acceleration), there is no resultant force on it (Newton's First Law, page 85). This means the forces are balanced and equal.
   $\therefore$ The frictional force $= 6$ N

b) When the applied force $= 10$ N, the frictional force is still 6 N.
   $\therefore$ The resultant (unbalanced) force $= (10 - 6)$ N $= \underline{4\ \text{N}}$

c) Formula first: Force $=$ mass $\times$ acceleration
   Then numbers: 4 N $= 2$ kg $\times$ acceleration
   $$\therefore \text{acceleration} = \frac{4}{2} = 2\ \text{m/s}^2$$

d) Formula first: $v = u + at$ (see page 137)
   Then numbers: $= 0 + 2 \times 10$
        velocity $= 20$ m/s

If you need to find the distance travelled from rest in 10 seconds, which equation of motion would you use next? (See page 137.)

Gravitational field strength *g*

The Earth is surrounded by an invisible *gravitational field*. This field exerts a force (called weight) on any mass which is in the field.

On the surface of the Earth, experiments show that this force is **9.8 N on every 1 kg** (see page 83). This is the **gravitational field strength, *g* $= 9.8$ N/kg.**

We now have 2 ways of thinking of *g*:

1. For an object in free fall, the acceleration *g* $= 9.8$ m/s^2
2. For any object, moving or at rest, the gravitational field strength *g* $= 9.8$ N/kg.

For *m* kg the force is *m* times as much,

$$\boxed{\therefore \textbf{Weight } = \textit{m} \times \textit{g} \textbf{ newtons}}$$

We usually take *g* $= 10$ m/s^2 $= 10$ N/kg (here on Earth).

On the Moon, *g* is only 1.6 N/kg. On Jupiter, *g* $= 26$ N/kg!

If your mass is 50 kg, what is your weight on Earth, and what would it be on the Moon, and on Jupiter?

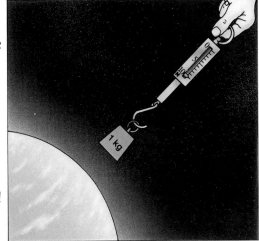

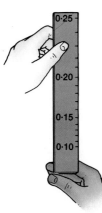

▷ Something to make – a reaction timer

You can test a friend's reaction time by holding the top of a 30 cm (1 foot) ruler so that it hangs vertically with your friend's fingers close to (but not touching) the zero mark at the bottom of the ruler. When you let go (with no warning), your friend has to grip the ruler as quickly as possible.

The slower your friend's reaction time, the farther the ruler will fall as it accelerates downwards (at 9.8 m/s^2).

The table below shows how to mark the ruler (or a strip of card). It also shows how far you would travel in a car at 50 km/h (30 m.p.h.) (that is, how far you would travel **before** your foot could reach the brake pedal to **start** to slow down).

At this distance (in cm) from the zero mark:	4.9	5.9	7.1	8.3	9.6	11.0	12.5	14.2	15.9	17.7	19.6	21.6	23.7	25.9	28.2	30.6
Mark this time (in seconds) on your ruler:	0.10	0.11	0.12	0.13	0.14	0.15	0.16	0.17	0.18	0.19	0.20	0.21	0.22	0.23	0.24	0.25
Distance travelled (in metres) at 50 km/h (30 m.p.h.)	1.4	1.5	1.7	1.8	1.9	2.1	2.2	2.4	2.5	2.6	2.8	2.9	3.1	3.2	3.3	3.5

In a real driving situation, most people's reaction time is over **three** times longer than this! (See also page 119.)

Summary

$$\text{Average speed (m/s)} = \frac{\text{distance travelled (metres)}}{\text{time taken (seconds)}}$$

$$\text{Acceleration} = \frac{\text{change in velocity (m/s)}}{\text{time taken for change (s)}}$$

Velocity (a vector quantity) is speed in a particular direction.

Acceleration is shown by the **slope** of a velocity–time graph.
Distance is shown by the **area** under a velocity–time graph.

For constant acceleration: $v = u + at$ $\qquad s = \frac{(u + v)}{2} t$

$$s = ut + \tfrac{1}{2}at^2 \qquad v^2 = u^2 + 2as$$

The acceleration of free fall (due to gravity) = 9.8 m/s^2 (10 m/s^2)
The horizontal and vertical motions of a body are independent.

Newton's Second Law: **Force = mass × acceleration**
(N)   (kg)   (m/s^2)

1 N = force which gives to a mass of 1 kg an acceleration of 1 m/s^2.
Weight = **mg** newtons.

▷ Questions

(Take $g = 10 \text{ m/s}^2 = 10 \text{ N/kg}$ where necessary)

1. a) A car travels 100 m in 5 seconds.
   What is its average speed?
   b) A car accelerates from 5 m/s to 25 m/s in 10 seconds. What is its acceleration?

2. Sketch a speed–time graph of you walking to a bus-stop, catching a bus, getting off it, and walking to school.

3. The table gives some data for a Ferrari racing car at the start of a Grand Prix race:

Speed (m/s)	0	10	20	29	37	50	59	64	65	65
Time (s)	0	1	2	3	4	6	8	10	12	14

   a) Plot a speed–time graph for the car.
   b) What was its acceleration at:
      i) 13 seconds? ii) 1 second?

4. Describe in as much detail as possible, the motion of a car which has this graph:

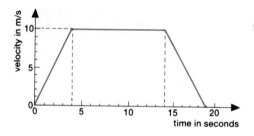

5. The Highway Code says that at 20 m/s, the reacting (thinking) distance of a driver is 14 m, and the brakes take 3 seconds to halt the car.
   a) What is the reaction time in this case?
   b) Plot a velocity–time graph for these data (assume the brakes are on steadily for 3 s).
   c) Use the graph to estimate the speed at which the driver would hit a wall 3 seconds after first noticing it.

6. Sketch a displacement–time graph for the car in question 4.

7. A sports car accelerates from rest at 4 m/s² for 10 seconds. Calculate a) the final velocity [use equation (1)] and b) the distance travelled.

8. Explain why parachutists and snowflakes do *not* fall with a constantly accelerating motion.

9. A sky-diver weighing 500 N falls through the air at a steady speed of 50 m/s.
   a) Draw a diagram of the forces on her.
   b) What is the air resistance on her (in N)?
   c) What is her mass?
   d) Sketch a velocity–time graph to show what happens when her parachute opens.

10. The Eiffel Tower is 300 m high. A boy at the top slips on a banana skin and falls over the side. How long has he got to live?

11. A ball is thrown upwards with a velocity of 10 m/s. a) How high does it go? b) How long is it in the air?

12. A film stunt man dives off a cliff 20 m high, so that initially his *vertical* velocity is zero but his horizontal velocity is 5 m/s. Find a) his vertical velocity as he hits the water b) the time taken and c) the horizontal distance travelled during this time.

13. A car travelling within the town speed limit, at 13 m/s, hits a brick wall. For a passenger without a seat belt, this is like falling from the top of a house of height *h* metres. Find the value of *h*. (Houses are about 7 m high.)

14. A mass of 4 kg is accelerated by a force of 20 N. What is the acceleration if there is no friction? If a frictional force of 8 N is acting, what is the acceleration?

15. A Saturn V Moon rocket has a mass at lift-off of 3.0×10^6 kg. The thrust at lift-off is 3.3×10^7 N. Find
    a) the weight of the rocket on Earth
    b) the resultant (unbalanced) force at lift-off
    c) the acceleration at lift-off
    d) the apparent weight of the rocket in orbit.

16. An astronaut has a mass of 100 kg. What is his weight a) on Earth b) on the Moon where the gravitational field strength = 1.6 N/kg?

Further questions on page 154.

▷ Physics at work: Sport

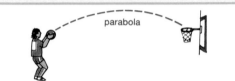

In many games, a ball is thrown or hit. The ball does not travel in a straight line, but in a **parabola** shape. (This is because gravity is pulling it downwards.)
To be more accurate, it is the centre of gravity of the ball that follows the parabola.

The centre of gravity (centre of mass) of a person is usually behind the navel, but it depends on the positions of the arms and legs.
When an acrobat or a diver jumps through the air, the centre of gravity moves in a smooth parabola, no matter how the person twists or spins.

The gymnast must keep her centre of mass directly over her hands, in order to balance.

In judo it's an advantage to have a low centre of mass, and stand with your legs apart, for stability.

Skiing is another sport where a low centre of mass is helpful.

Can you think of any others?

The long jump

Sprinting: The athlete accelerates because of the force F (the friction force, see also page 98).
His weight W is balanced by the reaction force R.
You can measure the acceleration of a runner by using ticker-tape (see page 133).

The take-off:
A final push to make R and F as large as possible.

The landing:
Now the forces F and R bring him to a stop.

His centre of gravity follows a parabola between take-off and landing.
The world record is about 9 m.

Using friction to turn corners

In many sports you have to turn corners.
This can only happen if there is an inwards force F provided by friction. This is a **centripetal force**.

The faster you go, and the sharper the bend, the bigger the force F needs to be (and the bigger the angle you lean over).
What happens if the friction F is not big enough (for example, if the cyclist meets ice or gravel)?

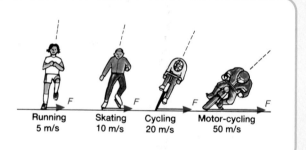

| Running 5 m/s | Skating 10 m/s | Cycling 20 m/s | Motor-cycling 50 m/s |

144

▷ Physics at work: Sport

Here is a velocity–time graph for a runner in the Olympic 100-metre sprint:

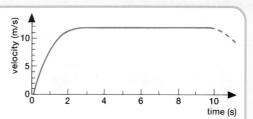

He accelerates to almost 12 m/s in about 2 seconds, and then travels at almost constant speed.

If his mass is 60 kg, then:

his kinetic energy $= \frac{1}{2} \times$ mass $\times$ speed squared (p. 119)
$$= \tfrac{1}{2} \times 60 \times 12 \times 12 = \underline{\text{about 4000 J}}$$

Power developed in his legs at start $= \dfrac{\text{energy converted}}{\text{time taken}} = \dfrac{4000\,\text{J}}{2\,\text{s}} = \underline{2000\,\text{W}}$

In fact, to produce this much power in his legs, his whole body produces at least twice as much – more than 4 kW.
This is equivalent to four one-bar electric fires – he gets hot!

He loses the surplus energy by radiation from his skin (20%), convection (20%), evaporation of sweat from his skin (40%), and respiration (breathing out hot air, 20%).

What was his initial acceleration? Why did he stop accelerating?

Analysing movement

One way to analyse an athlete's movement is to take a video or cine film (see page 203), and then play it back slowly. Another way is to use flashing lights, like the 'strobe' lamps in a disco:

A camera is used in the dark, with a flashing lamp called a **stroboscope**. In the photo, it was flashing 10 times in each second. How long was the shutter open? Where do his feet move fastest?

A large mass can be an advantage.

The Sumo wrestler needs a large mass and a low centre of mass so that he is not easily toppled.

When the shot-putter throws the shot, it is rather like firing a gun (page 101). The heavier she is, the less recoil, and the farther it goes.

There are examples of physics in other sports: car-racing (page 110), snooker (p. 147), shooting (p. 148), cycling (p. 130), football (p. 146).

chapter 19

MOMENTUM

The *momentum* of an object depends on its *mass* and its *velocity*.
In fact:

Momentum = mass × velocity	or	**Momentum = *mv***
(kg) (m/s)		

Momentum is a *vector* quantity. It is measured in units of **kg m/s**.
A bicycle of mass 10 kg moving at 5 m/s has a momentum of 50 kg m/s.

Consider a force *F* acting on a mass *m* for a time *t* so that it
accelerates from velocity *u* to velocity *v*.
The quantity (force × time) is called *impulse*. **Impulse = *F* × *t*.**

From page 132, acceleration $a = \dfrac{v - u}{t}$

∴ Newton's Second Law (page 140) is: $F = ma = m\left(\dfrac{v - u}{t}\right) = \dfrac{mv - mu}{t}$

or in words: **Force = $\dfrac{\textbf{change in momentum}}{\textbf{time taken}}$**

Multiplying both sides by time, we get:

Impulse = Force × time = change in momentum	or	***Ft = mv − mu***
(N) (s) (kg m/s)		

Example 1
Consider first a boy kicking a **stone** of mass 1 kg and accelerating it
from rest to 10 m/s. Because the stone is rigid, the force of his foot
acts for only $\frac{1}{100}$ second. Calculate this force.

Formula first: Force × time = *mv − mu*

Then numbers: Force × $\frac{1}{100}$ = (1 × 10) − (1 × 0)

Force × $\frac{1}{100}$ = 10

∴ Force = 1000 N   (painful!)

Example 2
Now consider the same boy kicking a **football** of the same mass
(1 kg) to give it the same speed (10 m/s) and therefore the same
momentum (10 kg m/s).
Because the ball is soft and he follows through with his foot, this
force is applied for $\frac{1}{10}$ second (10 times longer than before).

Using the same formula with time = $\frac{1}{10}$, gives force = 100 N
Which kick hurts less? Why?

The *longer* the time of a collision, the *smaller* the force.

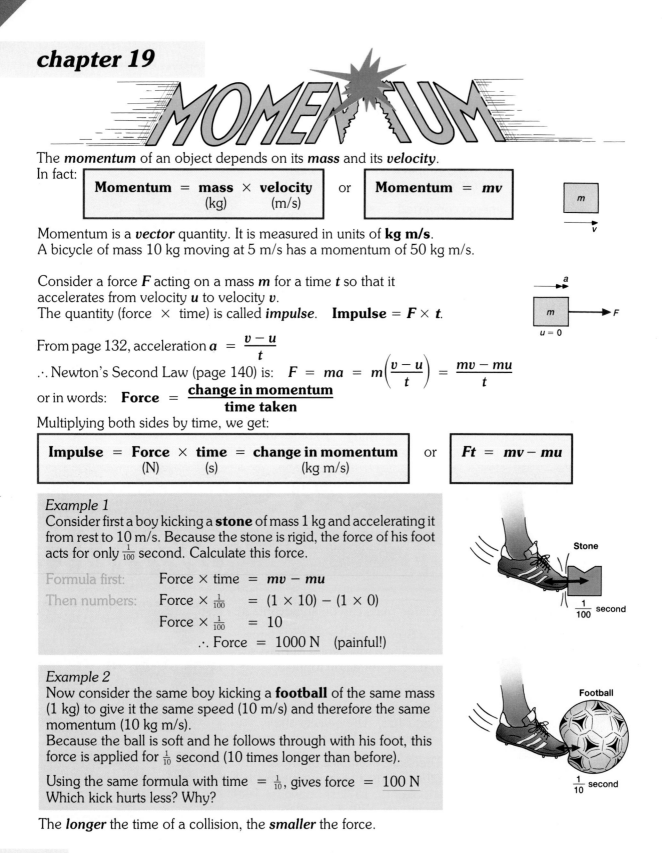

Stone

$\frac{1}{100}$ second

Football

$\frac{1}{10}$ second

▷ Collisions

Here are two balls rolling towards each other so that they collide:

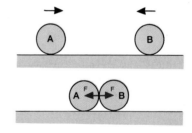

When they collide they exert a force **F** on each other for a short time **t** and so the momentum of each ball changes.
From Newton's Third Law (page 100), the forces are equal and opposite. Therefore (from the equation on the opposite page) the changes in momentum are equal and opposite.

That is, the momentum *gained* by one ball is equal to the momentum *lost* by the other ball.

This is the **Principle of Conservation of Momentum**:
When two or more bodies act on each other, their total momentum remains constant, providing there is no external force acting.

That is:

total momentum before collision	=	total momentum after collision

The same is true for an explosion (as an explosion is the opposite of a collision).

Example 3
A bullet, mass 10 g (0.01 kg), is fired into a block of wood, mass 390 g (0.39 kg), lying on a smooth surface. The wood then moves at a velocity of 10 m/s. a) What was the velocity of the bullet?
b) What is the kinetic energy before and after the collision?

a) Formula first: **total momentum before collision** = **total momentum after collision**

$$(\text{mass} \times \text{velocity})_{before} = (\text{mass} \times \text{velocity})_{after}$$

Then numbers:
$$0.01 \times V = (0.01 + 0.39) \times 10$$
$$0.01 \times V = 0.4 \times 10$$
$$V = \underline{400 \text{ m/s}}$$

b) Total KE *before* collision $= \frac{1}{2} \times \text{mass} \times \text{speed}^2$ (page 119)
$$= \frac{1}{2} \times 0.01 \times 400^2 = \underline{800 \text{ J}}$$

Total KE *after* collision $= \frac{1}{2} \times \text{mass} \times \text{speed}^2$
$$= \frac{1}{2} \times 0.4 \times 10^2 = \underline{20 \text{ J}}$$

The difference, 780 J, is converted to heat energy and sound energy during the collision.

Note: **momentum is *conserved* in a collision, but kinetic energy is not.**

More examples of momentum in collisions

In any collision, a force is exerted for a length of time. The examples on page 146 showed that the *longer the time* of a collision, the *smaller the force* exerted.

This idea is used to design safety into **cars.** For example, the front and back of a car is designed to crumple, in order to spread out the time of a collision, and so reduce the force on you. (See the photo on page 85.)

A **seat-belt** is designed to stretch slightly, to spread out the time of the crash even further, and so reduce the force on you to a safe level: A motor-cyclist's **safety helmet** is padded inside so as to extend the time of any collision.

In an opposite way, a **hammer** is designed from hard metal. This is so that the time of the collision is as small as possible, and so the force of the hammer on a nail is as large as possible.

Explosions

An explosion is the opposite of a collision – objects move apart instead of coming together. A **rocket** uses a controlled explosion. The rocket moves one way while the hot gases move the opposite way. (See also page 101.) The gain of momentum of the rocket is equal and opposite to the momentum of the hot gases that are ejected:

A **gun** firing a bullet is like a rocket ejecting fuel. The bullet gains momentum in one direction, while the gun recoils with momentum in the opposite direction.

Example 4
A bullet of mass 10 g (0.01 kg) is fired at 400 m/s from a rifle of mass 4 kg. What is v, the recoil velocity of the rifle? (See also page 101.)

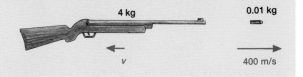

Remembering that momentum is a vector, let the positive direction be to the right.

Formula first:
$$\frac{\text{total momentum}}{\text{before explosion}} = \frac{\text{total momentum}}{\text{after explosion}}$$

Then numbers:
$$0 = (0.01 \times 400) - (4 \times v)$$
$$\therefore 4v = 0.01 \times 400$$
$$\therefore v = \underline{1 \text{ m/s}}$$

Summary

Momentum = mass × velocity. It is a vector quantity.
  (kg m/s)   (kg)   (m/s)

Newton's Second Law: $\text{Force} = \dfrac{\text{change in momentum}}{\text{time taken}} = $ **rate of change of momentum**

Impulse = Force × time = change in momentum $= mv - mu$
(Compare: Force × *distance* = change in KE $= \frac{1}{2}mv^2 - \frac{1}{2}mu^2$ from page 119)

Principle of Conservation of Momentum: **total momentum before** $=$ **total momentum after**
(if no external force is acting)   **collision (or explosion)**   **collision (or explosion)**

▷ **Questions**

1. Copy and complete:
   a) Momentum = ... × . . . . .
   b) It is a . . . . quantity.
   c) Its unit is . . . . .
   d) Newton's Second Law can be written as
   Force = mass × . . . . or
   Force = ... of change of . . . . .
   e) Impulse = ... × . . . . .
   f) Impulse = change in . . . . .
   g) The Principle of Conservation of Momentum states:
   h) . . . . is conserved in any collision but kinetic . . . . is not.

2. a) Explain the physics of this limerick:

   A dashing young footballer, Paul,
   Scored a goal with a one-kilo ball,
   But a similar kick
   To a one-kilo brick
   Made Paul bawl, and then fall, and then crawl.

   b) When you jump down from a table, why do you bend your legs rather than keep them rigid? Why do paratroopers roll on landing? Why would you prefer to fall on to a bed than on to concrete?
   c) Why are seat-belts designed to stretch in a collision? Why is the front of a car designed to collapse in a serious collision?
   d) Why does a cricket batsman, a tennis player or a golf player 'follow through'?
   e) Professor Messer invents a foam-rubber hammer ('to make less noise'). Will it work? Why?

3. A truck of mass 2 kg travels at 8 m/s towards a stationary truck of mass 6 kg. After colliding, the trucks link and move off together.
   a) What is their common velocity?
   b) What is the KE before and after the collision?
   c) Explain the apparent loss in energy.

4. In a sea battle, a cannonball of mass 30 kg was fired at 200 m/s from a cannon of mass 3000 kg. What was the recoil velocity of the gun?

5. A heavy car A, of mass 2000 kg, travelling at 10 m/s, has a head-on collision with a sports car B, of mass 500 kg. If both cars stop dead on colliding, what was the velocity of B?

6. A man wearing a bullet-proof vest stands still on roller skates. The total mass is 80 kg. A bullet of mass 20 grams is fired at 400 m/s. It is stopped by the vest and falls to the ground. What is then the velocity of the man? How does this compare with what you see in TV films?

7. A Saturn V Moon rocket burns fuel at the rate of 13 000 kg in each second. The exhaust gases rush out at 2500 m/s.
   a) What is the change in momentum of the fuel in each second?
   b) What is the thrust? (See also question 15, page 143.)

Further questions on page 155.

Further questions on mechanics

▷ Hooke's Law

1. Some students carry out an experiment to find out how a spring stretched when small loads were added to it.
   a) Draw a labelled diagram to show what is meant by the extension of the spring. [2]

The results of the experiment are shown in the table. One of the readings is incorrect.

Load (N)	0	2	4	6	8	10	12	14
Extension (mm)	0	16	32	58	64	80	96	112

   b) Use these results to plot a graph. [5 marks]
   c) Use your graph to find:
      i) the extension when the load is 3 N
      ii) the load which produces an extension of 40 mm. [4]
   d) Label the incorrect point on the graph with the letter 'E'. [1]
   (NEA)

2. Theory suggests that, when a beam is loaded at the centre, the deflection x is directly proportional to the cube of L, the distance between the supports, provided that the same force W is applied for each value of L.

In other words, $x = kL^3$, where k is a constant.

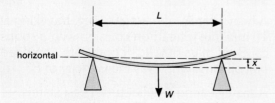

You are required to carry out an experiment to confirm that $x = kL^3$, using a metre rule as a beam. From initial checks you know that when a metre rule rests on edge supports 0.90 m apart, its centre is deflected by about 1 cm when a load of 8 N is hung from the centre. Describe:
   a) how you would set up the apparatus, [6]
   b) how you would make your measurements, [4]
   c) the number and range of measurements you would make, [5]
   d) how you would use your measurements to test the theory. [5] (SEG)

3. a) Explain briefly the meaning of each of the following terms: i) deformation ii) compression iii) tension. [6 marks]
   b) A perfectly straight plank was used to make a simple bridge over a stream as shown.

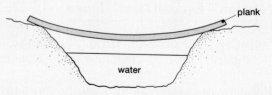

      i) Why does the plank sag slightly in the centre even before anyone uses the bridge?
      ii) When the plank is sagging, one part of it is longer and another part shorter than it is when the plank is straight. Copy the diagram and mark clearly that part of the plank:
         I which is longer,   II which is shorter,
         III which is under compression,
         IV which is under tension. [10]
   c) In an experiment a load is placed at the centre of the plank and the distance x that it is depressed at its centre is measured:

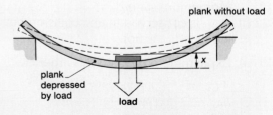

The table below shows some readings:

Load in newtons	Depression (caused by the load) in mm
200	10
400	20
600	30

      i) Show that these readings are consistent with the conclusion that the depression x is directly proportional to the **load** causing it.
      ii) Calculate the depression produced when a girl weighing 500 N, carrying a rucksack weighing 50 N, stands at the centre of the plank. [14] (NI)

150

▷ Pressure

4. a) Why is pressure more important than force when considering the damage which stiletto heels might cause to a floor? [2]

b) A girl has a mass of 50 kg. What is her weight? [1]

c) If she wears stiletto heels and the area of one of the heels is 0.25 cm², what pressure does she exert on the floor if she puts all her weight on one heel? [3]

d) An elephant has a mass of 2000 kg and each of its feet has an area of 500 cm². What pressure does it exert on the ground when standing on all four feet? [2]

e) How many similar elephants standing on top of this elephant would exert the same pressure on the ground as the girl? [1]
(MEG)

5. A rectangular block measures 8 cm by 5 cm by 4 cm, and has a mass of 1.2 kg.

a) i) If the gravitational field strength is 10 N/kg, what is the weight of the block?

ii) What is the area of the smallest face?

iii) What pressure (in N/cm²) will it exert if it is resting on a table on its smallest face?

iv) What is the least pressure the block could exert on the table?

b) i) What is the volume of the block?

ii) Calculate the density of the material from which the block is made. (NEA)

6. The diagram shows side and front views of a car tyre in contact with the road.

Side view of tyre   Front view

A tyre company stated that the area of the tyre in contact with the road was about the same as the area of the sole of one of your shoes.

a) Describe (with a diagram) how you would estimate the area of sole of your shoe. [3]

b) The car weighs 12 000 N. What is the force acting on one tyre if the weight of the car is evenly distributed amongst the tyres? [1]

c) If the area of contact of the tyre is 80 cm² (0.008 m²), calculate the pressure of the air in the tyre. [2] (NEA)

▷ Density

7. A piece of wood has a density of 0.6 g/cm³ and a volume of 25 cm³. The mass of the block in grams is:

A 1.5 **B** 15 **C** 41.6
D 150 **E** 1500 (NI)

8. You are given a bottle of milk and asked to find the density of the milk. How would you measure

a) the mass of the milk?

b) the volume of the milk?

c) Write down one likely source of error in your measurements. (NEA)

9. a) Here are some pieces of apparatus that a student could use to measure the volume of a stone.

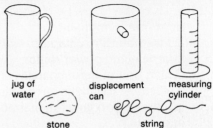

jug of water   displacement can   measuring cylinder
stone   string

Complete a list of instructions for this experiment. (The first one is: Tie the string around the stone.) [7 marks]

b) i) Why would this method **not** work for a lump of wood? [1]

ii) How could you find the **volume** of a lump of wood? [3]

c) You are given a rectangular block of wood. Describe, carefully, how you would measure the **length** of one of the sides using a metre rule. [5] (SEG)

10. A certain liquid (1,1,2,2-tetrabromoethane) has a density of 3.0 g/cm³. Diamonds have a density of 3.5 g/cm³. Sand has a density of 2.6 g/cm³.

a) Explain how diamonds could be separated from sand using this liquid. [2]

b) At the Mwadui diamond mine in Tanzania, diamonds are separated from sand by using a special mud. Complete the following sentence: The density of the mud must be less than . . g/cm³ but more than . . g/cm³. [1] (LEAG)

Further questions on mechanics

▷ **Moments and Machines**

11. A person pushes a door open by applying a force of 10 N at a point 0.8 m from the hinge of the door. The turning effect or moment of this force, in N m, is:
    A 0.08   **B** 8.0   **C** 10.8
    D 12.5   **E** 80.0   (NI)

12. A uniform metre rule of mass 100 g balances at the 40-cm mark when a mass X is placed at the 10-cm mark. What is the value of X?
    A 33.3 g   **B** 133.3 g   **C** 300 g
    D 400 g   **E** 1000 g   (WAEC)

13. a) What type of energy is converted to heat when: i) coal is burnt, ii) a moving object is stopped, iii) a current passes through a resistor? [3]
    b) Each of the following changes energy from one form to another.
    Bunsen burner   catapult   electric motor hydroelectric power station   weightlifter
    Which one of these changes energy from:
    i) electrical to kinetic   ii) elastic to kinetic
    iii) chemical to gravitational potential energy? [3]
    c) A pulley system can be used to lift a load of 400 N using a smaller force, called the effort, of 125 N as shown:

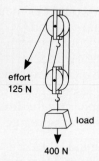

effort
125 N

load

400 N

The end of the rope has to be pulled through 2 m to lift the load vertically upwards by 0.5 m.
    i) How much work is done on the load in lifting it 0.5 m? [3]
    ii) How much work is done by the effort in pulling 2 m of rope through the pulley system? [3]
    iii) Calculate the efficiency of the system.
    iv) Suggest two reasons why the pulley system is not 100% efficient. [2] (NEA)

▷ **Energy**

14. Which list of energy sources, **A** to **D**, contains only non-renewable resources?
    A   wind, wave, nuclear
    B   hydroelectric, gas, oil
    C   coal, oil, geothermal
    D   gas, oil, nuclear   (NEA)

15. a) When a simple pendulum is oscillating, energy is changing from the form of . . . . energy to the form of . . . . energy and then back into the form of . . . . energy.
    b) In a gas fire, ... energy is converted into energy in the form of . . . . and . . . . .
    c) In a dynamo, ... energy is converted into . . . . energy.
    d) When a spring-driven toy motor-car is running, ... energy is converted into . . . . energy.
    e) In an explosion, ... energy is converted into kinetic energy and also into . . . . and . . . . .
    f) In order to propel an electric train, ... energy is converted into . . . . energy.
    g) In a battery, energy is stored in the form of . . . . energy ready to be released in the form of . . . . energy.
    h) When a catapult is used, ... energy is converted into . . . . energy.

16. The table lists three energy sources. The cost of each unit in which a source is bought and the number of megajoules of energy obtained from each unit are shown for each source:

Energy source	Cost of one unit	Megajoules per unit
Coal	£130	30 000
Gas	39p	105
Oil	20p	37

    a) Calculate how much energy in megajoules can be bought for £1 using
    i) coal   ii) gas   iii) oil. [6]
    b) Which energy source is the cheapest? [1]
    c) At current prices, about 70 MJ of electrical energy can be bought for £1. Suggest 2 reasons why electrical energy is more expensive than energy obtained directly from the sources above. [2] (NEA)

▷ Energy and power

17. To drag a 25 kg packing case across a floor needs a horizontal force of 100 N. How much work is done in moving it 2 metres?

A 50 J   **B** 100 J
C 200 J   **D** 500 J   (SEG)

18. a) State the Principle of Conservation of Energy.
b) Only a small amount of the chemical energy stored in the petrol of a car engine is changed into kinetic energy. Into which two other forms of energy is some of the chemical energy in the fuel *always* changed?
c) A motor-car engine uses fuel energy at a rate of 105 kW and converts it into useful power at a rate of 21 kW. Calculate the efficiency of the car.
d) Suggest one way in which friction is a nuisance in a motor car and say how its effect can be reduced.   (NI)

19. A worker on a building site raises a bucket full of cement at a slow steady speed, using a pulley as shown:

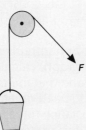

The weight of the bucket and cement is 200 N. The force **F** exerted by the worker is 210 N.
a) Why is **F** bigger than the weight of the bucket and cement?
The bucket is raised through a height of 4 m.
b) Through what distance does the worker pull the rope?
c) How much work is done on the bucket and cement?
d) What kind of energy is gained by the bucket?
e) How much work is done by the worker?
f) Where does the energy used by the worker come from?   (SEG)

20. The output power of a small car engine is 2000 W.
a) What does this statement mean?   [1]
b) How much work does the engine do in 30 seconds?   [2]
c) If the efficiency of the engine is 25%, how much energy is supplied to the engine in 30 seconds?   [2]   (SEG)

21. A motor is used on a building site to raise a block of stone. The weight of the block is 720 N and it is raised 20 m in 24 s. Calculate
a) the work done on the block,
b) the useful power supplied by the motor.

22. A boy of mass 50 kg races up a flight of 40 steps, each of height 15 cm, in 5 s. Calculate:
a) the work done   b) his average power.
c) Why is the energy he uses up greater than the calculated work done?   (WJEC)

23. A car is travelling at a steady speed of 15 m/s.
a) Calculate the distance moved in 10 s.   [4]
b) The total resisting force (friction and air resistance) is 800 N.
Calculate the work done by the car in 10 s in overcoming this resisting force.   [4]
c) Calculate the power developed at the driving wheels.   [4]   (NEA)

24. A car, moving along a level road, had a kinetic energy of 200 kJ. The car was put out of gear and the brakes were applied at the same time. The car stopped after travelling 100 m.
a) What was the kinetic energy of the car when it stopped? How much work was done to bring the car to rest?
b) What was the steady braking force which was applied to the car?
c) Total time needed to stop the car is 10 s. Calculate the average power at which the braking energy is converted.   (WJEC)

25. An electric pump whose efficiency is 60% raises water to a height of 15 m. If water is delivered at the rate of 360 kg per minute, what is the power rating of the pump? What is the energy lost by the pump?
(WAEC)

▷ Velocity and acceleration

26.

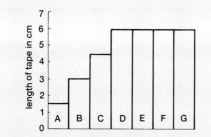

A tape attached to a trolley moving along a runway passes beneath a vibrator making 50 dots on the tape per second. The tape is then cut into successive lengths containing 10 spaces between the dots on each tape length. These are shown in the diagram.
a) What time does each tape length represent?
b) Calculate the average speed for length A.
c) Calculate the average speed for length B.
d) What is happening between A and D?
e) Describe briefly, with the aid of a diagram, how you would set up the trolley and runway apparatus to produce the type of motion shown by tapes A to D.
f) What is happening between D and G?
g) What is the total force acting on the trolley between D and G?
h) What is the total time from the start until the trolley reaches its maximum speed?

27. The velocity, v, of a car varies with time, t, according to the following table:

t (s)	0	5	10	15	20	25	30	35	40	45	50
v (m/s)	0	5	10	15	15	15	15	11	7.5	3.5	0

a) Plot a graph of v against t.
b) Describe the motion of the car during each of the following periods of time:
0–15 s   15–30 s   30–50 s
c) Calculate the acceleration of the car during the period 0–15 s.
d) How far did the car travel in the following periods of time?
0–15 s   15–30 s   0–30 s
e) State what forces would be acting on the car when it is travelling at constant velocity. What is the resultant of these forces equal to?

28. A man runs a race against a dog. Here is a graph showing how they moved:

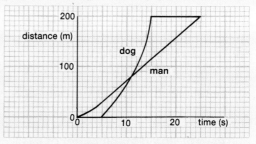

a) What was the distance for the race?
b) After how many seconds did the dog overtake the man?
c) How far from the start did the dog overtake the man?
d) What was the dog's time for the race?
e) Use the equation $v = \frac{d}{t}$ to calculate the average speed of the man.
f) After 8 seconds, is the speed of the man increasing, decreasing, staying the same?
g) What is the dog's speed after 18 s?   (NEA)

29. An underground train takes 1 minute to travel from one station to the next. The train accelerates from rest to a speed of 25 m/s in 20 s, travels at this constant speed for 30 s before coming to rest under a uniform braking force.
a) Draw a graph to represent the motion.
b) From the graph, calculate:
   i) The time that the train is travelling with a speed greater than 20 m/s,
   ii) the distance travelled by the train when it is moving at constant speed,
   iii) the retardation of the train,
   iv) the braking force applied to stop the train, given that the mass of the train is 100 000 kg.   [6]
c) How can you tell from the graph that the train travelled a greater distance in the last 30 s than in the first 30 s?   [1]   (WJEC)

30. A mass of 5 kg changes its velocity from 4 m/s to 24 m/s in 10 s. Calculate:
a) the acceleration,
b) the force needed for this acceleration,
c) the momentum at the start,
d) the kinetic energy at the start,
e) the kinetic energy after 2 s.

Force, acceleration and momentum

31. The table below is based on information from the Highway Code. The figures apply to a saloon car in good driving conditions.

Speed	Thinking distance	Braking distance
10 m/s	7 m	8 m
20 m/s	14 m	30 m
30 m/s	21 m	68 m

a) The thinking distances depend on the driver's reactions.
   i) How does the thinking distance depend on the car's speed? [2]
   ii) How much **thinking time** is the driver allowed according to the Highway Code? [2]
b) It takes the driver time to react (i.e. to register the need to brake) **and** apply the brakes. Once the brakes are applied it still takes the braking distance to stop the car.
   i) According to the table, what is the **total stopping distance** at 30 m/s? [1]
   ii) Two cars, one following the other, are both travelling at 30 m/s. Suggest, with reasons, a **minimum** safe distance separating the cars. [2]
   iii) Estimate a maximum safe driving speed if fog has reduced visibility to 40 m. Explain your reasoning. [2]
c) A car of mass 600 kg is travelling at 20 m/s.
   i) Calculate the kinetic energy of the car. [2]
   ii) What happens to this energy when the brakes are applied? [1]
   iii) What is the braking distance at this speed?
       Calculate the average braking force in bringing the car to rest from 20 m/s. [3]
d) If the speed of the car is doubled, the braking distance increases approximately four times. By energy considerations or otherwise explain this change. [2]
   (LEAG)

32.

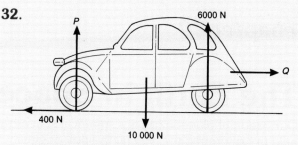

A front-wheel drive car is travelling at **constant velocity**. The forces acting on the car are shown in the diagram above. **Q** is the force of the air on the moving car. **P** is the total upward force on both front wheels.
a) Explain why i) **P** = 4000 N,
   ii) **Q** = 400 N. [2]
b) Calculate the mass of the car. [1]
c) The 400 N driving force to the left is suddenly doubled.
   i) Calculate the resultant force driving the car forward.
   ii) Calculate the acceleration of the car.
   iii) Draw a sketch graph showing how the velocity of the car changes with time. (Start your graph just before the driving force is doubled.) [5]
d) i) Passengers in a car are advised to wear a safety belt. Explain, in terms of Newton's Laws, how a safety belt can reduce injuries.
   ii) What other design feature in a car can offer protection in a crash? [4] (WJEC)

33. A trolley having a mass of 2 kg and velocity 4 m/s collides with another trolley having a mass of 1 kg which is at rest. If the two trolleys stick together after the impact, calculate
a) the total momentum before the collision,
b) the common velocity after the collision,
c) the kinetic energy before the collision,
d) the kinetic energy after the collision.
Why are the answers to (c) and (d) different?

Further reading
Physics Plus: *Karate; Physics and Sport; The Bicycle; Artificial Hip Joints* (CRAC)
Science of Movement, Hancock (Macmillan)
Physics of Human Movement, Page (Wheaton)

chapter 20

The Earth in Space

Our beautiful planet Earth is a spherical spaceship held in orbit round the Sun by the pull of gravity. The atmosphere and everything on Earth is held to it by gravity. This gravitational force also crushes the Earth itself so that the rocks inside are very dense.

Inside the Earth there are three main layers:
- The **core** is very hot metal, mostly iron and nickel. Most of it is liquid, but the centre is under such high pressure that it is solid. The Earth's magnetic field is probably due to currents in this metal core. The core is hot because it is radioactive (see page 340).
- The **mantle** is mostly solid rock but some of it is 'slushy' – a mixture of solid and hot molten (liquid) rock, rather like lava from a volcano. The mantle can move very slowly in giant convection currents.
- The **crust** is a very thin layer, like the shell on an egg. It is made up of huge '**plates**' of solid rock, some with a continent like America or Africa.

How do we know all this? We make deductions:

• Whenever there is an earthquake, the shock waves (seismic waves) travel into the Earth: The P-waves are *longitudinal* compression waves, like sound (see page 166). They travel fast through the mantle and the core. The S-waves are slower and *transverse* waves. They can only travel through a solid – so they cannot go through the liquid core.

Seismometer stations all over the world detect the waves. In the diagram, stations A, B, C and D all receive S-waves as well as P-waves. But stations X, Y and Z only receive P-waves. From this we know that the core is liquid, and its size can be calculated.

• Meteorites are rocks that sometimes arrive from space. They are the left-overs from when the Earth was formed, and often consist of iron and nickel.

• From measurements in astronomy we can find the average density of the whole Earth. It is 5500 kg/m^3. But the density of the surface rocks is only 2700 kg/m^3. So the centre of the Earth must be denser. Calculations show that iron has the right density.

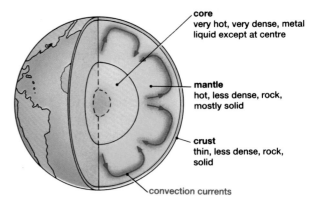

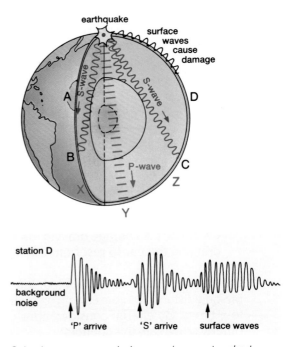

Seismic waves recorded on a seismometer chart

The changing Earth

Have you ever noticed how South America and Africa could fit together in a jig-saw puzzle? A map of the world looks as though these continents are fixed, but in fact they are moving!

Accurate measurements, by bouncing laser beams off satellites, show that they are moving apart at about 3 cm per year – about the speed that your finger-nails are growing!

Going backwards in time, they would have been touching about 200 million years ago, at the time of the dinosaurs: There is other evidence for this: the two continents have similar fossils in the rocks, and their age can be found by radioactive dating (see page 349).

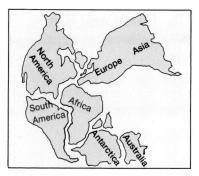

The jig-saw – 200 million years ago

The plate tectonics theory

It seems that each of the continents is on a huge 'plate' of rock, which moves slowly with the convection currents in the mantle. To see where these plates are, study the maps carefully:

What do you notice about the pattern of volcanoes and the pattern of earthquakes? We believe that these active areas are where one plate meets another.

Along other lines, called mid-ocean ridges, liquid rock or *magma* is found to be coming up through the ocean floor, to fill in gaps as the plates separate.

Notice that the most ancient rocks are usually well away from the active areas. The newest mountains are all in the green earthquake areas. These mountains are formed when the plates collide in slow motion, and force the rock upwards. The diagram below shows two ways in which the new mountains are formed:

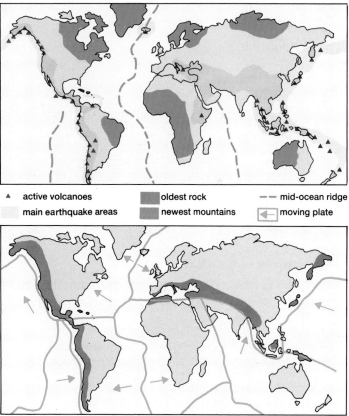

| ▲ active volcanoes | oldest rock | – – – mid-ocean ridge |
| main earthquake areas | newest mountains | moving plate |

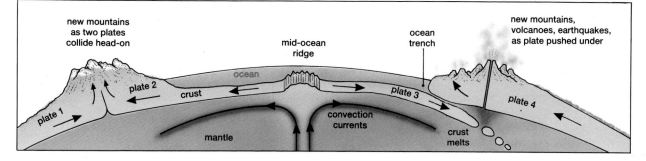

157

▷ Earth and Sun

The Earth moves round the Sun, in an *orbit*, which is (almost) a circle. The Earth is pulled into this orbit by a *centripetal force* (see page 86). This force is the gravitational force between the mass of the Sun and the mass of the Earth.

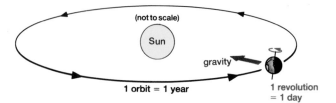

Experiment 20.1   A day
Use a white tennis ball as the Earth. Sketch the main continents on it, and mark the position of your school. Insert a knitting needle as its north–south axis. In a darkened room, use a lamp for the Sun.
Turn your 'Earth' slowly, so your 'school' is in daylight and then night.
For the real Earth, one revolution takes 24 hours, a complete day.
For other planets it is different.

The Sun 'rises' in the east. Which way should you rotate your 'Earth'?

Experiment 20.2   A year
Walk with your 'Earth' in a complete circle round the 'Sun'.
Mark out the 12 months of the year. For the real Earth, 1 complete orbit takes $365\frac{1}{4}$ days (this is why we need a leap year every 4 years).

Experiment 20.3   Polar and equatorial regions
Look carefully at the surface of your 'Earth' in the light from the 'Sun'.

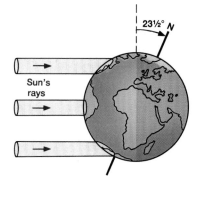

At the equator, where the Earth's surface is facing the Sun, it looks bright. But near the poles, where the Earth's surface is curved away from the Sun, it is less bright. Why is the Earth cold at the poles?

The seasons of the year

In fact, the Earth's axis is not at right-angles to the orbit as you might have assumed. *It is tilted* at an angle of $23\frac{1}{2}°$.

Experiment 20.4   Summer and winter
Repeat experiment 20.2 but with the axis *always tilted towards the same end of the classroom*.

Look carefully at how much of the Sun's energy arrives at the northern hemisphere at different times of the year. This explains three things:
• When the northern hemisphere is facing away from the Sun, it is winter. Six months later (half an orbit) it is facing towards the Sun (in summer). What happens in the southern hemisphere?
• The Sun appears higher in the sky in summer. It has a higher *elevation* above the horizon.
• Daylight lasts for more hours in summer.
Imagine you are at the North Pole. At what time of the year is it a) daylight for 24 hours?
b) night-time for 24 hours?

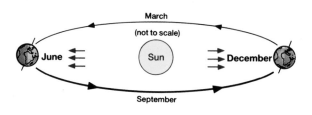

▷ Earth and Moon

The Moon is a stark and hostile world. Its mass is smaller than the Earth's, so its gravitational field is less. It is only one-sixth of the Earth's field (g = 1.6 kg/N, see page 141). How high could you jump on the Moon?

Because its gravity is less, all the molecules of gases have escaped, so it has no atmosphere. And because it has no atmosphere, any water has long since boiled away (see page 63).

The Moon moves in orbit round the Earth, held by the gravity pull between them. One orbit of the Moon takes 1 month (1 'moonth').

The Moon looks different to us at different times of the month. It has *phases*. The diagram shows the Moon at 8 different positions round the Earth with the 8 phases:

Earth-rise on the Moon

Experiment 20.5   Phases of the Moon
Use a lamp or an overhead projector for the Sun's rays, with a white ball for the Moon. Then stand in the centre (as the Earth) and look at the 'Moon' as it is moved round in orbit.

Experiment 20.6   Phases of the Moon
Observe the real Moon for a full month, weather permitting. Keep a diary of its appearance each evening, a) in writing and b) in sketches.

Try to find out when the next lunar eclipse will occur (see the diagrams on page 175).

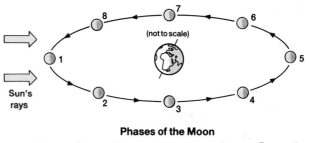

Phases of the Moon

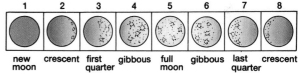

1	2	3	4	5	6	7	8
new moon	crescent	first quarter	gibbous	full moon	gibbous	last quarter	crescent

Tides

Just as the Earth pulls on the Moon to keep it in orbit, so the Moon pulls on the Earth (Newton's Third Law, see page 100).
This gravitational force also pulls on the Earth's seas so that the water piles up. For complex reasons, there are *two* bulges as shown:

As the Earth rotates, each part of the world gets two high tides each day, and two low tides.

The Sun also causes tides on Earth, but its effect is smaller. Sometimes its effect *adds* to the Moon's effect to give 'spring' high tides (at new moon and full moon). Sometimes its effect subtracts from the Moon's effect ('neap' tides).

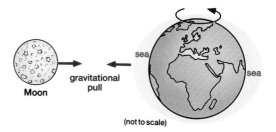

▷ The solar system

Our solar system consists of a central star (the Sun), surrounded by 9 planets, including Earth.

The first six planets are visible to the naked eye, and have been known since ancient times. The last three were discovered through telescopes.

The planets are held in their orbits by the gravitational pull of the Sun.

The planets vary enormously in their size (see below, with the Sun in yellow on the same scale).

They also vary in their distance from the Sun (see the diagram on the opposite page). This means that the conditions vary from planet to planet:

Saturn, photographed by the Voyager-2 space probe

1 Mercury

Mercury is closest to the Sun, and small for a planet (about the size of our Moon).
It has no atmosphere and is covered in craters.
The side facing the Sun is very hot, about 430 °C (that is hot enough to melt lead).

2 Venus

Venus is almost as big as the Earth, but very unpleasant.
It is covered in clouds of sulphuric acid, with an atmosphere of carbon dioxide at very high pressure.
Because of the CO_2 and the Greenhouse Effect (see page 54), it is even hotter than Mercury.

3 Earth

From space, Earth is a blue planet with swirls of cloud.
It is the only planet with water and oxygen and living things.
It is at the right distance from the Sun, with the right chemicals, to support life. Other stars may have planets with the same conditions.

4 Mars

Mars – the red planet – is a cold desert of red rocks, with huge mountains and canyons.
There is no life on Mars.
It has a thin atmosphere of carbon dioxide, and two small moons.
Between Mars and Jupiter there are thousands of rocks, called asteroids (see opposite).

5 Jupiter

Jupiter is the cold giant of the planets.
It has no solid surface, being mainly liquid hydrogen and helium, surrounded by gases and clouds.
The Great Red Spot is a giant storm, three times the size of Earth. Some of the 16 moons have volcanoes.

6 Saturn

Saturn is another 'gas giant', very like Jupiter.
The beautiful rings are not solid. They are made of billions of tiny chunks of ice, held in orbit by the pull of Saturn's gravity. As well as the rings, Saturn has more than 20 moons.

7 Uranus

Uranus is another giant planet. It looks pale green, with very faint rings and 15 moons.
It was discovered by William Herschel in 1781.
It is unusual because its axis is tilted right over, so that it is 'lying on its side' as it goes round the Sun.

8 Neptune

Neptune is the twin of Uranus. They are both about 4 times the size of Earth.
Neptune is bluish, due to a thick atmosphere of cold methane.
It has 8 moons, one with volcanoes.
The Great Dark Spot is a storm about the size of Earth.

9 Pluto

Pluto is the smallest of all, discovered in 1930.
We don't know much about it.
Pluto's orbit is not as circular as the others. Most of the time its orbit is outside Neptune's, but between 1979 and 1999 its orbit is *inside* Neptune's.
Some astronomers think there is a 'planet 10' beyond Pluto.

The Sun

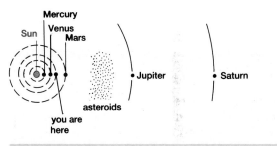

The distances to the planets, drawn to scale.
On this scale, the nearest star would be 1 km away!

Experiment 20.7   The solar system
Make a scale model of the solar system:
a) For the Sun use a grapefruit or cardboard disc of diameter 11 cm.

b) For the Earth, make a ball of plasticine 1 mm in diameter. From column **C** in the table below, make all the other planets to the same scale (e.g. Jupiter = 11 × Earth = 11 mm).

c) Hold your 'Earth' at a distance of 12 metres from the 'Sun'. From column **B** below, hold the other planets at the correct scale distances (e.g. Jupiter = 5 × Earth = 5 × 12 = 60 m).

You'll need to go on the playing field! On this scale, the nearest star would be another grapefruit about 3000 km away!

The birth of the solar system

Most astronomers believe that all the planets were formed at the same time as the Sun, about 4600 million years ago. A huge cloud of dust and gas in space shrank smaller and smaller (because of its own gravity), becoming hotter and hotter.

Eventually the hot centre became the Sun and the rest became the 9 planets, with some left-overs that formed the asteroids. Ice on the inner planets was boiled off by the Sun, to leave them rocky, and denser than the outer planets.

Here is a table of data about the solar system. Look down each column to see what patterns you can find (see also question 8, on page 165).

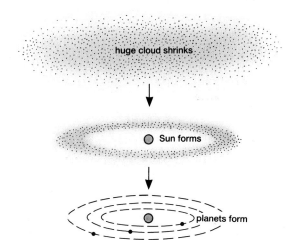

The planets	A Average distance from the Sun (millions of km)	B Average distance from the Sun (relative to Earth = 1)	C Diameter (Earth = 1)	D Density (kg/m³)	E Average temperature (°C)	F Mass (Earth = 1)	G Surface gravity (N/kg)	H Time for 1 orbit (years)	I Number of moons
1 Mercury	58	0.4	0.4	5500	+430 to −180	0.1	4	0.2	0
2 Venus	108	0.7	0.95	5200	+470	0.8	9	0.6	0
3 Earth	150	1.0	1.0	5500	+15	1.0	10	1.0	1
4 Mars	228	1.5	0.5	4000	−30	0.1	4	1.9	2
Asteroids									
5 Jupiter	778	5	11	1300	−150	320	26	12	16
6 Saturn	1427	9.5	9	700	−180	95	11	30	20 + rings
7 Uranus	2870	19	4	1300	−210	15	11	84	15 + rings
8 Neptune	4497	30	4	1700	−220	17	12	165	8
9 Pluto	5900	39	0.2	500	−230	0.002	4	248	1

▷ Our planet: the Earth

Our Earth is at the ideal distance from the Sun. If we were nearer, the seas would boil. Further away and they would be permanently solid ice.

Away from the Earth the universe is a hostile place, and being an astronaut is not easy. First he must escape the pull of gravity without his rockets blowing up. Once above the atmosphere he has no ozone layer to protect him from dangerous radiation (page 214). In space he must wear a spacesuit so that he can breathe, and so that his blood does not boil away (see page 63). He can be overheated by the Sun, or frozen by the coldness of space. Weightlessness makes it hard to drink and wastes his muscles due to lack of exercise. Distances are too vast to go much beyond the Moon, and on returning to Earth he is in danger of burning up like a meteorite!

Bruce McCandless in 'weightless' orbit near his spacecraft. He is using small jets of gas to manoeuvre.

Escaping from the Earth

Imagine firing a gun from the top of a very high mountain. The shell would fall back to the Earth, just like a football (see A). If the shell is fired faster, it falls further away (B and C). If the shell could be fired even faster (at 8 km/s, 25 times the speed of sound) then it would still fall towards Earth, but because of the curving of Earth, the shell would stay the same height above the ground (see D):

It is then a **satellite**, in orbit. Gravity gives it the **centripetal force** needed to keep it in orbit (p. 86).

An astronaut in orbit **feels** weightless, because the astronaut **and** his clothes **and** the spaceship are all falling at the same rate together. However to become truly weightless the astronaut would have to travel far away from the gravitational pull of the Earth and the Sun. The further the distance away, the smaller the gravitational pull.

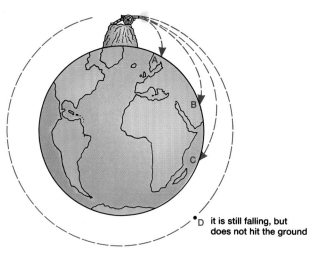

D it is still falling, but does not hit the ground

Is the Earth round?

Some people still believe that the Earth is flat! Here are some arguments for a round Earth:

Which of them do you find most convincing?

Which bits of evidence have you seen at **first hand**, and which are from secondary sources?

- when a ship moves away from you, in any direction, the hull always disappears first
- if you travel North, stars rise above the northern horizon and sink below the southern one
- during a lunar eclipse (page 175), the shadow of the Earth on the Moon is always circular
- a plane can continue travelling in the same direction, but arrives back at its starting point
- a photograph from a satellite in any position shows that the Earth looks round

▷ Our star: the Sun

The Sun is a star like the others that we see at night. Compared to other stars, the Sun is very ordinary and average.

The Sun is not burning like a fire – it is a huge controlled hydrogen bomb.
In the core, at a temperature of 14 million °C, hydrogen nuclei combine to form helium.
Each time this happens, some mass is lost and energy is released (see page 350).

The Sun is using up its mass at the rate of 4 million tonnes each second! It has been doing this for 4600 million years, and is now about half-way through its life-time!

The huge amount of energy is radiated out from the core. This radiation pressure **opposes** and balances the pull of gravity which is trying to crush the Sun.

When a star gets older it expands to become a **red giant**, but later gravity takes over and the star collapses down to form a **white dwarf**.

Sun-spots are slightly cooler areas on the Sun's surface – they are only 4000 °C!
Sun-spots come and go, being common every 11 years (the 'sun-spot cycle').
Sun-spots and **solar flares** affect the Earth, causing magnetic storms and radio interference. It is possible that the Sun's power may vary slightly from time to time, affecting the Earth's weather – perhaps even causing the Ice Ages.

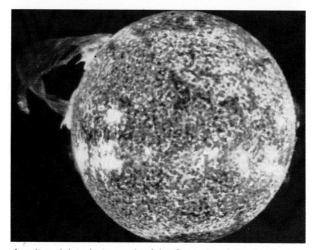

An ultra-violet photograph of the Sun, showing a huge prominence curving in the Sun's magnetic field

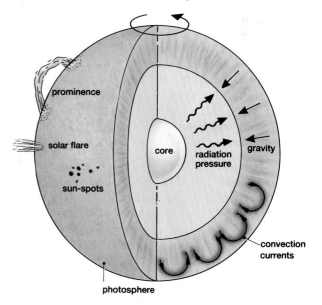

Twinkle, twinkle sunny star,
Now we know just what you are.
Nuclear fusion, burning brightly,
– Nice of you to switch off nightly!

▷ Our galaxy: the Milky Way

If you look carefully at the night sky, you can see a faint band of stars running across it. You are looking edge-on at our Galaxy.

It is a collection of about 100 000 million stars! Our Sun is just one of them, placed somewhere near the edge, in a **spiral arm**.

The Galaxy is huge. It takes 8 minutes for light to travel from the Sun to Earth; 4 years for light to travel from the nearest star; but 100 000 years for light to travel across our Galaxy!

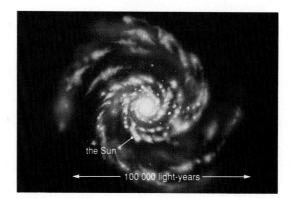

▷ The universe

The Milky Way is our galaxy, but it is not the only galaxy. Through telescopes we can see millions of other galaxies in the visible universe!

When we look at stars, we see them as they **were**, when the light left them. We are looking into the past. Some stars are 13 000 million light-years away, and so we see them as they were, 13 000 million years ago! So the universe is even older than this.

The expanding universe

Astronomers can analyse the spectrum of light from a star (see page 211). They have found that the light from other galaxies **always has a red-shift**. This is the same as the pitch of a police-car siren going lower as it races past. It is called a **Doppler** shift. It means that **all** the galaxies are moving **away** from us. In fact, the farthest galaxies are moving away fastest. This means the whole universe is **expanding**, just like the dots on a balloon move further apart as the balloon expands:

Going backwards in time, all the material in the universe was lumped together about 15 000 million years ago, when it was created in the '**Big Bang**'. The universe has been expanding ever since then, against the pull of gravity between the stars.
It is not clear if the universe will continue to expand, or stop and then contract to a 'Big Crash'.

The 'whirlpool' galaxy, 37 million light-years away. It is a 'spiral' galaxy, like ours, but others are different shapes.

Your place in the universe:

(each drawing is 1000 times wider than the one before it)

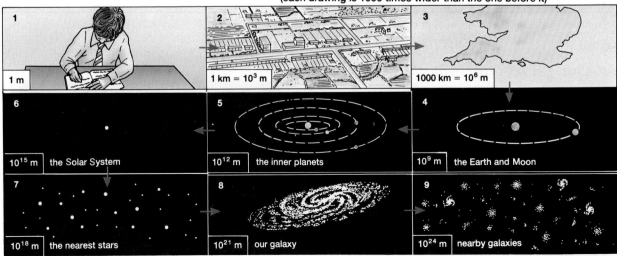

1 — 1 m	**2** — 1 km = 10^3 m	**3** — 1000 km = 10^6 m
6 — 10^{15} m — the Solar System	**5** — 10^{12} m — the inner planets	**4** — 10^9 m — the Earth and Moon
7 — 10^{18} m — the nearest stars	**8** — 10^{21} m — our galaxy	**9** — 10^{24} m — nearby galaxies

Summary

The Earth's continents move slowly on large 'plates' pushed by convection currents.
The solar system is the Sun and 9 planets.
The Sun is a star in our galaxy, the Milky Way.

The movement of the Earth and Moon explains: a day, a month, a year, the seasons, the Moon's phases and the tides.

▷ Questions

1. Copy and complete:
   a) The Earth has . . layers: . . . . , ... , ...
   b) The . . . . is made of large plates, which move on . . . . currents in the . . . . .
   c) When two plates collide, ... are formed.

2. The distance between South America and Africa is about 5000 km. If they were touching 200 million years ago, calculate their average speed of separation, in cm/year.

3. Copy and complete:
   a) A day is the time for:
   b) A month is the time for:
   c) A year is the time for:
   d) The Earth's axis is tilted at an angle of . . . and this causes our . . . . .
   e) 'Spring' tides occur when:
   f) The names of the planets, in order, are:
   g) The solar system was formed by:
   h) The Sun is an ordinary . . . . , part of our . . . . called the Milky Way.
   i) The universe has been . . . . since the time of the . . . . .

4. Make up a sentence to remind you of the first letter of each of the planets, in order. Whose sentence is easiest to memorise?

5. Using reference books, write a report on the human exploration of space, including some of the difficulties and dangers of space travel.

6. Look at page 160 and the table on page 161. Write two paragraphs about each planet to describe what you would see, and feel, if you visited each one in a spaceship.

7. On a sheet of graph paper, draw bar charts of:
   a) the diameter of each planet,
   b) the density of each planet.
   What pattern can you see?
   How does this fit with the 'nebula' cloud theory of the origin of the solar system?

8. Look at the table on page 161.
   a) Looking down column **B** and column **E**, what pattern do you notice? Why is this?
   b) Looking down columns **B** and **H**, what pattern do you see? Why is this?
   c) Looking down column **D**, what do you notice? List the names of the planets, divided into two clear groups.
   d) What do you notice from column **C**? Why do astronomers consider Pluto to be odd?
   e) From column **G**, how much would you weigh on i) Mercury ii) Jupiter?
   f) What is the connection between column **D** and columns **C** and **F**?
   g) For an asteroid, estimate i) the distance from the Sun, ii) the average temperature, iii) the length of its 'year'.
   h) How many orbits or 'years' has Uranus completed since the Battle of Hastings?

9. From reference books, write a paragraph about each of the following: a) red giants, b) white dwarfs, c) supernovas, d) neutron stars (pulsars), e) black holes, f) quasars.

10. Estimate the possibility of ET (extra-terrestrial) life elsewhere in our galaxy, if there are:
    – 100 000 000 000 stars in our galaxy,
    – 1 in 100 are suitable stars with planets,
    – 1 in 10 planets are at a suitable distance from the star.

11. Using the table on page 161:
    a) Sketch a graph of distance from the Sun, d, (column **B**) against periodic time, T, of the planet (column **H**). Is it a straight line?
    b) Now try d^3 against T^2.
    What conclusion can you make (see page 364)? (This is Kepler's Law.)

12. Professor Messer dreamed he was living on an asteroid. Why was his dream a nightmare?

chapter 21

Waves Waves Waves

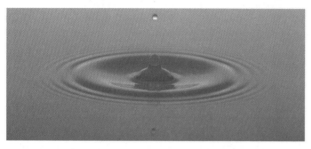

Energy is often moved from one place to another by waves.

You can see this if you throw a stone into a pond – waves spread out from the splash and soon the water at the edge of the pond moves as it gets some of the energy of the splash.

There are **two** kinds of waves:

▷ Transverse waves

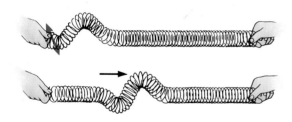

Experiment 21.1
To see a transverse wave, stretch a long 'slinky' spring along a smooth bench or floor and then give one end a quick wiggle **at right angles** to the spring. What happens?

You can see that energy is moving down the spring although each bit of the spring is only vibrating. It is vibrating at right angles (**transversely**) to the spring.

In this chapter we will be investigating transverse water waves. Light waves are also transverse waves.

▷ Longitudinal waves

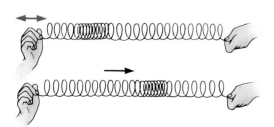

Experiment 21.2
Stretch the same 'slinky' spring along the floor and this time give it a quick wiggle to and fro **along the length** of the spring. What happens?

If you look closely you can see that again, energy is passing down the spring although each bit of the spring is only vibrating (or 'oscillating'). It is vibrating lengthways or **longitudinally**.

We will investigate longitudinal waves in chapter 29 (page 220). Sound waves are longitudinal waves.

▷ The ripple tank

We can study transverse waves by using water waves in a ripple tank.

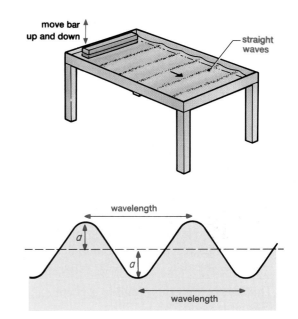

As the waves move down the ripple tank, which way do the pieces of cork move?
Are these transverse or longitudinal waves?

Here is a side view of a transverse water wave:

The high parts are called **crests** and the low parts are called **troughs**.
The distance marked **a** is called the **amplitude**. It is the height of a crest above the average water level (or the depth of a trough below it).

The distance between two successive crests (or troughs) is called the **wavelength**.

If the wooden bar (or the pieces of cork) in the experiment moves up and down through two complete vibrations in each second, then the **frequency** of the vibrations is two cycles per second (this is also written as 2 hertz or 2 Hz).

A wave which vibrates only up and down like this is said to be **polarised** vertically.

You should be able to see that:
a) the speed stays the same and
b) when the frequency increases, the wavelength decreases:

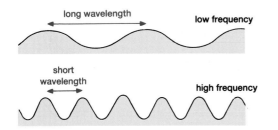

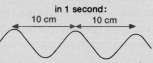
In fact, for any wave:

wave speed = frequency × wavelength
(m/s)   (Hz)   (m)

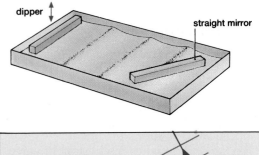

▷ Reflection and Refraction

Experiment 21.5   Reflection at a straight mirror
Place a straight metal or plastic barrier in the tank
so that it reflects the waves (like a mirror).
Send single waves across the ripple tank and look
carefully for the waves reflected off the 'mirror'.

What do you notice about the angle at which the
wave hits the mirror and the angle at which the
reflected wave leaves the mirror?
What do you know about the angles when light is
reflected from a mirror? (See page 179.)

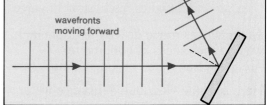

Experiment 21.6   Reflection at a curved mirror
Place a curved barrier in the ripple tank. Again
send single waves across the tank and look carefully for the reflected waves. What do you see?

The waves are **converged** to a point called the **focus**:
(See also page 183.)

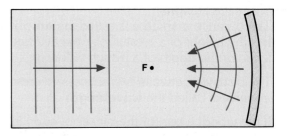

Experiment 21.7
Repeat experiment 21.6 with the curved mirror
turned round the other way.

Can you see that the waves are spread out or
diverged by the mirror? (See also page 183.)

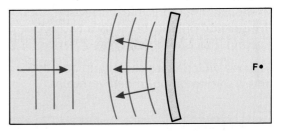

Experiment 21.8   Refraction
Place a thick sheet of glass in the tank and adjust
the depth so that the water is very shallow over
the glass sheet.
Again send waves across the tank and look at the
waves when they travel through the shallow water.

What do you see?   What happens to:
a) the speed and b) the wavelength of the waves
when they move into the shallow water?

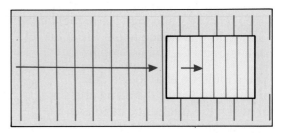

Experiment 21.9   Refraction
Now place the sheet of glass at an angle and repeat
the experiment.

What happens to the **direction** of the waves as they
slow down in the shallow water?

When waves change their speed and direction like
this, we say they have been **refracted**.
Light waves can also be refracted (see page 186).

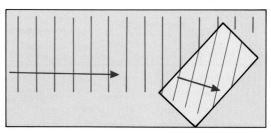

▷ Diffraction

Do waves change shape as they go through a gap in a wall?

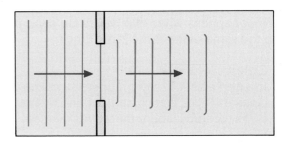

Experiment 21.10   Wide gap
Use a ripple tank to send straight waves across water to a wide gap between two barriers.

Look at the waves after they have passed through the gap. Are they still perfectly straight?

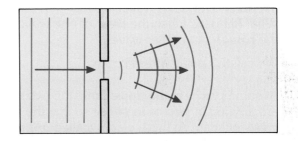

Experiment 21.11   Narrow gap
Now send straight waves through a narrow gap.

What happens to the shape of the waves?

In the first diagram, with a wide gap, the waves travel almost straight on (*rectilinear propagation*). There is a clear shadow behind each barrier.

In the second diagram, the waves spread out after passing through the gap. This is called *diffraction*.

Diffraction is most obvious when the *width* of the gap is about the same size as the *wavelength* of the waves.

The photograph shows some sea waves passing through into a harbour. If you look carefully you can see the waves spreading out in the harbour:

Sound waves are diffracted through doorways, because the wavelengths of sounds are about the same size as doorways. This is one reason why you can hear people without seeing them.

Light is seen to be diffracted only if it passes through a very narrow slit. This shows us that *light is a wave motion*, with a very small wavelength.

Radio and TV waves can also be diffracted. (Your teacher may be able to show you this with 3 cm radar waves.)

This affects the lives of people living near hills:

Radio waves (longer wavelength) can diffract round the hill to be received by the house, but TV waves (short wavelength) do not bend as much. The house will have poor TV reception.

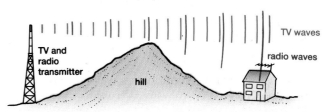

▷ Interference of water waves

What happens when two waves are **superposed** – in the same place at the same time?

> **Experiment 21.12**
> In a ripple tank, use two dippers to make two circular waves that spread out and overlap. Look carefully where the waves overlap.

What happens when the **crest** of one wave overlaps the **crest** of another wave?
What happens when the **crest** of one wave overlaps the **trough** of another wave?

We call this effect **interference**.

At some places in the ripple tank, the two waves are arriving in step or **in phase** so that the two crests overlap. At these places (called **antinodes**) the waves add up to give a bigger wave.
This is called **constructive interference**.

At other places in the ripple tank, the waves are exactly out of step (out of phase or in **antiphase**). At these places (called **nodes**) one wave is trying to push the water up while the other wave is trying to push it down and so the water does not move. This cancelling effect is called **destructive interference**.

The photograph shows two water waves interfering: The sources of the waves (the dippers) are marked S_1 and S_2. Look at the photograph carefully:

Close to the dippers you can see the circular waves spreading out and then overlapping.

Can you see that there are lines of places where there is constructive interference (marked C).
In between, there are other lines where the water is not moving (destructive interference, marked D).

Interference can occur with light waves and with sound waves.
Two sound waves in antiphase can add up to give silence! This idea is sometimes used to reduce the noise level of machinery.

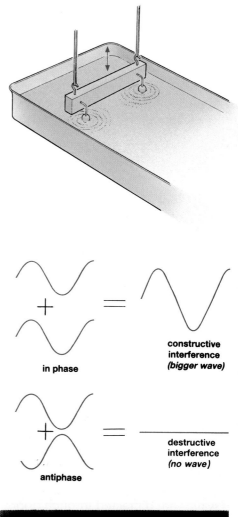

in phase

constructive
interference
(bigger wave)

antiphase

destructive
interference
(no wave)

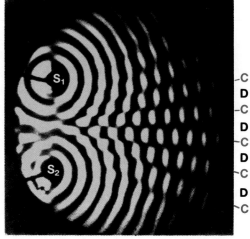

Top view of a ripple tank

170

Summary

Waves move energy from one place to another (although each part of the wave only **vibrates** to and fro). In **transverse** waves (like water waves and light waves) this vibration is at right angles to the direction of the wave. In **longitudinal** waves (like sound waves) the vibration is along the direction of the waves.

For **all** waves: speed = frequency × wavelength

Waves can be reflected and refracted.

Waves can be diffracted through a narrow opening.

Two or more waves can interfere constructively or destructively.

▷ Questions

1. Copy out and complete:
   a) If the vibrations of a wave are at right angles to the direction of the wave, it is called a . . . . wave. If the vibrations of a wave are along the direction of the wave, it is called a . . . . wave. A water wave is a . . . . wave and a sound wave is a . . . . wave.
   b) The wavelength of a wave is the . . . . between two successive . . . . (or . . . .). The frequency of a vibration is measured in . . . . or . . . . .
   c) The speed of a wave (measured in . . . .) is equal to the frequency (measured in . . . .) multiplied by the . . . . (measured in . . . .).

2. a) When waves travel through a narrow gap in a wall, they are . . . . . This effect is most obvious when the . . . . of the gap is similar to the . . . . .
   b) If two waves overlap and are in step (in . . . .) then they interfere to give a . . . . wave (. . . . interference).
   c) If two equal waves overlap and are exactly out of step (in . . . .) then they interfere to give . . . . wave (. . . . interference).

3. A certain sound wave has a frequency of 170 hertz (cycles per second) and a wavelength of 2 metres. What is the speed of sound?

4. A certain radio wave has a frequency of 200 000 hertz (cycles per second) and a wavelength of 1500 metres. What is the speed of radio waves?

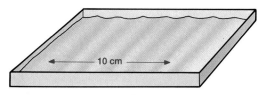

5. The diagram shows some waves in a ripple tank travelling at a speed of 6 cm/s. What is a) the wavelength b) the frequency?

6. The diagram shows straight waves approaching a straight barrier at an angle of 45°. Draw a diagram to show how the waves are reflected from the 'mirror'.

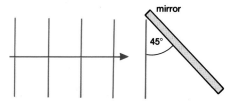

7. A vibrating source **S** produces circular water waves near a straight reflector:

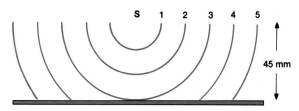

   a) Copy and complete the diagram to show how crests 4 and 5 are reflected.
   b) What is the wavelength of these waves?
   c) If their speed is 60 mm/s, what is the frequency?

Further questions on page 234.

chapter 22

Visible light is a form of energy that we can detect with our eyes.

Objects that produce their own visible light are called **luminous sources** (for example, the Sun, other stars, electric lamps, television screens, glow-worms).

Other objects are **illuminated** by this light and **reflect** it into our eyes (for example, this page and most objects in this room, and the Moon).

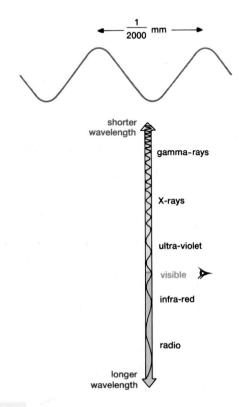

▷ Light waves

Light is a wave motion, rather like the water waves that you may have seen on a pond or in a ripple tank (see page 167).

The light that we can see has a **wavelength** of only about $\frac{1}{2000}$ of a millimetre!

Some waves have a wavelength even shorter than this, but then we cannot see them – for example, ultra-violet waves, X-rays, and gamma-rays.

Other waves have a wavelength longer than visible light. Again we cannot see them – for example, infra-red (heat rays), and radio waves.

All these waves are called **electromagnetic waves** (see page 212).
In a vacuum, they all travel at the same speed.

▷ The speed of light

Light travels at a very high speed – about a million times faster than the speed of sound.
This is fast enough to travel 8 times round the Earth in one second or fast enough to cross this room in about one-hundred-millionth of a second.
The speed of light in air is 300 million metres **per second**.

A racing car travelling non-stop would take about 100 years to travel from the Sun to the Earth.
Light does the journey in about 8 minutes!
This means that you see the Sun as it was 8 minutes ago.

Other stars are much farther away.
From some stars, the light has taken 1000 million years to reach us, so we are seeing them as they were 1000 million years ago.
How long would you have to wait to see them as they are now?

A **light year** is a distance – how far is it?

▷ Rays of light

For experiments with visible light, we often use a very narrow beam of light called a **ray**.

Experiment 22.1
Use a **raybox** to shine a narrow beam of light across a piece of paper.

What do you notice about the edges of the beam of light?

Why can't you see round corners?

Light travels in straight lines.

Sometimes the word **rectilinear** is used because it means 'in a straight line'.

Mary had a little lamp
It shone so straight for her.
The rays of light so straight and true
Were rectilinear.

Lasers produce a narrow beam of light:

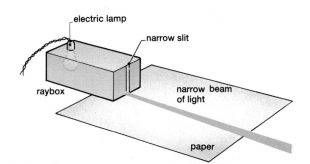

173

▷ Shadows

Shadows are formed when some rays of light continue to travel in straight lines, while other rays are stopped by an object.

The kind of shadow that is formed depends on the size of the source of light:

1 Shadow due to a small source of light

Experiment 22.2
Hold a table-tennis ball between a **small** electric lamp and a white screen.

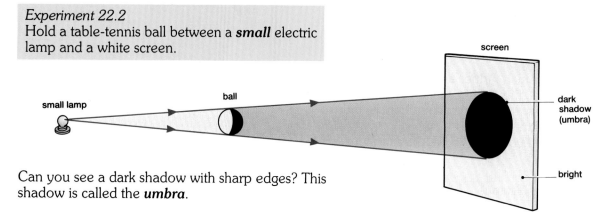

Can you see a dark shadow with sharp edges? This shadow is called the **umbra**.

2 Shadow due to a large source of light

Experiment 22.3
Change the small lamp for a **large** pearl lamp.

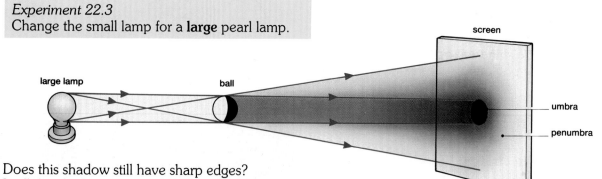

Does this shadow still have sharp edges?
Is all the shadow equally dark?

The dark centre of the shadow is still called the umbra. No light from the lamp reaches the umbra.
It is surrounded by a fuzzy partial shadow or **penumbra**.
The penumbra varies in brightness – from very dark near the umbra (where not much light can reach it) to very bright at the outer part where more light can reach.

Which kind of shadow is formed on a clear sunny day?
Which kind of shadow is formed on a cloudy day?

How many things can you deduce from this photo?

▷ Eclipses

Eclipses occur because the Moon and the Earth cast large shadows.

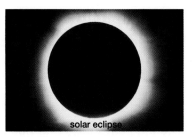
solar eclipse

1 Eclipse of the Sun (solar eclipse)

This happens when the Moon comes between the Sun and the Earth, so that the Earth (in the shadow) darkens during the day.
At the time of total eclipse, only the flames of the outer edge of the Sun can be seen.

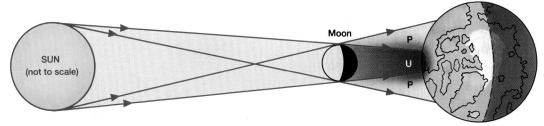

Experiment 22.4
Set up a model of an eclipse of the Sun (in a darkened room) using a large pearl lamp for the Sun, a table-tennis ball for the Moon and a white football for the Earth.

Total eclipses of the Sun do not occur often – the next to be seen in Britain will be on 11th August 1999 and then 7th October 2135.

2 Eclipse of the Moon (lunar eclipse)

This happens when the Moon moves into the Earth's shadow. When the Moon is in the Earth's umbra, no sunlight can reach the Moon and it is seen to go dark.

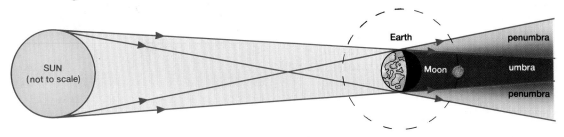

Experiment 22.5
Use the same apparatus as in experiment 22.4 to show an eclipse of the Moon.

Eclipses of the Moon can be seen much more often than eclipses of the Sun. Try to find out the date of the next eclipse of the Moon to be seen from where you live.
How long does it take for the Moon to move once round the Earth?

partial lunar eclipse

▷ The pinhole camera

A simple pinhole camera consists of a closed box with a screen at one end and a small pinhole at the other end.
It uses the fact that the light going in through the pinhole travels in straight lines to the screen.

The diagram shows one way to make a pinhole camera.

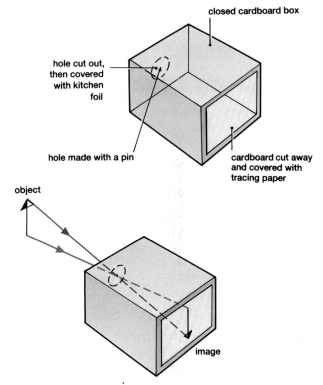

Experiment 22.6
Use a pinhole camera in a darkened room and point the pinhole at a lamp or at the outside world through a gap in the curtains.

What do you see on the screen?

The picture that you see is called an *image*. In this case, because the picture is formed on a screen, it is called a *real* image.
In the pinhole camera the image is *inverted* – it is turned upside-down and turned left-to-right.
The diagram shows just two of the many rays of light going through the pinhole.

Experiment 22.7
Look carefully at the image on your screen and notice how sharp and dim it is.
Then use a pencil to make the pinhole larger.

Is the image sharper or more blurred?
Is the image dimmer or brighter?
Why does this happen?

This kind of camera is *not* often used to take a photograph because to get a sharp picture, a small pinhole must be used and this means a long exposure time for the film. (See also page 201.)

A pinhole camera gives an inverted image

Improve your eye-sight without spectacles

If you normally wear glasses you can also make things look sharper by looking through a small pinhole in a piece of card.
In this way you are converting your eye (which is like a lens camera, see page 200) to a pinhole camera.

Summary

Shadows and eclipses occur because light travels in straight lines.
A small source of light causes a sharp shadow, called the umbra.
A large source of light causes a blurred shadow with two parts, the umbra and the penumbra.

In a pinhole camera, the image is real and inverted.
A small hole gives a sharp but dim image.
A large hole gives a bright but blurred image.
Visible light travels at the same high speed as gamma-rays, X-rays, ultra-violet, infra-red and radio waves.

▷ Questions

1. Copy out and complete:
   a) Light travels in . . . . lines.
   b) A shadow formed by a small source of light has . . . . edges and is called an . . . . .
      A large source of light causes a shadow with two parts called the . . . . and the . . . . .
   c) In an eclipse of the Sun, the . . . . comes between the Sun and the . . . . .
      In an eclipse of the Moon, the . . . . comes between the . . . . and the . . . . .
   d) In a pinhole camera, the image is . . . . and . . . . .
      If the pinhole is made larger, the image becomes . . . . and . . . . .

2. How could a detective tell from a photograph, without seeing the sky, whether it was a bright clear day or cloudy?

3. *Professor Messer thinks he's clever,*
   Put him right or he's gone for ever:

4. A laser beam of light can be bounced off the Moon (from a special mirror left there by astronauts). The light travels there and back in 2.6 seconds. What is the distance to the Moon?
   (Speed of light = 300 000 000 m/s = 3×10^8 m/s)

5. Draw a diagram to explain an eclipse of the Sun. Label all parts of the diagram, including the umbra and penumbra.
   Why is your diagram not drawn to scale?

6. Draw a diagram of a pinhole camera with two rays passing from the object through the pin-hole. How can you make the image
   a) sharper b) brighter c) larger?

7. Professor Messer fell asleep and dreamed that light no longer travelled in straight lines. Why was his dream a nightmare?

chapter 23

REFLECTION ᴎOITƆƎ⅃ᖴƎЯ

Have you ever reflected sunlight from a mirror to dazzle someone or to flash messages to a friend?

Experiment 23.1 Investigating reflection
Get hold of a **plane** (flat) glass mirror and look at it carefully.
Where is the light reflected from – the glass at the front of the mirror or the 'silvering' at the back?

Place the plane mirror on a sheet of white paper, holding it vertical with plasticine or a block of wood.

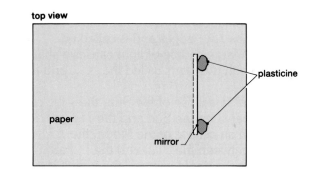

Draw a line along the **back** of the mirror (where the light is reflected).

Use a raybox to shine a ray of light on to the mirror. This is called the **incident ray**.

Can you see the reflected ray of light?

Mark the position of each ray with crosses as shown in the diagram.

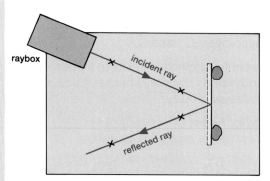

Now take off all the apparatus and use a ruler to join up the crosses, along the lines of the rays of light.

Use a protractor or a set-square to draw a line at right angles (90°) to the mirror, as shown. This line is called the **normal**.

Now use a protractor to measure the angles on each side of the normal.
What do you find?

Now try the experiment again with different angles.

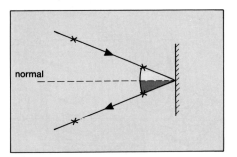

What do you find each time?

The diagram shows the correct scientific words:

The two rays and the normal line are **all in the same plane** (the plane of this piece of paper).

The **law of reflection**:

| the angle of incidence | = | the angle of reflection |

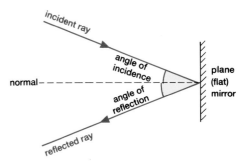

Water waves are reflected in the same way as light waves (see the ripple tank experiments on page 168).

If the angle of incidence is 45°, what is the angle of reflection and what is the angle turned by the ray of light?

The periscope

Periscopes are used in submarines so that people below the surface can see what is happening above the surface.

The diagram shows how light is reflected by the two mirrors.

Remember that the angle of incidence = the angle of reflection and the rays should be turned through 90° each time. At what angle should each mirror be placed?

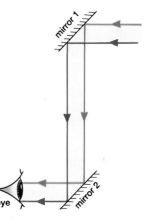

Building a periscope
You can use two handbag mirrors and a plastic bottle (like a washing-up-liquid bottle) to build a periscope which you can use to see over people's heads or to spy round corners.

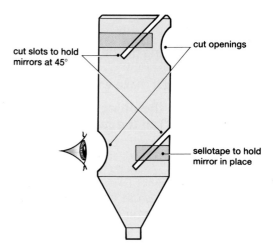

▷ The image in a plane mirror

When you look into a plane (flat) mirror, you can see a picture or *image* of yourself. Whereabouts do you think this image is — on the mirror or behind the mirror?

Experiment 23.2   To find out where the image is
Use a vertical sheet of clear glass as a mirror and place a lighted candle (or Bunsen burner) on the bench in front of it. (Take care.)

Then move an unlighted candle or Bunsen burner (of the same height) into a position behind the mirror so that it *appears* to be burning.

When the image of the flame appears to be on the unlighted candle, we know where the image is — it must be in the same position as the unlighted candle.

Measure the distance of the image *from the mirror* and the distance of the lighted candle (called the object) *from the mirror*. What do you notice?

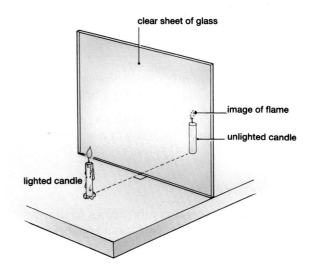

The image is as far behind the mirror as the object is in front.

This means that if your face is 30 cm in front of a mirror, then your image is 30 cm behind the mirror.

The diagram shows how the reflected rays appear to come from the image:

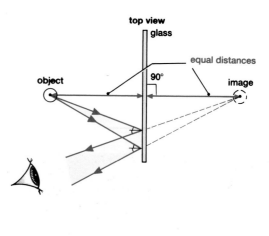

Imagine a line joining the object to the image in this experiment.
What angle would it make with the mirror?

The line joining the object to the image is at right angles (90°) to the mirror.

'Imagic' tricks

Magicians sometimes use images in their tricks.
If you put your finger on the unlighted candle in this experiment, it looks as if your finger is burning!

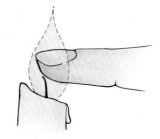

Another trick is to fix the unlighted candle in a tall beaker which you then fill with water so that it appears to burn underwater. Can you think of some other tricks like these?

Virtual images

If you go behind the mirror to look for the image, you cannot see it.

This *virtual* image only *appears* to be behind the mirror (because the rays of light are reflected to your eye as though they came from behind the mirror). This virtual image cannot be formed on a screen (as a real image can, see page 176).

Look at the *size* of the real flame compared with its virtual image.
What do you notice?

The image in a plane mirror is virtual and is the same size as the object.

Experiment 23.3
Look into a mirror and hold your *right* ear.
Which ear is your image holding?

Write your name on a piece of paper and pin it to your chest as a label. What do you notice when you look in the mirror?
This is an example of *lateral inversion*.

The image in a plane mirror is laterally inverted.

A tattooist once wrote on my throat
A short but mysterious note
And so to inspect it
I tried to reflect it
And found I'd got TAOAHT on my throat!

Regular and diffuse reflection

We have been looking at *regular* reflection from mirrors and other shiny flat surfaces.
Other surfaces, like this sheet of paper, also reflect light but because the surface is rough, the light is reflected off in all directions.
This is *diffuse* reflection.

If a surface does not reflect any light at all, it looks dull black, like this black ink.

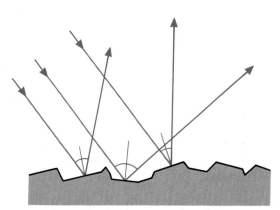

Diffuse reflection from a rough surface

Summary

The law of reflection:

the angle of incidence = the angle of reflection

The image in a plane mirror is:

a) as far behind the mirror as the object is in front, with the line joining the object and the image at right angles (90°) to the mirror
b) the same size as the object
c) virtual
d) laterally inverted.

▷ Questions

1. The angle between an incident ray and the mirror is 30°.
   a) What is the angle of incidence?
   b) What is the angle of reflection?
   c) What is the total angle turned by the ray?

2. Complete this verse:

   A periscope's a useful thing,
   As everyone agrees.
   It has . . mirrors fixed inside,
   At . . degrees.

3. A boy with a mouth 5 cm wide stands 2 m from a plane mirror. Where is his image and how wide is the image of his mouth?
   He walks towards the mirror at 1 m/s. At what speed does his image approach him?

4. **AMBULANCE** Explain where you would see this sign and explain why it is written like this.
   Write STOP as it should be written on the front of a police car.

5. In experiment 23.2 you can often see *two* images, close together. Why is this?

6. Explain how to read the following message which was found on some blotting paper:

7. Explain what is happening in the photo.

8. Draw a labelled ray diagram to show this:

   A periscope builder called Fred,
   Fixed two mirrors to the head of his bed.
   The light was reflected,
   And so he inspected,
   The hair on top of his head.

9. Professor Messer fell asleep and dreamed that all objects stopped reflecting light. Why was his dream a nightmare?

10. Explain, with a diagram, how 2 plane mirrors could be fixed at a dangerous T-junction so that motorists approaching the junction could see all other traffic.

11. An object O is placed near a plane mirror:

    Draw it full size, with two rays from O to the mirror. Working carefully, draw in the two reflected rays and the position of the image.

12. A room with only one window is dark even at mid-day. Explain how you could use wall-paper, furnishings and mirrors to brighten the room.

Further questions on page 234.

chapter 24

CURVED MIRRORS

Have you ever looked into a spoon to see an image of your face? Have you noticed that the front and back of a spoon give different images?

Two kinds of curved mirror:
A mirror that curves in (like a cave) is called a *concave* mirror.
A mirror that curves outwards is called a *convex* mirror.

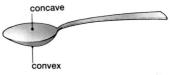

Experiment 24.1 A concave mirror
Use a raybox (on a sheet of white paper) to produce several rays of light to shine on to a *concave* mirror. It is better if you can adjust your raybox to make parallel rays of light.

Can you see that the rays coming off the concave mirror are getting closer together – we say they are *converging*. (See also page 168.)
A concave mirror is a converging mirror.

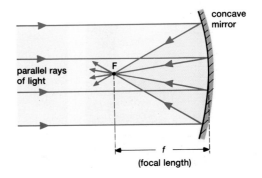

If the rays of light from the raybox are *parallel* rays, the rays reflected off the mirror converge to one point called the *principal focus* (marked F on the diagram).

Parallel rays of light are reflected through the principal focus of a concave mirror.

The distance from F to the mirror is called the *focal length* of the mirror.

Experiment 24.2 A convex mirror
Repeat the experiment, with a *convex* mirror.

Can you see that the light rays are spread out or *diverged* by this mirror?

A convex mirror is a diverging mirror.
Parallel rays of light are reflected so that they appear to come from the principal focus (F) of a convex mirror.

Where have you seen curved mirrors used?

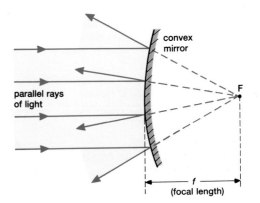

▷ Uses of concave reflectors

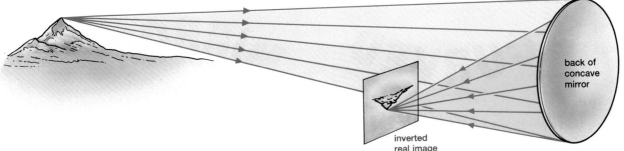

Concave mirrors are used to collect light energy. They are also used to collect sound, heat radiation, radar, and TV signals.

This idea is used in TV satellite dishes, solar ovens (see page 54), telescopes (see page 207), and radar receivers (see page 215).

In an opposite way, energy can be sent out in a beam from a concave mirror. This is used in car headlamps (see Q 4 opposite), film projectors (see page 208), and electric fires (see page 252).

If you move *close* to a concave mirror, then you will see a magnified image of yourself. This image is upright and virtual as well as magnified. This is a shaving mirror or make-up mirror:

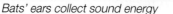

Satellite 'dishes' collect energy for TV sets   *Bats' ears collect sound energy*   *Concave mirrors magnify (if you are close)*

▷ Uses of convex mirrors

Convex mirrors always produce virtual, upright images. The image is always 'diminished' (it is smaller than the object).

Convex mirrors are useful if you want a wide field of view, such as in a car driving mirror or a shop security mirror:

A hairy young student called Raven
Used a diverging mirror for shaving,
His image was diminished,
And when he had finished,
His shaving looked more like engraving!

184

Summary

Concave mirrors **converge** parallel rays of light towards a principal focus in front of the mirror.

For distant objects, concave mirrors produce inverted real images. For a near object, the image is upright and magnified and virtual.

Convex mirrors **diverge** parallel rays of light away from the principal focus which is behind the mirror.
The image in a convex mirror is always erect, virtual and diminished.
Convex mirrors have a wide field of view.

▷ Questions

1. Copy and complete:
   a) A concave mirror . . . . rays of light, whereas a convex mirror . . . . rays of light.
   b) Parallel rays of light are reflected by a concave mirror to a point called the . . . . The focal length is the distance from the . . . . to the mirror.
   c) For a convex mirror, parallel rays of light are reflected . . . . a point called the . . . .

2. The diagram shows a dish-aerial which is used to receive television signals from a satellite.

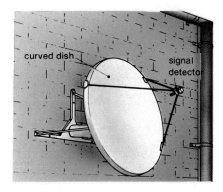

The signal detector (aerial) is fixed in front of the curved dish.
   a) What is the purpose of the dish?
   b) Should it be concave or convex?
   c) Where should the aerial be positioned to receive the strongest possible signal?
   d) Explain what change you would expect in the signal if a larger dish was used.

Further questions on page 234.

3. Draw up two lists headed 'Concave mirrors' and 'Convex mirrors' from the following: shaving mirror, car headlight mirror, searchlight mirror, driving mirror, dentist's inspection mirror, torch mirror, projector lamp mirror, staircase mirror on a double-decker bus, make-up mirror, reflecting telescope mirror, solar furnace mirror, satellite TV dish, shop security mirror.

4. A car headlight bulb contains two filaments labelled A and B in the diagram. Filament B is at the principal focus of the concave mirror. Copy the diagram and draw in the rays of light from each filament. Why is this arrangement used in car headlights?

5. *Professor Messer's in a dither,*
   He's seeing things in his driving mirror:

chapter 25

REFRACTION

Why do swimming pools look shallower than they really are? Is this pencil really broken?

Before we can answer questions like these, we must find out what happens when light travels from one substance into another.

Experiment 25.1
Place a glass block on a sheet of white paper and draw a line all round it. Use a raybox to shine a ray of light from the air into the glass.

Look carefully for the ray inside the glass and where it comes out again.
Has the ray of light travelled straight into the glass or has it been bent?

Mark the incident ray and the emergent ray with crosses as shown in the diagram.
Then take off all the apparatus and use a ruler to join the crosses and re-construct the path of the ray of light.

You can see that the ray was bent or **refracted** where it entered the glass and where it left the glass.

Use a protractor or a set-square to draw lines at right angles (90°) to the glass surface, as shown. Each of these lines is called a **normal** (N).

Where is the light bent towards a normal and where is it bent away from a normal?

Rays of light travelling from air into glass are bent or refracted *towards* the normal.
Rays of light travelling from glass into air are refracted *away from* the normal.

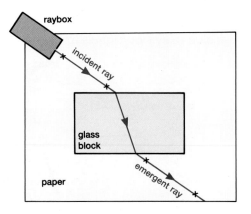

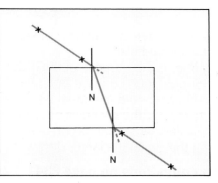

Here are some of the scientific words used to describe experiment 25.1.

Angle *i* is called the **angle of incidence**.
Angle *r* is called the **angle of refraction**.

Changing speed

We saw on page 173 that the speed of light in air is about 300 000 000 metres/second.
But light travels more **slowly** in glass – in fact only two-thirds as fast.
This is the reason why light is refracted – in the same way as a fast car travelling on the road is 'refracted' if it moves at an angle on to thick mud.
The mud slows down one side of the car, and the path of the car bends.
The more it is slowed, the more it bends.

Look again at the ripple tank experiment 21.9 on page 168. This shows how waves are refracted when they slow down.

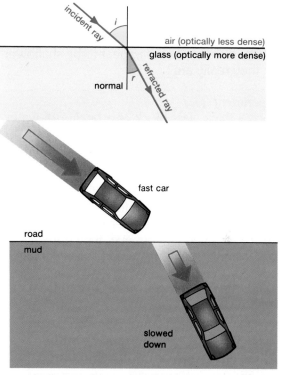

A motorist drove down a straight country road,
When the kerb gave the steering a twitch,
One wheel was slowed by the edge of the road,
And refracted him into a ditch.

Refractive index

A number called the **refractive index** of a substance is the ratio of the two speeds:

$$\text{Refractive index of a substance} = \frac{\text{speed of light in air}}{\text{speed of light in substance}}$$

The refractive index of most kinds of glass is about 1.5.
The refractive index of water is less (1.33) which means that the light is not bent as much when it enters water, because it is not slowed as much.

Experiment 25.2 (more difficult)
In experiment 25.1, measure *i*, the angle of incidence, and *r*, the angle of refraction.
Then use your calculator or a book of mathematical tables to find 'sine *i*' and 'sine *r*'.
The refractive index of the glass can be found from another formula, called **Snell's Law**:

$$\text{Refractive index} = \frac{\text{sine } i}{\text{sine } r}$$

Substance	Refractive index	Speed of light (m/s)
Air	1.0	300 000 000
Water	1.33	225 000 000
Perspex	1.5	200 000 000
Glass	About 1.5	200 000 000
Diamond	2.4	120 000 000

▷ Real and apparent depth

Have you noticed that swimming pools and buckets of water always appear to be shallower than they really are?

> *Experiment 25.3*
> Place a thick glass block or a beaker of water on top of some writing.

What do you notice?

The apparent depth is *less* than the real depth because rays of light are refracted away from the normal as they leave the glass or the water. When they are received by your eye, they *appear* to come from the point marked I (a virtual image):

The apparent depth can be calculated from the formula:

$$\frac{\textbf{Real depth}}{\textbf{Apparent depth}} = \frac{\textbf{refractive index}}{\textbf{of the substance}}$$

This means that water is really 1.33 times as deep as it appears to be; and glass is about 1.5 times as deep as it appears to be.

> *Experiment 25.4   An 'imagic' trick*
> Put a coin in the bottom of an empty cup and position your head so that the coin is just out of sight.

How can you bring it into view without moving your head or the cup or the coin (and without using a mirror)?

Simply pour in water. Rays of light from the coin are refracted away from the normal and into your eye so that you can see it. Its apparent position is higher than its real position.

> *Experiment 25.5   The 'broken' pencil*
> Hold a pencil at an angle in a beaker of water.

What do you see? (See page 186.)

Use the diagram to explain why the pencil appears to be broken.

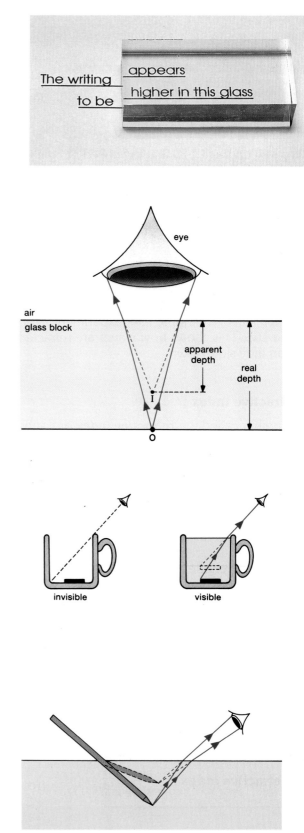

▷ Total internal reflection

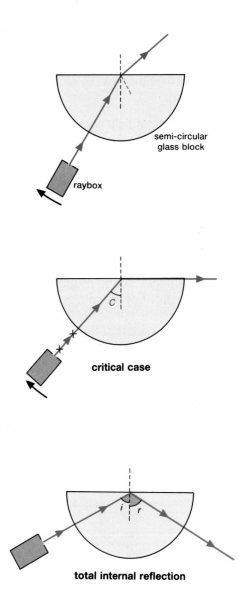

semi-circular
glass block

raybox

critical case

total internal reflection

Can you see the ray of light refracted away from the normal as it comes out of the glass block?
Can you also see that only a small amount of light is reflected back?

Now move the raybox round so that the ray of light still goes straight in to the mid-point, but with the angle of incidence increased until the refracted ray comes out *along the edge of the block*:

When this happens, the angle of incidence in the glass is called the *critical angle* (*C*).

Mark the ray with crosses so that you can take the apparatus off and re-construct the diagram.
Use a protractor to measure the critical angle (*C*).
Is it the same size as the angle in the diagram?

Now move the raybox to increase the angle of incidence so that it is *greater* than the critical angle.
What happens?

We call this *total internal reflection*: 'total' because *all* the light is reflected and 'internal' because it only takes place *inside* the more dense substance.

As with ordinary reflection, angle *i* = angle *r*.

Total internal reflection takes place only
when i) the rays are travelling in a dense medium towards a less-dense medium
and ii) the angle of incidence is greater than the critical angle.

The critical angle is about 42° for glass, and 49° for water.

More difficult: using the formula on page 187 and referring to the middle diagram above:

$$\text{refractive index} = \frac{\text{sine } 90°}{\text{sine } C}$$

∴ $$\boxed{\textbf{refractive index} = \frac{1}{\textbf{sine } C}}$$

▷ Uses of total internal reflection: totally reflecting prisms

Experiment 25.7
Put a prism of glass with angles measuring
45°–45°–90° on a sheet of white paper.
Use a raybox to shine in a ray of light as shown
in the diagram:

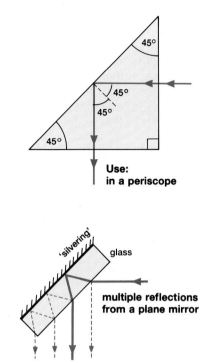

Use:
in a periscope

Notice that the ray of light goes straight through
the first surface. Inside the glass it meets the
second surface at an angle of incidence of 45°.
This is **greater** than the critical angle of 42° and so
total internal reflection takes place.
The long side of the prism acts as a mirror and
turns the ray of light through 90°.

In periscopes, prisms are better than ordinary plane
mirrors (page 179). This is because ordinary mirrors
give multiple images due to multiple reflection in
the glass at the front of the mirror. This is shown
in the diagram:
Two other reasons are that the silvering on a mirror
is easily damaged and that a mirror reflects less
light than a totally reflecting prism.

'silvering'   glass

multiple reflections
from a plane mirror

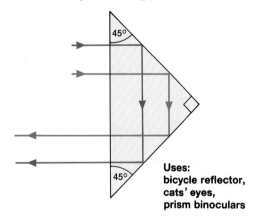

Uses:
bicycle reflector,
cats' eyes,
prism binoculars

Experiment 25.8
Use a 45°–45°–90° prism in a different way, as
shown in the diagram:

Use your raybox to show that the rays are totally
reflected. This is because the angle of incidence
(45°) is greater than the critical angle (42°).

The rays of light are turned through 180° so that
they return to the direction they came from.

Bicycle reflectors and "cats' eyes" reflectors on
roads are built like this so that they reflect back
the light from car headlights.

Diamonds are cut in a similar way to make use of
total internal reflection.

Notice that the two rays of light are inverted
when they come out of this prism.
This is used in prism binoculars (see page 206)
where prisms are used to invert the image and
shorten the binoculars.

▷ More examples of total reflection

The light-guide (or optical fibre)

Experiment 25.9
Shine a lamp into one end of a curved glass or perspex rod.

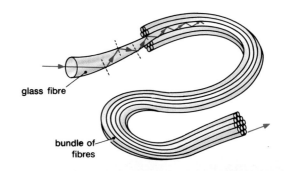

glass fibre

bundle of fibres

What do you see at the other end?

This is possible because of total internal reflection of the rays of light inside the rod (as long as the angle of incidence is greater than the critical angle).

If the rod is thin and flexible it can even be tied in knots or pushed down your throat to take photographs of the inside of your lungs!

Light guides are sometimes used to illuminate the dials on radios and in cars.
The streams of water in illuminated fountains also act as light-guides.

See also the examples on page 194.

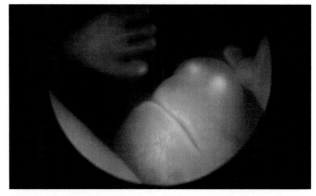

Fibre optic view of a 10-week-old human embryo

Mirages

Sometimes, on a hot day, you might see what look like pools of water on a long stretch of road.

This happens because there are hot layers of air near the hot road and cooler (denser) layers of air higher up.

A ray of light is gradually refracted more and more towards the horizontal. Eventually it meets a hot layer near the ground at an angle greater than the critical angle and total internal reflection takes place.

A mirage: Concorde and its reflection

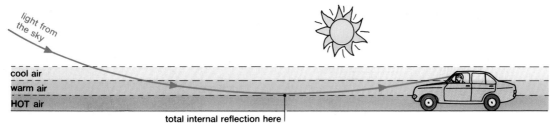

light from the sky

cool air
warm air
HOT air

total internal reflection here
looks like a pool of water

A fish's-eye view

A fish or a diver under water can *see* everything above the surface, but their view is squeezed into a cone with an angle of 98° (twice the critical angle for water).

Outside this cone, the surface looks like a silvered mirror and reflects the light from objects inside the pond (by total internal reflection).

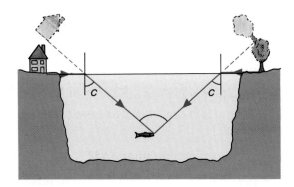

Summary

Rays of light travelling into a 'denser' medium (e.g. from air into glass) are refracted **towards** the normal; when leaving the denser medium (e.g. glass to air) the rays are refracted **away from** the normal. This happens because light travels slower in glass than in air.

The apparent depth of an object is less than its real depth. Total internal reflection takes place if
i) the rays are travelling in a dense medium towards a less-dense medium **and**
ii) the angle of incidence is greater than the critical angle **C**.
    (**C** = 42° for glass, **C** = 49° for water.)

$$\text{Refractive index of a substance} = \frac{\text{speed of light in air}}{\text{speed of light in substance}} = \frac{\text{real depth}}{\text{apparent depth}} = \frac{\text{sine } i}{\text{sine } r} = \frac{1}{\text{sine } C}$$

▷ Questions

1. Copy and complete:
    a) A ray of light travelling from air into glass is . . . . or bent . . . . the normal. A ray of light travelling from water to air is . . . . or bent . . . . the normal.
    b) The speed of light in water is . . . . than the speed of light in air.
    c) A swimming pool looks . . . . than it really is, because light from the bottom is refracted . . . . the normal on passing into air.
    d) Total internal reflection takes place in a glass prism if the angle of incidence in the . . . . is . . . . than the . . . . angle. The angles of a totally reflecting prism are . . degrees . . degrees . . degrees.

2. Copy and complete the diagrams:

3. Explain with the aid of diagrams:
    a) A swimming pool appears to be shallower than its real depth.
    b) Diving for a coin is more difficult than it seems to be from the side of the pool.
    c) A hunter who is spear-fishing does not aim his spear at where the fish appears to be.
    d) In a photograph of a man rowing a boat, the oars seem to be broken.

4. Explain with the aid of diagrams:
    a) To a diver under water, most of the surface looks silvery. Bubbles of air rising from the diver look silvery.
    b) Light can be shone in inaccessible places by means of a 'light guide'.

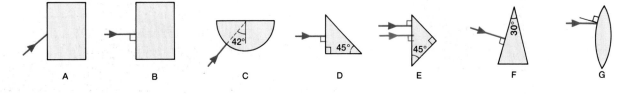

5. Explain with diagrams, how 45°–45°–90° prisms can be used to reflect light through a) 90° b) 180°.

6. The diagram shows a bicycle reflector made of red transparent plastic:

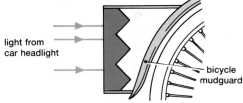

light from car headlight

bicycle mudguard

   a) Copy and complete the diagram.
   b) What does the car driver see?
   c) Why is the reflected light red?

7. Draw a diagram showing a periscope made with 45°–45°–90° prisms. What are the advantages over ordinary plane mirrors?

8. Attic windows, roof-lights on caravans and the covers over fluorescent lights are sometimes made of ribbed plastic or glass:

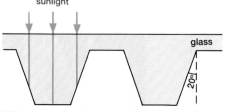

sunlight

glass

20°

   a) Where is the light totally internally reflected and where is it refracted?
   b) Copy and complete the diagram carefully.
   c) Why is the glass made like this?
   d) What would happen if the glass was reversed?

9. Professor Messer goes out of his depth again:

10. Professor Messer fell asleep and dreamed that light could not be refracted by any substance. Why was his dream a nightmare?

11. a) From the diagram, calculate the angle of incidence and the angle of refraction.
    b) What is the refractive index of substance X?

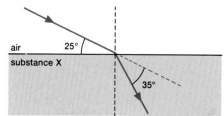

air   25°

substance X

35°

12. Look at the table on page 187 showing the speeds of light. Which substance refracts light: a) the most? b) the least?
    c) If light travels from perspex to water, will it bend towards or away from the normal?

13. Here are some angles of incidence *i* with the corresponding angles of refraction *r* for glass:

i	0	10	20	30	40	50	60	70	80	90
r	0	7	13	19	25	30	35	39	41	42

   a) Plot a graph of angle *i* against angle *r*.
   b) What angle of incidence will give an angle of refraction of 10°?
   c) What angle of refraction would you get for an angle of incidence of 36°?

14. A swimming pool appears to be 3 m deep. If the refractive index of water is $\frac{4}{3}$, what is the real depth?

Further questions on page 235.

THE WATER LOOKS SHALLOW AND I'VE GOT MY FISHING BOOTS ON!

WHAT WENT WRONG?

▷ Physics at work: Fibre optics

An optical fibre is a very narrow *light-guide* (see page 191). It is as thin as a human hair and as flexible. It has a very narrow core made of very pure glass. This glass is so clear that if the seas were made of it, you could see the sea-bed easily.

The core is surrounded by a 'cladding', also made of a very pure glass. However, this glass has a slightly lower refractive index (see page 187). Because the cladding (like air) has a lower refractive index than the core, the light rays inside the core are *totally internally reflected*.

The edges of the core act like a perfect mirror and the light rays can travel a long way inside the glass. Optical fibres can be as long as 200 km.

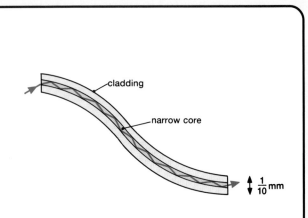

Communications

Optical fibres are being used to replace the copper wires in the telephone system.
The telephone conversations are sent down the optical fibre by switching on and off a LED (Light-Emitting Diode, page 316) or a laser (see opposite page). Infra-red rays are often used.
At the other end, the signals are converted back to electricity by a *photo-diode* which acts rather like an LDR (Light-Dependent Resistor, page 317).

Over 10 000 telephone conversations can be carried at the same time by one fibre! Fibre optics can also be used to send data from one computer to another, or to carry up to ten TV channels.

Optical fibres have many advantages over copper cables. They are thinner, cheaper, carry more signals, have no cross-talk between two telephone conversations and are almost impossible to 'bug'.

Optical fibres in the eye of a needle

Medicine

Doctors use optical fibres to look at the inside of people's lungs and stomachs:

A large number of optical fibres are held together in a bundle which goes down the patient's throat. Light is sent down some of the fibres to illuminate the stomach.
Reflected light comes back up the bundle of fibres to form an image of bright and dark spots of light.

If an ulcer is found, a laser beam can be sent down the fibres to burn and seal the ulcer neatly.

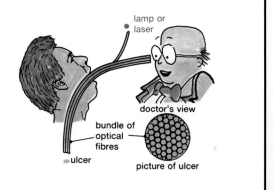

▷ Physics at work: Lasers

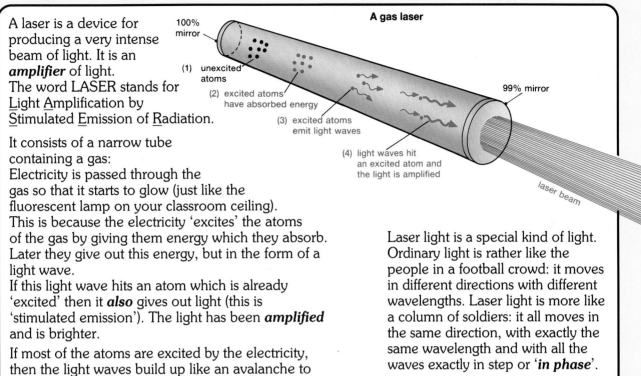

A gas laser

100% mirror
(1) unexcited atoms
(2) excited atoms have absorbed energy
(3) excited atoms emit light waves
(4) light waves hit an excited atom and the light is amplified
99% mirror
laser beam

A laser is a device for producing a very intense beam of light. It is an *amplifier* of light.
The word LASER stands for Light Amplification by Stimulated Emission of Radiation.

It consists of a narrow tube containing a gas:
Electricity is passed through the gas so that it starts to glow (just like the fluorescent lamp on your classroom ceiling).
This is because the electricity 'excites' the atoms of the gas by giving them energy which they absorb. Later they give out this energy, but in the form of a light wave.
If this light wave hits an atom which is already 'excited' then it *also* gives out light (this is 'stimulated emission'). The light has been *amplified* and is brighter.

If most of the atoms are excited by the electricity, then the light waves build up like an avalanche to make a very bright light. The mirrors help by reflecting the light up and down the tube many times. One of the mirrors is not a perfect reflector so some of the light escapes. This is the laser beam.

Laser light is a special kind of light. Ordinary light is rather like the people in a football crowd: it moves in different directions with different wavelengths. Laser light is more like a column of soldiers: it all moves in the same direction, with exactly the same wavelength and with all the waves exactly in step or '*in phase*'.

Lasers can be made from solids and liquids as well as gases. Some are smaller than a pencil, others are as big as a room.

Uses of lasers

Because laser beams are perfectly straight, they are used by surveyors and engineers to ensure that roads, pipes, railway lines, oil tankers and buildings are built accurately.

Laser beams are very intense (*never* look up the beam) and they can melt or even vaporise substances. Laser beams are used to 'drill' holes and 'cut' sheets of metal. They do this more quickly and accurately than metal drills and saws. The cloth for your jeans was probably cut by a laser.

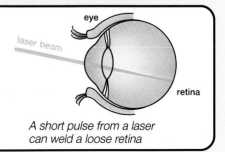

Drilling a hole with a laser beam.
The metal is melted then vaporised.

Lasers are used to weld pieces of metal together. The laser melts the metal, which mixes and later solidifies.
Lasers are used by doctors to 'weld' skin. For example, if the retina in a person's eye becomes loose, a laser is used to 'weld' it back on. Optical fibres can be used to guide the laser beam (see opposite page).

Lasers are also used in 'compact disc' players (see page 306) and laser-TVs can project big TV pictures on to a wall.

eye
laser beam
retina

A short pulse from a laser can weld a loose retina

chapter 26

LENSES

Lenses are very useful – in fact, you are using the lens in your eye to focus on these words now.

Two kinds of lenses

A lens which is thicker at the centre than at the edges is called a *convex* lens. A lens which is thinner at its centre is called a *concave* lens.

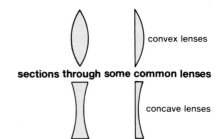

sections through some common lenses

Experiment 26.1   A convex lens
Use a raybox (on a sheet of white paper) to produce several rays of light to shine on to a convex lens. It is better if you can adjust your raybox to make *parallel* rays of light.

Can you see that the rays after the convex lens are getting closer together? We say they are *converging*.
A convex lens is a converging lens.

If the rays of light from the raybox are *parallel* rays, the rays refracted through the lens converge to one point called the *principal focus* (marked F on the diagram).

Parallel rays of light are refracted through the principal focus of a convex lens.
The distance from F to the centre of the lens is called the *focal length* of the lens.

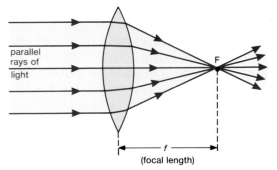

Convex (converging) lens
(used like this as a burning glass or the objective lens of a telescope)

Experiment 26.2   A concave lens
Repeat the experiment, but with a concave lens.

Can you see that the light rays are spread out or *diverged* by this lens?

A concave lens is a diverging lens.

Parallel rays are diverged so that they appear to come from the principal focus (F) of a concave lens.

Objects always look *smaller* through a concave lens.

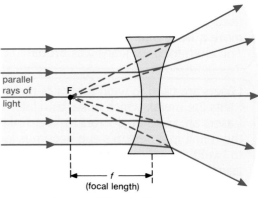

Concave (diverging) lens

▷ Forming images

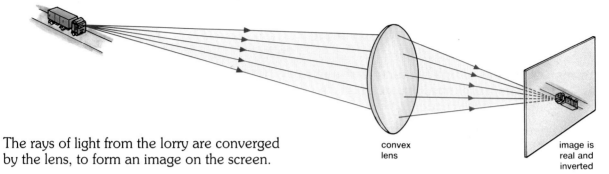

The rays of light from the lorry are converged by the lens, to form an image on the screen.

convex lens

image is real and inverted

This image is:
– **inverted** (upside-down)
– **real** (because rays of light go through it, and the image is shown on the screen).

A camera uses a lens like this (see page 200):

As the lorry moves nearer to the lens, the image moves *away from* the lens.
A camera would have to be re-focussed by moving the film away from the lens.

Images can be **magnified** larger than the object (that is, the magnification is greater than 1) or **diminished** smaller than the object (the magnification is less than 1).
What does 'magnification = 1' mean?

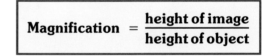

$$\text{Magnification} = \frac{\text{height of image}}{\text{height of object}}$$

Experiment 26.3 Measuring the focal length
Hold a convex (converging) lens so that the parallel rays of light from a distant house or tree are focussed on a piece of paper.

The piece of paper must then be at the principal focus F. The distance from the (inverted) image to the centre of the lens is the focal length f.

Do this experiment first with a fat lens, and then with a thin lens. What do you find?

PARALLEL RAYS

FOCAL LENGTH
f

A fat lens is a strong lens, with a short focal length.

A thin lens is a weak lens, with a long focal length.

▷ Ray diagrams for convex (converging) lenses

To draw accurate ray diagrams you can use two constructions:

Construction 1: parallel rays of light are refracted through the principal focus F. (See experiment 26.1.)
It is also useful to mark points called '2F' at twice the distance.

Construction 2: rays of light passing through the centre of the lens travel straight on. This is true for a thin lens because, at the centre, its sides are parallel.

The image depends on where the object is placed:

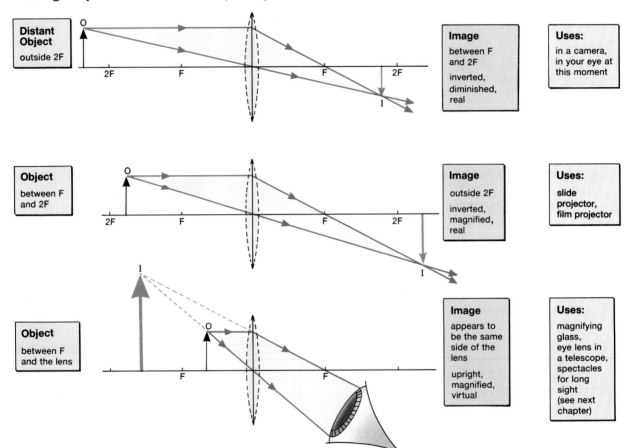

	Image	Uses:
Distant Object outside 2F	between F and 2F inverted, diminished, real	in a camera, in your eye at this moment
Object between F and 2F	outside 2F inverted, magnified, real	slide projector, film projector
Object between F and the lens	appears to be the same side of the lens upright, magnified, virtual	magnifying glass, eye lens in a telescope, spectacles for long sight (see next chapter)

The last diagram is particularly important – it shows how a magnified *virtual* image is seen through a magnifying glass (as in the photograph).

Professor Messer gets in₁s,
Things go wrong when ᵏᵉˢ ᵍᵘᵉˢˢᵉˢ.
As you will see, he's n₉ₕₜ.ᵇʳⁱᵍʰᵗ,
It's up to you to put

▷ Questions

1. Copy out and complete:
   a) A convex lens . . . . rays of light, whereas a concave lens . . . . rays of light.
   b) Parallel rays of light are refracted by a convex lens to a point called the . . . . The focal length is the distance from the . . . . to the lens.
   c) A fat convex lens is a . . . . lens, with a . . . . focal length.
   d) The image in a convex lens depends upon the distance of the . . . . from the lens.
   e) For a concave lens, parallel rays of light are refracted . . . . a point called the . . . . The image in a concave lens is always . . . . than the object.

2. Copy and complete these ray diagrams:

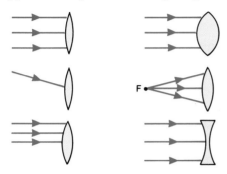

3. a) Describe an experiment to measure the focal length of a convex lens.
   b) Your eye contains a convex lens – why is it unwise to look at the Sun?
   c) Why is it unwise to leave glass bottles in a forest?

4. *Detective Messer looks for clues*
   But hasn't a clue which lens to choose:

5. We can imagine that a lens is made up of many prisms:

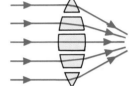

Use the diagram to explain why a convex lens converges rays of light.
Draw a similar diagram for a diverging lens, showing clearly what happens at each surface.

6. An object 4 cm high is placed 15 cm from a convex lens of focal length 5 cm. Draw a ray diagram on graph paper (full size or $\frac{1}{2}$ scale) to find the position, size and nature of the image.

7. Professor Messer did an experiment with a convex lens. He put an object at different distances (25 cm, 30 cm, 40 cm, 60 cm, 120 cm) from the lens. In each case he measured the distance of the image from the lens. His results were: 100 cm, 24 cm, 60 cm, 30 cm, 40 cm. Unfortunately his results are written in the wrong order.
   a) Rewrite the image distances in the correct order.
   b) Plot a graph of object distance : image distance.
   c) What would be the image distance if the object distance was 90 cm?
   d) Which of the object distances gives the biggest image?
   e) When the object distance equals the image distance, they are each at twice the focal length from the lens. What is the focal length of this lens?

Further questions on page 235.

chapter 27

 ptical instruments

First we will consider optical instruments that form **real** images:
the camera, your eye, the film projector.
Then we will consider optical instruments that form **virtual** images
for your eye to look at: microscopes and telescopes.

▷ The lens camera

A camera consists of a light-tight box with a convex (converging)
lens at one end and the film at the other end.

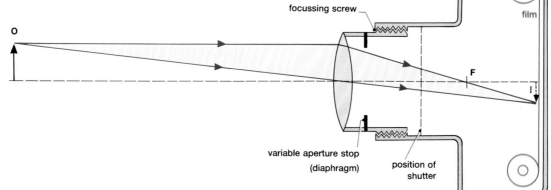

The image on the film is small, and upside-down.
It is a **real** image, because the light-rays actually pass through it.

Experiment 27.1
Convert the pinhole camera (described on page 176)
into a lens camera by enlarging the hole at the front of
the box and holding a lens over the hole (choose a
lens with a focal length about as long as the box).
Adjust the position of the lens for either near or far objects
to make a sharp image on the screen.

I knew a famous photographer chap,
He snapped everyone in the town,
But his clients weren't happy,
– their tempers would snap,
When he made them all stand
upside down.

Is the image erect (upright) or inverted (upside-down)?
If the object is coloured, is the image coloured?
Why is the image brighter than in a pinhole camera?

▶ Operating a camera

1 Focussing

A camera is focussed by moving the lens. If the object moves *nearer* the lens, its image moves farther away (see page 197). So to keep the image in focus on the film, the lens must be moved farther *out*.

Night scene
1 second at f/2, fast film

2 The shutter

The amount of light entering the camera can be controlled by the length of time that the shutter is open. Fast moving objects will appear blurred unless the exposure time is very short (perhaps only $\frac{1}{500}$ second).

Moving object
1/500th second at f/4, fast film

3 The aperture stop or diaphragm

The amount of light entering the camera can also be controlled by varying the size of the hole in the diaphragm which is just behind the lens. To take a photograph in dim light, a large hole is needed.

The control which varies the size of the hole is marked in *f*-numbers. A value of *f*/8 means that the diameter of the hole is $\frac{1}{8}$th of the focal length. A value of *f*/11 is a *smaller* opening.

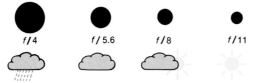

f/4   f/5.6   f/8   f/11

Still object; detail important
1/15th second at f/8, slow film

The size of this hole also controls the *depth of focus*. A large depth of focus means that both near and far objects will appear to be in focus at the same time. This is obtained by having a *small* hole in the diaphragm.

4 The film

The film is coated with chemicals that are sensitive to light. The chemicals are changed by the light of the image. A 'fast' film is more sensitive to light than a 'slow' film. Fast film can be used in dim light or when a fast shutter speed is needed to 'freeze' a moving object. However, a fast film may not give the quality of a slow film.

Large depth of field
1/15th second at f/16, slow film

Setting the controls

The shutter time and aperture that you should use depend upon:
a) the brightness of the object,
b) the sensitivity or 'speed' of the film,
c) the kind of effect that you want.

Look at the values given with each photograph:
Can you decide why the photographer chose the values in each case?

Small depth of field
1/1000th second at f/2, slow film

▷ The human eye

Study the different parts of the diagram of your eye:

The light enters your eye through the transparent **cornea**, passes through the **lens** and is focussed on the **retina**.
The retina is sensitive to light and sends messages to your brain by way of the **optic nerve**.

The **iris** changes in size to vary the amount of light that enters through the **pupil**.
When is your pupil small and when is it large?

Notice that the image on your retina is inverted (see also the ray diagram on page 198). Although the image is inverted, your brain has learned to interpret this correctly.

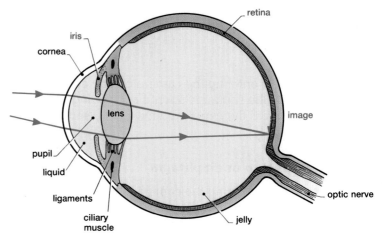

top view of your right eye

Focussing your eye

Most of the bending of the light rays is done by the curved cornea, but your lens can change its shape to alter its focal length slightly.
To focus distant objects the lens is **thin** (with the **ciliary muscles** relaxed).

To focus on a near object (like this book) the ciliary muscles tighten (rather like the draw-strings on a duffel bag). This allows the lens to become fatter (with a shorter focal length) and so it can focus the light from near objects.

Experiment 27.2   Find your eye's blind spot
Cover your **left** eye, hold the page at arm's length and keep staring at Professor Messer's magic wand. Then move the page slowly towards your eye until the rabbit disappears.

This happens because the light rays from the rabbit are then arriving at a place where there is a gap in your retina. This gap is where the optic nerve leaves your eye.

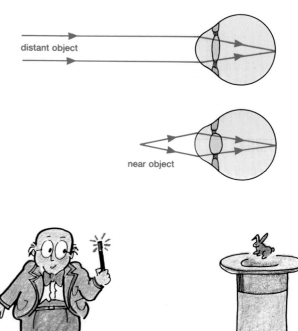

Experiment 27.3   Binocular vision
Hold a pencil in one hand and close one eye.
Then, with another pencil in your other hand,
try to make the two pencil points touch.

Now try this with both eyes open.
Which is easier? Why is this?

Having two eyes open allows you to see an object
from two slightly different angles and so you can
judge distances more accurately.

Experiment 27.4   Persistence of vision
Look at the little man at the bottom corner of
this page and the next few pages.
What appears to happen if you flick over the
corners of the pages quickly?

The effect of the image on your retina lasts for
about $\frac{1}{10}$th of a second – so that you can appear
to see an image after it has disappeared.
This effect is used in television and in the cinema
where you are shown about 25 pictures every
second, each picture changing slightly from the
one before.

Experiment 27.5   Optical illusions
Judge each of the diagrams with your eyes and
then answer each of the questions by measuring
with a ruler.

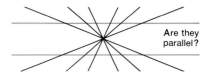

Are they parallel?

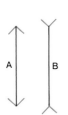

Which is longer,
A or B?

Which is greater,
the height of the hat or
the width of the brim?

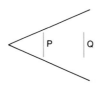

Which is longer, P or Q?

All equal in length or
one shorter?

▷ Defects of vision

People with normal vision can focus clearly on very distant objects at 'infinity'. We say their *far point* is at infinity.

People with normal vision can also focus clearly on near objects such as this book. The closest point at which they can see an object clearly is called the *near point*. The near point for adults is often about 25 cm from the eye (less for teenagers).

> Experiment 27.6
> Hold a book at arm's length and move it closer to find the nearest distance that you can focus it clearly without straining your eyes.
> What is the distance to your near point?

Long sight

A person who can see distant objects clearly but cannot focus near objects is said to be *long-sighted*.

This is because his eye-ball is too short or his eye-lens is too thin even though his ciliary muscles are fully squeezed.

This means that rays of light from a close object are focussed towards a point *behind* his retina.

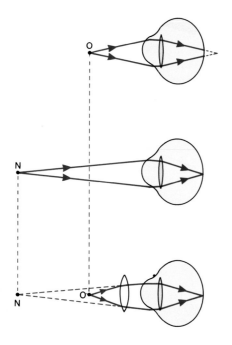

Rays of light from this person's near point N can just be focussed by his eye.

This person should wear a *convex* (*converging*) spectacle lens. Then rays of light from a close object are converged so that, as far as his eye is concerned, the rays appear to come from his near point N. Then they can be focussed by his eye.

Did you hear about the optician who fell into his lens-grinding machine and made a spectacle of himself?

Short sight

A person who can see near objects clearly but cannot focus distant objects is **short-sighted**.

This is because her eye-ball is too long or her eye-lens is too strong. (The lens is too thick even though the ciliary muscles are relaxed.)

This means that rays of light from a distant object (at 'infinity') are focussed **in front of** her retina:

Rays of light from this person's far point F can just be focussed:

This person should wear a **concave** (**diverging**) spectacle lens.
Then rays of light from a distant object are diverged so that they appear to come from her far point F and so they can be focussed by her eye:

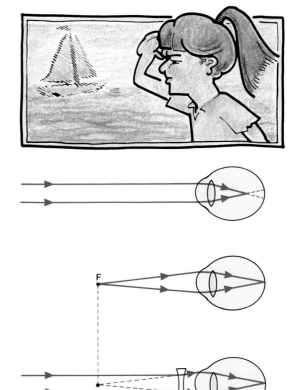

Other defects

In older people, the eye lens may become cloudy (a cataract), or the pressure in the eye may build up too high (glaucoma). The retina may become detached (see page 195).

A short-sighted teacher called Rose,
Confused the optician she chose,
And so in her specs,
Each lens was convex,
And her far point was the end of her nose.

Experiment 27.7   Colour-blindness
If some specially coloured charts are available, your teacher may be able to test you for colour-blindness (the commonest kind is when red and green look the same to the colour-blind person).

Things aren't always what they seem, Is the Professor in a dream?

▷ Telescopes

The astronomical refracting telescope

A simple refracting telescope contains 2 lenses:
1. an **objective lens** which has a **long** focal length
2. an **eye lens** which has a **short** focal length.

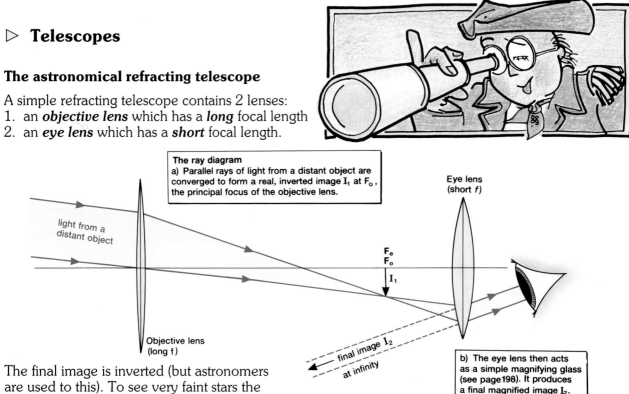

> **The ray diagram**
> a) Parallel rays of light from a distant object are converged to form a real, inverted image I_1 at F_o, the principal focus of the objective lens.

Eye lens (short *f*)

F_e
F_o
I_1

light from a distant object

Objective lens (long *f*)

final image I_2 at infinity

> b) The eye lens then acts as a simple magnifying glass (see page 198). It produces a final magnified image I_2.

The final image is inverted (but astronomers are used to this). To see very faint stars the first lens must be as wide as possible, to collect as much light as possible.

Experiment 27.8   Making a refracting telescope
Use two lenses, of focal lengths 20 cm and 5 cm, with a half-metre rule and some plasticine, as shown in the diagram. (Can you remember how to find the focal length of each lens? See page 197.)

Adjust the distance between the lenses until the image is in sharp focus.

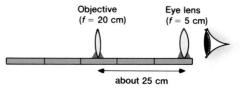

Objective (*f* = 20 cm)   Eye lens (*f* = 5 cm)

about 25 cm

Is the image magnified?
Is the image erect or inverted?

Prism binoculars

Binoculars are made of two refracting telescopes side by side, one for each eye.

In experiment 25.8 on page 190, we saw that $45° - 45° - 90°$ prisms can use total internal reflection to **invert** rays of light.

By including two of these prisms in each telescope, the final image I_2 is turned the right way up, and the telescope can be made shorter and more compact.

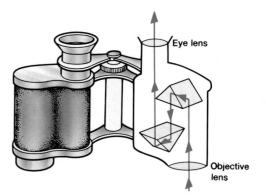

Eye lens

Objective lens

Reflecting telescopes

The *reflecting* telescope (invented by Sir Isaac Newton) has a large concave mirror to collect the light and to converge it towards a small plane mirror.

The plane mirror reflects the light sideways to an eye lens. This acts as a magnifying glass in the same way as the eye lens on the opposite page.

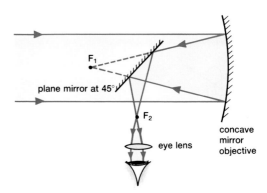

Large telescopes

Large lenses tend to sag, so all the very large telescopes are the reflecting type because mirrors can be supported all over the rear surface.
The largest objective lens in a telescope is 1 metre in diameter; the largest mirror is 6 metres in diameter.
This means that the larger reflecting telescopes can collect more light and let astronomers see fainter stars in the sky. Some telescopes are so sensitive they could detect a candle at 20 000 km!

Radio telescopes are the reflecting kind:

We have already looked at other examples of reflecting telescopes – the solar furnace (page 54), and the satellite TV dish (page 184).

A radio telescope

▷ The microscope

A microscope uses two lenses to magnify very small objects.

Light from a mirror shines through the object and into the objective lens. This produces a large image, just like a slide projector (see the next page).

You look at this image through an *eye lens*, which acts as a simple magnifying glass (see page 198). This produces a final magnified image for your eye to look at. The image is upside-down.

The image can appear 1000 times bigger than the object.

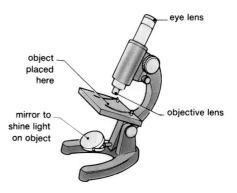

▷ The projector (for slides or for ciné film)

A projector contains a **lamp** and a **concave mirror** to make the image brighter. The lamp is placed so that the rays of light are reflected back along their own path.

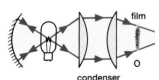
concave mirror   lamp

To give a brighter picture, a **condenser** is included. It is usually made of two plano-convex lenses, as shown.

Now the light is converging towards the screen and the film is illuminated both brightly and evenly.

The light is then scattered by the film and focussed by a convex (converging) **projection lens** on to the screen:

condenser

O   projection lens

I

large image on screen

The film O is placed between F and 2F of the projection lens (see the middle diagram on page 198), so that the image I is real, inverted and magnified. The film is put in the projector upside-down so that the picture is seen the right way up.

How is the projector focussed?
Why does a projector usually contain a fan?

Summary

In a camera, the real, inverted image is focussed by moving the lens.
The brightness of the picture depends on:
a) the exposure time
b) the size of the aperture (*f*-number)
c) the 'speed' of the film.

A projector produces a real, inverted, magnified image on the screen.

In your eye, the real, inverted image is focussed by changing the shape of the lens.

With long sight, a person cannot see near objects. It is corrected by a convex (converging) spectacle lens.

With short sight, a person cannot see distant objects. It is corrected by a concave (diverging) spectacle lens.

▷ Questions

1. Comparing the eye and a camera, make lists of a) similarities and b) differences.

2. Draw a large labelled diagram of a human eye. Explain the function of a) ciliary muscles b) the iris c) the retina. Where is the blind spot? Why are two eyes better than one?

3. Use a ruler to check the illusions on page 203. In order to look slimmer, should you wear clothes with vertical or horizontal stripes? What optical illusions are used by artists?

4. Among animals, the 'hunters' usually have their eyes facing forward at the front of their heads, whereas the 'hunted' usually have eyes at the sides of their head. Why is this?

5. In a certain murder investigation, it was important to discover whether the victim was long-sighted or short-sighted. How could a detective decide by examining the spectacles?

6. With the aid of ray diagrams, explain what is wrong with the eyes of a person with a) long sight b) short sight. Use ray diagrams to explain how each fault can be corrected.

7. The diagram shows possible settings for a camera using a fast film. Analyse the data.

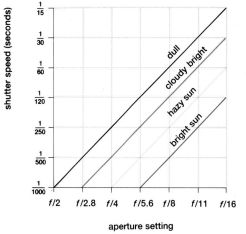

Explain, with reasons, the settings of shutter and aperture you would use for these photos:
a) a speeding car in bright sun
b) a rabbit in a wood
c) a friend in hazy sunlight with a mountain behind her
d) the friend's head without the background.

8. Professor Messer fell asleep and dreamed that light could not be refracted.
Why was his dream a nightmare?

9. Another night, Professor Messer dreamed that none of the optical instruments in this chapter (except the eye) had been invented. In what ways would our lives be different if this was so?

10. Discuss the physics of this cartoon:

Further questions on page 235.

chapter 28

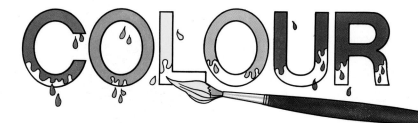

What causes the colours in a rainbow or in a diamond? To answer this, we must investigate the **spectrum** of white light.

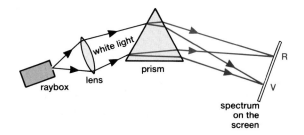

Experiment 28.1   Newton's experiment
Use a raybox (or a slide projector) with a narrow slit. Shine a ray of white light through a glass **prism** and on to a white screen, as shown:

What do you see on the screen?
How many different colours can you count?

It seems that white light is really a mixture of several colours and can be split up by a prism. We say that the white light from the raybox has been **dispersed** by the prism to form a **visible spectrum.**
The colours, in order, are **R**ed, **O**range, **Y**ellow, **G**reen, **B**lue, **I**ndigo, **V**iolet. (Remember them by forming them into a boy's name: **ROY G. BIV.**)

The colour that is **deviated** (bent) least by the prism is red; violet is deviated through a larger angle, as shown in the diagram.

A name you should remember
As long as you may live.
The colours of the spectrum,
Are known by **ROY G. BIV**.

Experiment 28.2   An almost pure spectrum
In that experiment, the spectrum was 'impure' because the colours overlapped each other on the screen. To form a 'pure' spectrum, a lens must be used to focus the rays.
Use a narrow slit and move the screen (or the lens) until the colours do not overlap and the spectrum is (almost) pure.

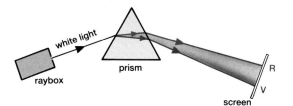

The colours of a rainbow or a diamond or a cut-glass necklace are caused by the refraction and dispersion of white light into a spectrum.

If white light is composed of the colours **ROY G. BIV**, then we should be able to get white light by adding these colours together.

Spin a colour wheel quickly so that the colours appear to be mixed together.

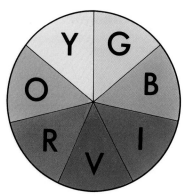

Why does it appear to be white?
What do we really mean by white light?

How does a prism disperse white light into a spectrum?

Different colours of light have different wave-lengths (rather like waves at sea have a different wavelength from ripples on a pond – see page 167).

All colours of light travel at the same speed in a vacuum. When they enter a transparent substance like glass, they all slow down but by *different* amounts.
Because they slow down, they are refracted (see page 187) but because they slow down by *different* amounts, different colours are refracted through *different* angles.

Violet (the shortest wavelength) is slowed down the most and so is refracted through the largest angle. Red (a longer wavelength) slows down less and so is deviated through a smaller angle.

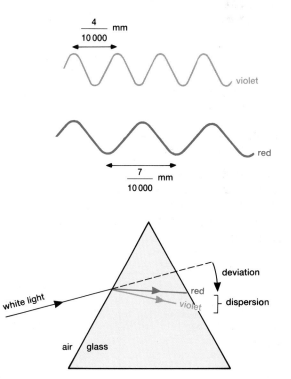

The full electro-magnetic spectrum

The visible spectrum is only a small part of a much larger spectrum containing many other wavelengths which we cannot see with our eyes (see also page 172).

The full electromagnetic spectrum is shown in more detail on the next four pages.
Study each part of it carefully.

The visible spectrum

▷ The electromagnetic spectrum

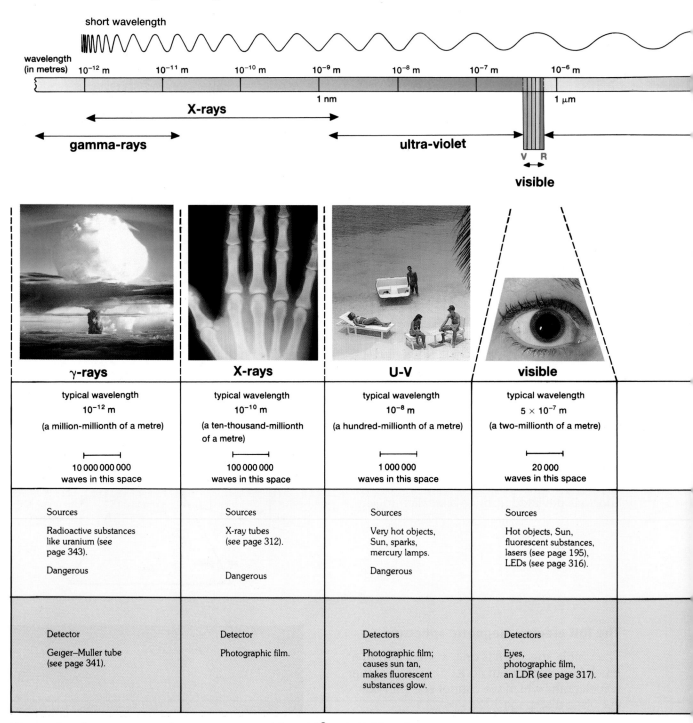

short wavelength

wavelength (in metres) 10^{-12} m 10^{-11} m 10^{-10} m 10^{-9} m 10^{-8} m 10^{-7} m 10^{-6} m

1 nm   1 μm

X-rays

gamma-rays

ultra-violet

V   R

visible

γ-rays	X-rays	U-V	visible	
typical wavelength 10^{-12} m (a million-millionth of a metre)	typical wavelength 10^{-10} m (a ten-thousand-millionth of a metre)	typical wavelength 10^{-8} m (a hundred-millionth of a metre)	typical wavelength 5×10^{-7} m (a two-millionth of a metre)	
⊢——⊣ 10 000 000 000 waves in this space	⊢——⊣ 100 000 000 waves in this space	⊢——⊣ 1 000 000 waves in this space	⊢——⊣ 20 000 waves in this space	
Sources Radioactive substances like uranium (see page 343). Dangerous	Sources X-ray tubes (see page 312). Dangerous	Sources Very hot objects, Sun, sparks, mercury lamps. Dangerous	Sources Hot objects, Sun, fluorescent substances, lasers (see page 195), LEDs (see page 316).	
Detector Geiger–Muller tube (see page 341).	Detector Photographic film.	Detectors Photographic film; causes sun tan, makes fluorescent substances glow.	Detectors Eyes, photographic film, an LDR (see page 317).	

All travel at the same speed, in a vacuum, of 3×10^8 m/s (300 million metres per second).

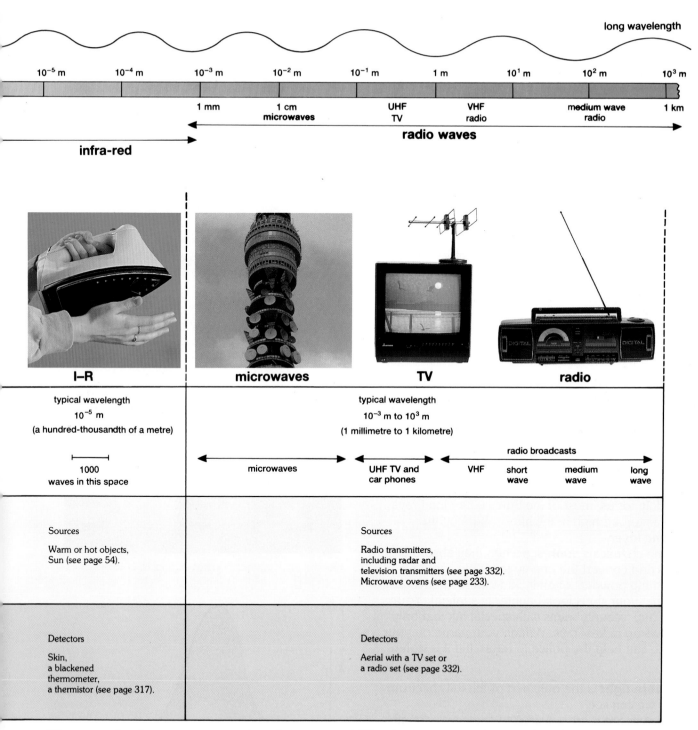

long wavelength

| 10^{-5} m | 10^{-4} m | 10^{-3} m | 10^{-2} m | 10^{-1} m | 1 m | 10^1 m | 10^2 m | 10^3 m |

1 mm   1 cm microwaves   UHF TV   VHF radio   medium wave radio   1 km

radio waves

infra-red

I–R	**microwaves**	**TV**	**radio**
typical wavelength 10^{-5} m (a hundred-thousandth of a metre)	typical wavelength 10^{-3} m to 10^3 m (1 millimetre to 1 kilometre)		
1000 waves in this space	microwaves UHF TV and car phones radio broadcasts: VHF short wave medium wave long wave		
Sources Warm or hot objects, Sun (see page 54).	Sources Radio transmitters, including radar and television transmitters (see page 332). Microwave ovens (see page 233).		
Detectors Skin, a blackened thermometer, a thermistor (see page 317).	Detectors Aerial with a TV set or a radio set (see page 332).		

Wave speed = frequency × wavelength (see page 167).   Each section is discussed on the next two pages:

▷ The electromagnetic spectrum

Each part of the full electromagnetic spectrum (see the previous page) has its own properties and uses.

Gamma-rays have a very short wavelength and are very penetrating. They are produced by radioactive substances and are dangerous to humans unless used very carefully.
Gamma-rays can be used to sterilise food so that it does not rot so quickly (see page 349).
They can be used to check that two pieces of metal have welded together properly.

X-rays are very like gamma-rays, but are produced by an X-ray tube (see page 312). Doctors and dentists use X-rays to check bones and teeth.

Like gamma-rays, X-rays are used by engineers to check welds and metal joints.
In factories, X-rays are used to check that meat and vegetables do not have metal or stones in them.

Ultra-violet rays can also be dangerous to us. Hot objects like the Sun produce ultra-violet rays, and so do the electric arcs used in electric welding:

Small amounts of ultra-violet rays are good for us, producing vitamin D in our skin. Ultra-violet rays also give us a suntan. Large amounts of ultra-violet are bad for our *eyes* and cause skin cancer. Luckily for us, most of the Sun's ultra-violet rays are absorbed high in the atmosphere by the *ozone layer*.
Some chemicals *fluoresce* when they absorb U-V rays and convert the energy to visible light rays. Washing powders often include these chemicals so that your shirts look brighter in sunlight. You can buy 'security' pens with special ink that only shows up in U-V rays. Writing your name on your radio can help the police to return it if it is stolen.

Visible light is the only part of the full spectrum that we can see.
Our eyes are more sensitive to some wavelengths than to others, as the graph shows:

Your eye is most sensitive to green-yellow light.

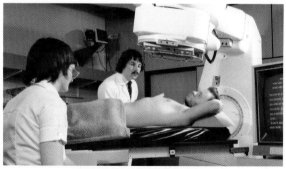
A carefully-controlled beam of gamma-rays is used to kill cancer cells. This is radiotherapy.

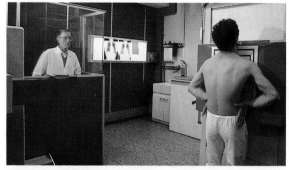
Having an X-ray (the operator at a safe distance)

Welders must protect themselves against U-V rays

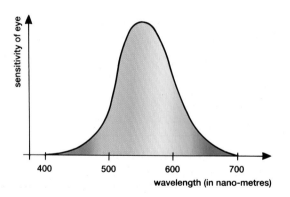

Infra-red rays are given out by warm objects. In fact, *every* object that has a temperature above absolute zero gives out infra-red waves – including you!
Because of this, fire-fighters can use infra-red viewers to search for unconscious people in smoke-filled buildings, and to search for survivors trapped alive beneath buildings by earthquakes.

Special photographs taken with infra-red rays are called **thermographs**. They can help doctors to detect circulation problems (where the skin is cooler, so the thermograph is bluer) and arthritis and cancer (where the skin is warmer).

Astronomers take infra-red photos to get data about the temperatures of planets and stars. Burglar alarms are designed to detect the infra-red rays from an intruder.
Other uses of I-R rays are shown on page 56.

Radio waves have a longer wavelength.
There are several kinds of radio waves:

Microwaves are the shortest of the radio waves. They are used in microwave ovens (see page 233). They are also used for communicating with satellites (see page 333) and for radar:

UHF waves (Ultra High Frequency) are used to transmit TV programmes to your home.

VHF waves (Very High Frequency) are used to transmit local radio programmes, and local police and ambulance messages.

Medium wave and **long wave** radio are used to transmit over long distances. Because they have long wavelengths, they can diffract round the curve of the Earth and round hills (see page 169).

Properties of all electromagnetic waves:

1 They transfer energy from one place to another.
2 They are all transverse waves.
3 They can all be reflected, refracted, diffracted.
4 They can all travel through a vacuum.
5 They all travel at 300 000 000 m/s in a vacuum.
6 Wave speed = frequency × wavelength.
7 The shorter the wavelength (the higher the frequency), the more dangerous they are.

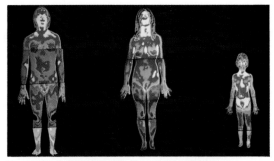

These thermographs are coloured so that white/red = most radiation; blue = least

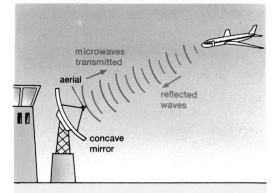

Radar uses reflected microwaves to detect planes

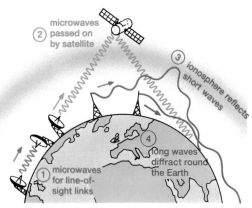

Communicating with other parts of the Earth

▷ Subtraction of colour (absorption)

Within the visible part of the electromagnetic spectrum, colour plays an important part in our lives and in our environment. Colour mixing is important to artists and to textile designers.

Filters

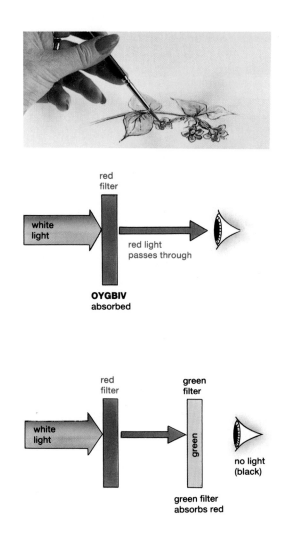

Experiment 28.4
Look through a red plastic filter. What happens to white light as it passes through the red filter?

Remember white light is really a mixture of several colours (ROY G. BIV).
Which one of these colours passes through a red filter? Which colours are *absorbed* (and subtracted) by a red filter?

What happens as white light passes through a *blue* filter?

Experiment 28.5   Two filters
Place a red filter and a green filter together and look through them.

What happens to the white light? Why?

Try the same experiment with red and blue filters and then green and blue filters. (In practice most filters are not perfect, so some light may get through both.)

Coloured objects in white light

When white light shines on this white paper, the white light is reflected to your eye. When white light shines on this black ink, all the light is absorbed and none is reflected.

What happens when white light shines on this red ink (or red paint or red cloth)?

White light (really ROY G. BIV) shines on the red ink which absorbs OYGBIV leaving only red to be reflected to your eye and so the ink appears red.
(In practice, red paint usually reflects small amounts of other colours as well.)

What happens when white light shines on to *blue* ink?

What happens if you mix red, green and blue inks together? Why?

Coloured objects viewed through filters

Experiment 28.6
Look at white paper through a green filter.
Look at a green book through a green filter.
What do you see?

Look at a red book through a green filter.

What do you see?
Explain why this is so.

Use other filters and brightly coloured objects
to fill in a copy of this table:

Colour of object	Filter	Appearance
White	Green	Green
Green	Green	Green
Red	Green	Black
Blue	Red	

Look at the cartoon through a red filter and then
through a green filter. Explain what you see.

Coloured objects in coloured light

Experiment 28.7
Place a red filter over a raybox or a torch and shine the
red light on to a bright blue object (in a dark room).

What do you see?
Explain why this is so.

Use different colours and fill in a table like the one above.

Look at the cartoon in red light and in green light and
explain what you see.

Mixing paints

When an artist mixes two paints, the final colour is the one
which is not absorbed by either of the paints.

Use the diagrams to explain why yellow + blue = green.

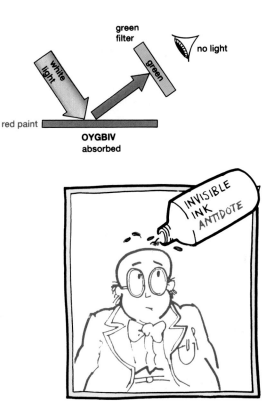

red paint — OYGBIV absorbed

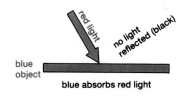

blue object — blue absorbs red light — no light reflected (black)

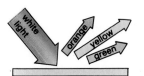

yellow paint
(absorbs R · · · **BIV**)

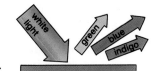

blue paint
(absorbs **ROY** · · · V)

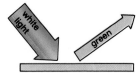

appears green
(**ROY · BIV** absorbed)

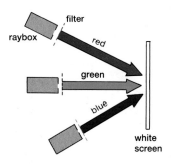

▷ Addition of colours

Experiment 28.8   Primary colours
Use 3 rayboxes (or slide projectors) fitted with a red, a green and a blue filter and shine them on to the same white screen so that the colours overlap and add together.

If you adjust the brightness or alter the distances correctly, you can make the three colours add up to make the screen appear white. Because they add together to make white light, **red, green and blue are called the primary colours of light**.

In your eye there are three types of colour-sensitive cells, called **cones**. One type of cone detects red light, one type detects green light, and the third type responds to blue light.

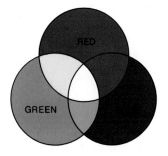

Experiment 28.9   Secondary colours
Switch on the lamps in pairs so that you see the result of adding *two* primary colours. These colours are called *secondary* colours.

The 3 secondary colours are **yellow**; a purple colour called **magenta**; and a greeny-blue colour called **cyan**.

Switch on all 3 lights and put your hand in front of the screen to cause shadows. Can you explain what you see?

Look closely at the screen of a colour TV set (see page 307). Can you see the red, green and blue dots? Why are these three colours used? Which dots must light up to make the TV screen look yellow?

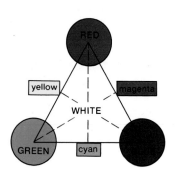

The colour-triangle diagram shows the primary and secondary colours and also pairs of **complementary** colours which add together to form white, e.g. red + cyan = white
green + magenta = white.
Can you name two more colours that are complementary?

Summary

The spectrum of white light can be remembered by ROY G. BIV.
Red has the longest wavelength and is deviated least by a prism.

The full electromagnetic spectrum in order of increasing wavelength is: gamma-rays, X-rays, ultra-violet, visible light, infra-red, radio waves.

Subtraction of colours (absorption): filters, paints, pigments and inks subtract colours, e.g. a blue flower absorbs ROYG, , IV and reflects blue light. In red light it looks black.

Addition of colours: the three primary colours (R,G,B) add to give white.
Secondary colours are yellow, magenta, cyan.

▷ Questions

1. Copy out and complete:
   a) White light is composed of . . . . colours, in order: red, ... , ... , ... , ... , ... , ... , which form the visible . . . . . The colour with the longest wavelength is . . . . . The colour deviated through the largest angle by a prism is . . . . .
   b) The full electromagnetic spectrum, in order, is: gamma-rays, ... , ... , ... , ... , ... The section with the longest wavelength is . . . . .
   c) The three primary colours of lights are . . . . , ... , ... and when added together they give . . . . light.

2. The diagram shows white light being dispersed by a prism:

   a) What colour would you see i) at X ii) at Y?
   b) What may be detected i) above X ii) below Y?

3. Name a type of electromagnetic wave which:
   a) can pass through metals,
   b) can cause a suntan,
   c) is used for radar,
   d) is diffracted round hills,
   e) is emitted by warm objects.

4. On the chart on pages 212–13, find the wavelength of your favourite radio station.

5. From the eye-sensitivity graph on page 214, which wavelength has the same sensitivity as red light of wavelength 650 nm?
On that graph, where is a) I-R b) U-V?

6. In the radar diagram on page 215, the wave takes 0.0001 s to travel to the plane and back. How far away is the plane?

7. Draw a scale diagram of the Earth (radius 6400 km) and a 'geostationary' satellite moving over a fixed point with the Earth, in an orbit of radius 42 000 km. Show that three satellites spaced out in this orbit can cover communications over the whole Earth. The microwave path from Britain to USA via such a satellite is about 90 000 km. Calculate the time taken for the waves to travel. Would you notice this in a telephone conversation?

8. What colour would a green book look:
   a) in white light?   b) in green light?
   c) in red light?   d) through a red filter?

9. Copy out and complete the table:

Colours of lights added together	Appearance
Red + Green + Blue	
Red + Green	
Red + Blue	
Blue + Green	
Blue + Yellow	

10. Why does red lipstick look unpleasant under yellow street lighting?

11. Write essays or discuss the following topics:
   a) The use of coloured spot lights for different stage effects.
   b) The choice of colours for road signs, advertisements, clothes and rooms.
   c) The use of colour in language, e.g. feeling blue, red with rage, green beginner, etc.

12. *His face is red, he's feeling blue,*
Can't see his green mistake – can you?

Further questions on page 236.

chapter 29

SOUND!

What is sound caused by?

> *Experiment 29.1*
> Hold down one end of a ruler firmly to the bench.
> Flick the other end to make a sound.

Look at the end of the ruler. What is it doing?
Does it make any sound when it has stopped
vibrating?

> *Experiment 29.2*
> Place your fingertips against the front of your
> throat. What can you feel when you make a noise?

> *Experiment 29.3*
> Get a tuning fork and bang it on a cork to make it
> vibrate (it vibrates like two rulers fastened
> together).   What can you hear?

Can you see that the ends are vibrating? If not,
touch the ends to the surface of some water in a
beaker. What do you see?

These experiments show that **sound is caused by
vibrations** and is a form of kinetic energy (see
page 114).
How are these vibrations caused in
a) a drum?   b) a guitar?

How are the vibrations caused by a gramophone
record?

> *Experiment 29.4*
> Look at an old gramophone record through a
> magnifying glass. Can you see the wobbly grooves?

> Push a needle through a piece of card and then
> put the point of the needle on to the record as it
> revolves on a turntable.

What can you hear?
What are the wobbly grooves doing to the needle
and card?

How do the vibrations travel to your ear?

Experiment 29.5
Stretch a long 'slinky' spring along a smooth bench and vibrate one end to and fro along the length of the spring, to send a longitudinal wave down the spring (see page 166 again).

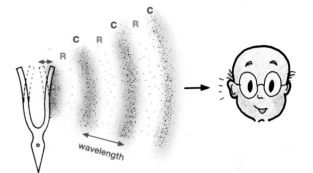

If you look closely at the spring you can see that, at any instant, some parts of the spring are pushed closer together (**compression**) and some parts are pulled farther apart (**decompression** or **rarefaction**).

It is the same with a sound wave in air. In some places the **molecules** of air are pushed together at a slightly higher pressure (compression) and in some places the molecules are farther apart at a slightly lower pressure (rarefaction).

These compressions and rarefactions shoot out across the room to your ear, travelling at the speed of sound. Molecules of air do not travel across the room – they just vibrate to and fro.

The **wavelength** of the sound is the distance between two successive compressions (or rarefactions).

For sound waves, like all other waves (page 167),

speed	**=**	**frequency**	**×**	**wavelength**
(m/s)		(Hz)		(m)

What happens if there are no molecules?

Experiment 29.6
Hang an electric bell inside a jar connected to a vacuum pump.
Switch on the bell. Can you hear it?

Start the pump to take the air molecules out of the jar. What happens to the sound of the bell?

Sound cannot travel through a vacuum, because there are no molecules to pass on the vibrations. Why can't we hear the sound of the explosions on the Sun?

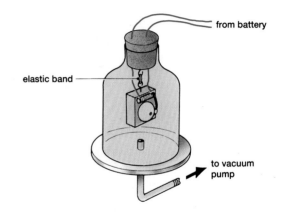

▷ Reflection of sound

Have you ever heard an *echo*?

Can sound be reflected by walls?
Why do noises sound louder in a room than in the open air?

You may have seen that some of the compressions on the slinky spring (in experiment 29.5) were reflected back down the spring. In the same way, compressions and rarefactions are reflected back from the walls of a room.

If the distance is long enough you may hear a clear echo.

Echo-sounding

Ships can use echoes to find the depth of the sea.

Example
A ship sends out a sound wave and receives an echo after 1 second. If the speed of sound in water is 1500 m/s, how deep is the water?

Time for sound to reach the bottom $= \frac{1}{2}$ second (and $\frac{1}{2}$ second to return)

$\therefore$ Depth of water $= \frac{1}{2} \times 1500$

$= 750$ metres

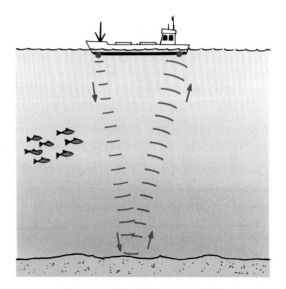

Fishing boats use this 'sonar' to detect shoals of fish. If the shoal of fish in the diagram swim under the boat, how will the captain know?

This echo-sounder uses a very high frequency sound. This **ultrasound** is so high that we cannot hear it. Because ultrasound has a high frequency, it has a short wavelength. This means that it can be sent out as a narrow beam (if the waves are longer, they spread out more due to *diffraction*, see page 169).

Bats also use ultrasound and listen to the echoes to 'see' their surroundings (see page 232).

Geologists also use echo-sounding on land. By hitting the ground hard (or using explosives) and detecting the echo, the geologist can get information about the different layers of rocks.

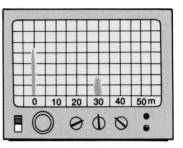

The sonar display on the captain's oscilloscope. It shows the transmitted pulse and the echo.

▷ Measuring the speed of sound

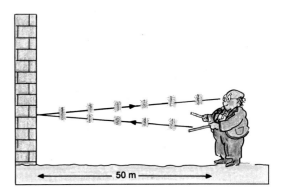

Experiment 29.7   Outdoors

Stand a measured 50 metres from a large wall. Clap, or bang sticks together, and listen to the echo. Then try to clap in an even rhythm of clap–echo–clap–echo–clap . . . while a friend times 100 of your claps with a stopwatch.

During the time from one clap to the next clap, the sound would have time to go to the wall and back, **twice** – that is a distance of 200 metres. In the time of 100 claps, the sound would travel $200 \times 100 = 20\,000$ metres.

$$\therefore \text{Speed} = \frac{\text{distance travelled}}{\text{time taken}} = \frac{20\,000 \text{ metres}}{\text{time in seconds}}$$

Experiment 29.8   Indoors

An alternative method, for indoors, is to use sound switches with an electronic timer:

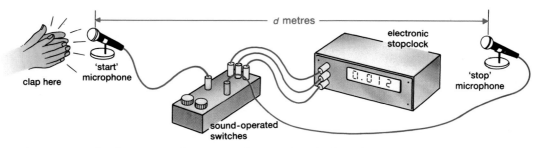

The sound arriving at the first microphone switches **on** the clock and the same sound arriving at the second microphone switches it **off**. The clock shows the time for the sound to travel the distance **d**.

$$\therefore \text{Speed of sound} = \frac{\text{distance } (d)}{\text{time taken}}$$

At 0 °C the speed of sound in air is 331 m/s (1200 km/h or 740 m.p.h.). At room temperature, its speed in air is faster, about 340 m/s.

If sound travels 1 kilometre in 3 seconds, how can you use lightning and thunder to find your distance from a storm?

Experiment 29.9

Measure the speed of sound in wood by clamping the microphones (of experiment 29.8) face down to a wooden bench. What do you find?

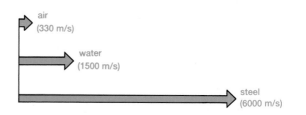

Sound travels faster through solids than through gases (or liquids) as you can see in the diagram. This is because the molecules are more tightly linked together in a solid.

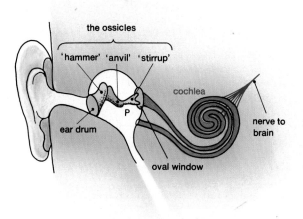

the ossicles
'hammer' 'anvil' 'stirrup'
cochlea
ear drum
nerve to brain
P
oval window

▷ The ear

Sound waves are collected by your outer ear and passed in to the **eardrum** which is made to vibrate by the compressions and rarefactions.

These vibrations are passed to the **oval window** by three bones (called the **hammer, anvil** and **stirrup**) which act as a lever (with the pivot at point P). This means that they magnify the force of the vibrations (see page 125).
Also, the oval window has a smaller area than the eardrum, so this increases the pressure on the oval window and on the liquid in the **cochlea** (see page 91).

The vibrations of the liquid in the cochlea affect thousands of **nerves** which send messages to your brain. This allows you to recognise the sound.

What range of **frequencies** can you hear?

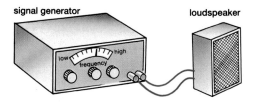

signal generator   loudspeaker

low   high
frequency

Doctor: Have your ears been checked lately?
Professor Messer: No, they've always been a pinkish colour.

Experiment 29.10
Connect a loudspeaker to a signal generator (which produces different frequencies as the pointer is moved to different positions).

Turn the pointer to lower and lower frequencies. What happens to the **pitch** of the note from the loudspeaker?
What is the lowest frequency you can hear?
How can you make this a fair test?
What is the highest frequency you can hear?
Does this vary from one person to another?

Young children can hear sounds as low as 20 hertz (20 cycles per second) and as high as 20 000 hertz, but as you get older, this range becomes less.

About 20% of the population has some kind of hearing defect. This may be due to old age, or an infection in the ear, or damage to the cochlea by very loud noises (for example, in a disco).

The graph shows a boy's **audiogram**.

This shows that he has a hearing loss at high frequencies (above 2000 Hz). His hearing loss is about 60 decibels (see also page 230).
He should wear a hearing aid which is adjusted to boost the high frequency sounds.

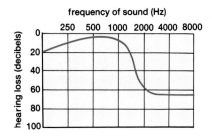

frequency of sound (Hz)
250 500 1000 2000 4000 8000
hearing loss (decibels)
0
20
40
60
80
100

▷ Resonance

A swing with someone sitting on it has a certain **natural frequency** of vibration. If you are asked to push the swing to make it go higher, you will obviously push it each time it comes near you. That is, the frequency of your pushes will be the **same** as the natural frequency of the swing. Then the swing vibrates with a large amplitude.

This is an example of **resonance**. We say the swing is **resonating**. If you push at a different frequency, it will not swing as high (and you might hurt your hand).

Resonance occurs when:

the applied frequency of the pushes	=	**the natural frequency of the object**

A short-sighted singer called Groat, Could do wonderful things with his throat.

At his specs he aimed sound, And with resonance found, That they cracked when he sang the right note.

Experiment 29.11
Hang a weight from a length of string (like a pendulum) and then use a straw to make it swing with a large amplitude,
a) by giving it one strong tap
b) by giving it lots of little taps at just the right frequency.

Which way gives the bigger swings?

In a similar way, if soldiers march in step over a bridge, they can make the bridge vibrate so much at its natural frequency that it may break.

A large bridge in America once fell down because it resonated with vibrations caused by the wind.

Have you heard windows or seats on a bus start to rattle (as they resonate) when the engine speeds up to a certain frequency?

If a singer sings near a wine glass with a frequency equal to the natural frequency of the glass, it may resonate so strongly that the glass breaks.

▷ Pitch, loudness and quality of musical notes

The notes from a musical instrument can vary in three ways:

1. in *pitch*
2. in *loudness*
3. in *quality* or *tone*.

What do these three things depend on?

1 Pitch

We have already seen (in experiment 29.10) that
the pitch of a note depends on the frequency.

Experiment 29.12
An oscilloscope (a kind of television set, see page 310) is very useful for showing vibrations. Connect a microphone to an oscilloscope and hold a tuning fork nearby.

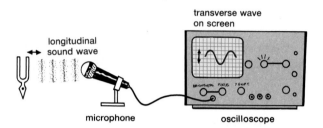

microphone   oscilloscope

When the oscilloscope is correctly adjusted, it shows on the screen a **transverse** wave of the **same frequency** as the **longitudinal** sound wave from the tuning fork.

Experiment 29.13
Look at the waves on the screen caused by a low-pitched tuning fork and a high-pitched tuning fork. What do you notice?

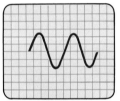

Low pitch
(low frequency)
long wavelength

High pitch
(high frequency)
short wavelength

2 Loudness

Experiment 29.14
Whistle a soft note into the microphone and then the same note but louder. Look at the waves on the screen. What do you notice?

The loudness depends upon the amplitude of the wave.
A wave with a larger amplitude contains more energy.

The loudness of a sound also depends on how much air is made to vibrate.

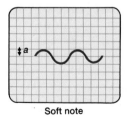
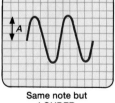

Soft note

Same note but
LOUDER

Experiment 29.15
Repeat experiment 29.4 with the gramophone record, using a small piece of card and then a large piece of card. Which sounds louder?

3 Quality (or tone)

If a violin and a piano play the same note (at the same pitch and the same loudness), you can still tell them apart because the notes have a different *tone* or *quality*.

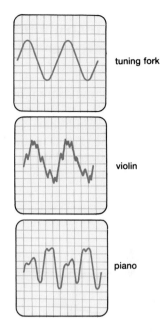

tuning fork

violin

piano

> *Experiment 29.16*
> Play different musical instruments in front of a microphone connected to an oscilloscope.
> Play the same note on each instrument and sketch the *waveform* that you see.

What do you notice?

The quality of a sound depends upon the waveform.

What causes these different waveforms?

A tuning fork produces a pure note with only one frequency. Other instruments usually produce many frequencies or *harmonics* at the same time.
All the harmonics add together to give a complicated waveform:

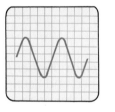 + 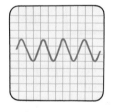 =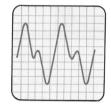

If this wave   is added to   this wave of **twice**   the result is this waveform
(the first harmonic)   the frequency
                (the second harmonic)

If more harmonics are added in, different waveforms are produced. Different instruments give different harmonics and different waveforms depending on the shape and size of the instrument, and so they give different sounds.

An electronic *synthesizer* allows you to choose which frequencies you want to mix together.
You can use it to imitate other musical instruments, or you can create entirely new sounds:

> *Experiment 29.17*
> Look at the oscilloscope while you sing or hum into the microphone. Try to keep the same pitch and loudness but change the shape of your mouth with your lips and your tongue. What do you notice?

▷ Musical instruments

1 Pipes

The diagrams show some pipe instruments.
These make a sound because the air inside the pipe is made to vibrate and resonate at its natural frequency.

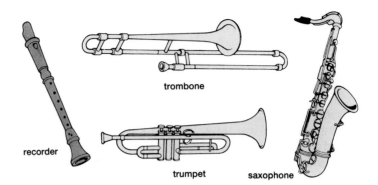

recorder

trombone

trumpet

saxophone

How does the natural frequency (pitch) depend upon the length of the pipe?

Experiment 29.18
Flatten the end of a drinking straw and then trim the corners with scissors as in the diagram. Put the trimmed end in your mouth and make a steady note by blowing.

Now cut bits off the other end to shorten the pipe and listen to the note. What happens to the pitch (frequency) of the note?

A shorter pipe produces a higher pitch (higher frequency).
The sound is produced by the trimmed end acting as a *reed.*
Which of the instruments at the top of the page uses a reed?

Experiment 29.19
Set up a row of test-tubes and fill them with different amounts of water, so that the length of air is different in each tube.

Blow gently across the top edge of each tube to make different notes. See if you can adjust the levels to make a musical scale.

Which tube produces the highest frequency (pitch)?
Which of the instruments at the top of the page produces sound by blowing across an edge like this?

How are the vibrations produced in a trumpet and a trombone?

Why do different pipes have a different tone (quality)?

2 Strings

The diagrams show some stringed instruments. They make sounds because each string vibrates at its own natural frequency (pitch).

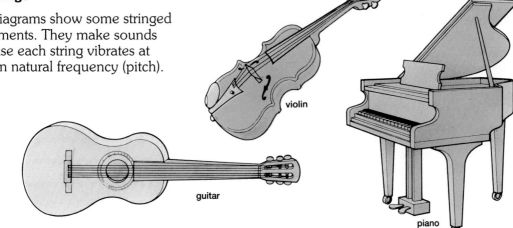

violin

guitar

piano

What does the natural (resonant) frequency of a string depend on?

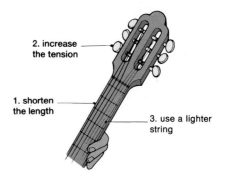

Experiment 29.20
Use a guitar (or a sonometer, a kind of one-string guitar).

a) Pluck a string and listen to the note. Then use your finger to shorten the length and pluck it again. What happens to the pitch (frequency)?
b) Use a screw at the end to increase the tension. What happens to the pitch (frequency)?
c) Pluck a light thin string and then a thick heavy string (with about the same tension and length). What do you find?

The natural frequency (or pitch) of a string can be **increased** by
– **shortening** the length
– **increasing** the tension
– using a **lighter** string.

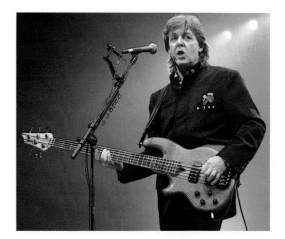

2. increase the tension

1. shorten the length

3. use a lighter string

Of the three instruments at the top of the page, which makes the strings start to vibrate by
– plucking?
– scraping?
– hitting?

Why do different stringed instruments have a different quality (tone)?

Professor Messer: If you drop a piano down a mineshaft, what key does it play in?
Mrs Messer: A flat minor?

▷ Noise

Noise is any sound that we do not like.
Noise can be irritating, and a loud noise can do permanent damage to your hearing.
The chart shows some common noise levels, measured in **decibels** (dB), using a sound-level meter:

Workers in noisy factories must wear ear-protectors to muffle the noise.
In Britain the law limits the maximum noise dose to 90 dB for an 8-hour working day.
A level of 100 dB is allowed for only 48 minutes.
However, personal stereo headphones and discos often have dangerous noise levels of over 100 dB!

Here is a simple test: if you have to shout to be heard by someone at arm's length, then your environment is dangerously noisy and you should leave. Once damaged by noise, your ears cannot be mended – the nerves in your cochlea are dead.

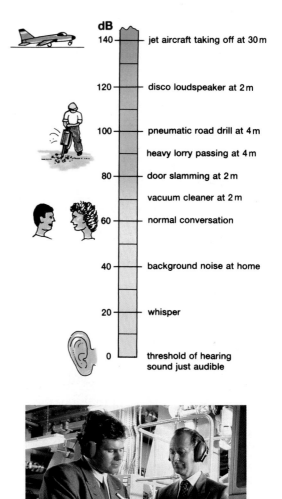

Reverberation

Echoes can be a nuisance. In an empty room or hall, the reflected sound can take a long time to die away. A long **reverberation time** makes it difficult for you to hear someone speaking.

The **acoustics** of a hall or a factory can be improved by using curtains, carpets and soft materials to absorb the sound. Concert halls are very carefully designed with absorbers to ensure the right amount of reverberation.

Measuring a high noise-level with a sound-meter

Summary

Sound is caused by vibrations and cannot travel through a vacuum.
A sound wave consists of compressions and rarefactions of the air.
Speed of sound = frequency × wavelength

Echoes are caused by the reflection of sound.
Resonance occurs when the applied frequency equals the natural frequency of the object.

The pitch of a note depends on the frequency.
Loudness depends on the amplitude.
Quality depends on the waveform.

The natural frequency of a pipe is increased as the pipe is shortened.
The natural frequency of a string is increased as the length is shortened; as the tension is increased; as the mass of the string is decreased.

▷ Questions

1.
a) Sound is caused by . . . . .
b) A sound wave consists of places at higher pressure (called . . . .) and places of . . . . pressure (called . . . .).
c) Wave speed (in metres per second) equals frequency (in . . . .) multiplied by . . . . (in . . . .).
d) Sound cannot travel through a . . . . .
e) Echoes are caused by the . . . . of sound.
f) The speed of sound in a solid is . . . . than the speed of sound in air.
g) Resonance occurs when the applied . . . . of the pushes equals the natural . . . . of the object.
h) Pitch depends on . . . . .
Loudness depends on . . . . .
Quality depends on . . . . .
i) If the length of a pipe is decreased, its natural frequency is . . . . .
j) The natural frequency of a string can be increased by . . . . the length, ... the weight of the string or . . . . the tension.

2. The speed of sound is 340 m/s. If thunder is heard 20 seconds after lightning, how far away is the storm? What must you assume?

3. What is the wavelength of a sound wave of frequency 100 Hz?
(Speed of sound = 340 m/s.)

4.
a) What are the highest and lowest frequencies that the human ear can detect?
b) What are the shortest and longest wavelengths that the human ear can detect?

5. Draw a line graph of distance travelled against time (up to 5 seconds) for sound in air.
Use your graph to find the distance of the lifeboat crew from a sinking ship if they see a distress rocket and then hear it after 4.2 s.

6. Explain how echoes are used:
a) by a ship to find the depth of the water
b) by geologists looking for oil
c) by bats and dolphins
d) by blind people with special equipment.

7. A man fires a gun and hears the echo from a cliff after 4 seconds. How far away is the cliff? (Speed of sound = 340 m/s.)

8. A sonar pulse sent out by a boat arrives back after 3 seconds. If the speed of sound in water is 1500 m/s, how deep is the water?

9. A girl stands 90 metres from a wall and claps her hands to hear clap–echo–clap–echo at an even steady rate of 1 clap per second. What result does she get for the speed of sound?

10. Explain, with diagrams, what is meant by:
a) amplitude   b) wavelength
c) frequency   d) quality of a note
e) resonance.

11. The diagram shows a graph of a sound wave given by a tuning fork.

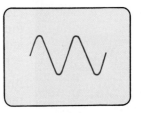

Copy this and then draw graphs to show:
a) a sound wave of higher pitch but the same loudness
b) a sound wave of the same pitch but louder
c) a sound wave of the same pitch but given by a different instrument.

12. Explain three methods by which the note from a guitar may be lowered in pitch.

13. A sound-level meter is used to find the noise level (in decibels, dB) at different distances from a disco loudspeaker:

Distance (m)	1	2	4	8	16
Noise level (dB)	130	120	110	100	90

Plot a graph and use it to find the noise level at 6 m. Is this distance safe for your ears?

14. A man is kidnapped, blindfolded and imprisoned in a room. How could he tell if he was in   a) a town?   b) the country?   c) a bare room?   d) a furnished room?

Further questions on page 237.

▷ Physics at work: Ultrasonic echoes

We cannot hear sounds which have frequencies above 20 000 Hz, but many animals can. Dogs will respond to an ultrasonic whistle even though it seems silent to us.

Bats and dolphins use ultrasound frequencies of about 150 000 Hz to 'see' by listening to the echoes. Echo-sounders on ships (see page 222) also use a very high frequency.

There is a reason for using a very high frequency: a high frequency means a **short** wavelength. To see this we can use the formula: **speed = frequency × wavelength** (see page 221) to calculate the wavelength. The speed of sound in water is 1500 m/s, so an **audible** sound of frequency 1500 Hz has a wavelength of 1 metre. With this long wavelength it is difficult to detect small objects (because the wave diffracts and bends round them, see page 169). However if the sound has an ultrasonic frequency of 150 000 Hz, the wavelength is only 0.01 m (= 1 cm). With this shorter wavelength the sound is reflected back from smaller objects and so the echo can be timed (and the distance to the object calculated).

An ultrasound beam is also much more directional and can be aimed rather like a torch. This has been used in ultrasonic spectacles for blind people. The spectacles have a transmitter and a receiver. The receiver produces a high or low sound in the person's ear depending on whether the object causing the echo is near or far.

Echo sounding is also used to detect flaws inside pieces of metal. A transmitter sends out pulses of ultrasound and a receiver picks up the echoes from different parts of the metal and shows the results on a CRO (cathode ray oscilloscope, page 310):

Pulse A is the transmitted pulse; pulse B has been reflected by the flaw (b) in the metal; pulse C is the echo from the end (c) of the metal.

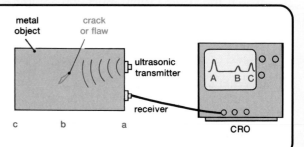

Ultrasonic echoes are also used in medicine (instead of X-rays, which can be dangerous).

Unborn babies can be seen by moving an ultrasonic transmitter/receiver across the mother's stomach.

Different tissues (skin, muscle, bone) reflect the sound waves to produce a lot of echoes. The machine uses these echoes to build up a picture on a TV screen:

Ultrasonic beams can also be used by dentists – the vibrations can shake dirt and harmful plaque off your teeth. This ultrasonic cleaning is also used in industry.

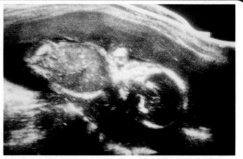

An ultrasound picture of an unborn baby (19 weeks after conception)

▷ Physics at work: Electromagnetic waves

Electromagnetic waves can have very short or very long wavelengths (see page 212). Different wavelengths have different properties and different uses. Different wavelengths give us different information about a hand:

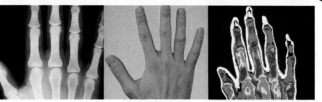

X-rays transmitted   Visible light reflected   Infra-red rays emitted

X-rays
X-rays are used in medicine (see page 312).

They can also be used in detective work. This old painting has been photographed in visible light and then using X-rays:

You can see that the artist changed his mind and then painted the head in a new position.

Ultra-violet
Ultra-violet waves can be used in detective work, in *forensic science*. The photograph shows a cheque photographed in ultra-violet – you can see it is a forgery.

Some chemicals glow or *fluoresce* in U-V rays. Washing powders include these chemicals to make clothes look brighter in sunlight.

In visible light, and in ultra-violet:

Infra-red
Some uses of infra-red waves are shown on pages 56 and 215. Satellite photos using I-R can tell the difference between fields of healthy and diseased crops.

Missiles can follow the infra-red from the hot engine of an aircraft. A remote control for your TV sends commands in infra-red rays.

Microwaves
A microwave oven uses radio waves to cook food very quickly. The microwaves are produced by a 'magnetron' and guided to a metal stirrer which reflects the waves into different parts of the oven. Microwaves are reflected by metal but absorbed by food.

The waves have a wavelength of 12 cm at a frequency of 2500 MHz. At this frequency the electromagnetic waves are absorbed by water molecules in the food, heating them up and so cooking the food. The food turns on a turntable so it is cooked evenly.
The door has a wire mesh over the window (to reflect the microwaves back inside). The door must have a safety switch to turn off the microwaves if you open the door – otherwise you could cook your fingers!

Microwaves are also used for communicating with satellites (page 333).

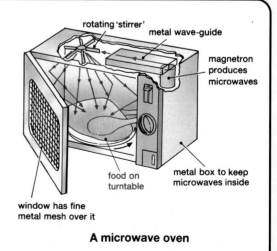

rotating 'stirrer'
metal wave-guide
magnetron produces microwaves
metal box to keep microwaves inside
food on turntable
window has fine metal mesh over it

A microwave oven

Further questions on waves: light and sound

▷ Waves

1. Which one of the following is a longitudinal wave?
   A water wave   B light wave
   C sound wave   D microwave   (NEA)

2. A siren emits a note which has a frequency of 400 hertz and a wavelength of 0.8 m. The sound will travel at a speed of
   A 0.002 m/s   B 320 m/s
   C 400.8 m/s   D 500 m/s   (SEG)

3. a) Here is an incomplete diagram which shows three successive straight waves A, B, and C on water, as they are being reflected at a straight barrier XY. Wave-front A is just about to be reflected while B and C have already been partly reflected:

   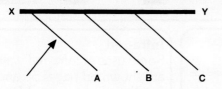

   Copy and complete the diagram showing the positions of the reflected parts of the wave-fronts B and C.   [4 marks]

   b) Here is an incomplete diagram which shows a circular wave-front originating at O just before reflection at a straight barrier AB:

   Copy and complete the diagram by drawing the wave-front as it would be just after reflection is complete. Indicate on the diagram where the reflected wave-front appears to come from.   [4]

   c) If the wavelength of an incident wave is 1.5 cm and the frequency of the source at O is 10 Hz, calculate
   i) the wavelength of the reflected wave,   [1]
   ii) the speed of the waves over the water.   [3] (NEA)

▷ Mirrors

4. The diagram below shows the plan view of an object O in front of a plane mirror.

   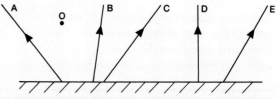

   Which one of the reflected rays of light appears to come from the *image* of O?   (NI)

5. The image of an object formed by a plane mirror is
   A virtual   B magnified   C real
   D diminished   E upside-down   (NI)

6. The diagram shows two rays of light shining on to a mirror from a torch labelled T.

   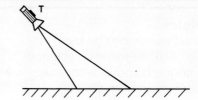

   a) Copy the diagram and:
   i) draw the two rays of light after they have been reflected by the mirror.   [3]
   ii) draw the **image** of the torch in the correct position (label this **I**).   [1]

   b) Could you use a mirror on its own to project an image on to a screen? Explain.   [2] (SEG)

7. a) The diagram shows a lamp bulb placed at the principal focus of a concave mirror. Copy it and draw **two** rays of light which leave the lamp and reflect from the mirror.   [2]

   b) Give two examples of devices which use concave reflectors for electromagnetic waves other than light. In each case, state which part of the electromagnetic spectrum is involved.   [4] (NEA)

▷ Refraction

8. A ray of light is shone on to a glass block at point A.

Some of the light is reflected and the rest is refracted.

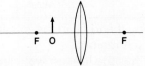

a) Copy and draw on the diagram the **paths** of the light as it leaves A. [3]

b) Light is a wave motion. What happens to the wavelength of the light as it is refracted by the glass? [1] (MEG)

9. a) The diagram shows a ray of red light incident on a glass prism. Copy and complete the diagram to show the path of the light through and out of the prism.

b) If white light were used instead of red light on this prism, what difference, if any, would you notice?

c) The diagram shows a ray of red light incident on a different prism. Copy and complete the diagram to show the path of the light through and out of the prism.

d) If white light were used instead of red light on this prism, what difference, if any, would you notice? (MEG)

10. An image of an object is created using four different devices. Copy the table below and place a tick where the word correctly compares the image with the object. The first column has been completed as an example.

	Simple camera	Magnifying glass	Projection lens	Plane mirror
real	✓			
virtual				
upright				
inverted	✓			
larger				
same size				
smaller	✓			

(NEA)

11. The diagram shows an object O placed between the principal focus F and the optical centre of a converging (convex) lens.

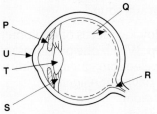

a) By drawing two rays of light from the object, copy and complete the diagram to show the image of O formed by the lens.

b) The converging lens is now held near a window and an image of some distant houses is formed on a screen.
  i) State three properties of the image on the screen.
  ii) The distance between the lens and the screen is 0.15 m. State and explain what can be deduced from this information. (NEA)

12. Here is a simple diagram of the eye.

a) Which is i) the retina ii) the cornea? [2]

b) A person enters a brightly lit room from a dark corridor.
  i) State the effect on the pupil of the eye.
  ii) How does this affect the amount of light entering the eye? [2]

c) A colour television screen consists of many thousands of dots arranged in threes as shown below. When the dots are lit, one produces red light, one produces green light and the third produces blue light.

  i) Which dots are lit if the screen is A) magenta B) yellow C) white? [3]
  ii) Explain how it might be possible to make the screen orange. [1] (NEA)

Further questions on waves: light and sound

▷ The spectrum and colour

13. The diagram shows the electromagnetic spectrum:

Radio		Visible	Ultra-violet		γ-rays

a) Copy it and fill in the missing names.
b) Which region:
   i) has the longest wavelength?
   ii) has the highest frequency?
   iii) causes a suntan?
   iv) is used in burglar alarms?
c) Some washing powders contain a chemical which is sensitive to ultra-violet radiation. State and explain what you see when clothes washed in such a powder are put in sunlight. (NEA)

14. An explosion emits light, infra-red and sound waves. Which waves will be received first by a person who is some distance away?
A infra-red and sound
B all three received together
C infra-red and light
D light and sound (LEAG)

15. The following pieces of equipment are used on a ship. Which one produces waves which are **not** electromagnetic in nature?
A the navigation lights
B the radar
C the siren
D the radio-transmitter
E the microwave oven (MEG)

16. A blue tee-shirt has a red 2 and a white 3 sewn on it.

Complete the table to show the appearance of the shirt under the given lighting conditions. [4]

Lighting	Colour of shirt	Colour of number 2	Colour of number 3	Number seen on the shirt
daylight	blue	red	white	23
red	black	red	red	23
blue				
green				

(WJEC)

▷ Sound

17. How much time is there between a person firing a starting pistol and being able to hear the echo from a wall 150 m away? The speed of sound is 300 m/s.
A 0.5 s   **B** 1.0 s
C 1.5 s   **D** 3.0 s   (SEG)

18. A wave source of frequency 1000 Hz emits waves of wavelength 0.1 m. How long does it take for the waves to travel 2500 m?
A 4 s   **B** 25 s   **C** 40 s
D 100 s   **E** 250 s   (MEG)

19. Here is an account of a student's experiment to find a value for the speed of sound in air.

We went to the school field well away from any buildings. We measured the length of the school field. Some of us stayed at one end of the field and some went to the far end of the field and fired a starting pistol. The time for the sound to travel across the field was measured. We then changed positions and fired the pistol from the other end of the field.

a) Why is there a need to be a long way from school buildings? [1]
b) Name a suitable instrument for measuring the length of the field. [1]
c) Here are four distances. Choose the **one** which would be most suitable in this experiment. [1]
  2 m   20 m   200 m   20 km
d) Here are four timing devices. Choose the **one** which would be most suitable. [1]
  pendulum, stopwatch, clock, ticker-timer
  Give a reason for your choice. [1]
e) Here are four times. Choose the **one** which is most likely to be recorded on the timing device. [1]
  0.07 s   0.7 s   7.0 s   7.0 minutes
f) Explain why the experiment is repeated from the other end of the field. [2] (SEG)

20. On a hot day it may be necessary to retune a guitar when it is taken from outside into a cold room. Explain
a) why the guitar would be out of tune,
b) how the instrument could be retuned.
(WJEC)

▷ Sound (contd)

21. Which one of the following waves is **longitudinal**?

A waves on a guitar string produced by plucking the string
B sound waves in air from a trumpet
C radio waves from a BBC transmitter
D waves produced in a slinky spring by flicking it from side to side
E ultra-violet waves from a suntan lamp

(NI)

22. a) State whether an ultrasonic wave is transverse or longitudinal. [1]

b) Bats emit ultrasonic waves of frequency 6.6×10^4 Hz. Calculate the wavelength of these waves. (The speed of ultrasound in air is 330 m/s.) [2]

c) A bat flies past a hole which is 0.10 m deep, as shown in the diagram.

i) Write down the difference in the distances travelled by sounds reflected from the cave wall and from the back of the hole. [1]

ii) Calculate the time difference which the bat must be able to detect if it is to be able to 'see' that there is a hole. [2]

d) The bat flies into a cave filled with marsh gas (methane) in which the speed of sound is 430 m/s. In what way will the bat be confused about the size of the cave?

[2] (LEAG)

23. a) Noise can be regarded as pollution. How is this problem overcome in:
   i) a house built next to a motorway? [1]
   ii) a motor car? [1]

b) Explain why discos could create hearing problems in later life for teenagers. [2]

c) i) By means of a diagram and notes explain how the depth of the sea can be found using sound waves. [3]

ii) Give **one** example where sound waves are used by geologists. [1]

(WJEC)

24. In an experiment Colin studies sound waves. He sets up a loudspeaker to produce sound as shown below:

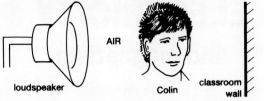

Colin adjusts the signal to the loudspeaker to give a sound of frequency 200 Hz.

a) What happens to the air in between Colin and the loudspeaker? [2]

b) Explain how Colin receives sound in **both** ears. [2]

c) State a way in which the waves along a stretched spring and the sound waves from the loudspeaker are similar. State **two** ways in which they are different. [3] (LEAG)

25. The diagram shows a fishing boat using sonar to detect a shoal of fish.

A short pulse of sound waves is emitted from the boat, and the echo from the shoal is detected $\frac{1}{10}$ s later. The sound waves travel through sea-water at 1500 m/s.

a) How far has the pulse travelled in $\frac{1}{10}$ s? [2]

b) How far below the boat is the shoal of fish? [2]

c) The reflected pulse lasts longer than the emitted pulse. Suggest a reason for this. [2] (MEG)

Further reading

Physics Plus: *Applications of Ultrasonics*; *Optical Fibres*; *Light and the Motor Car* (CRAC)

A Chronology of Science and Discovery – Asimov (Grafton)

Telecommunications – McGill (Stanley Thornes)

ELECTRICITY
and magnetism

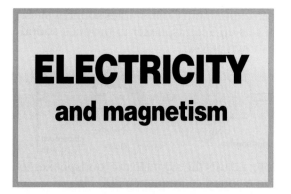

The photographs show some of the many uses of electricity and magnetism.

What difference would it make to your life if electricity and magnetism had not been discovered?

chapter 30

STATIC ELECTRICITY

Experiment 30.1
Rub a plastic pen or comb on your sleeve and then hold it near some tiny pieces of paper.

What happens?

Experiment 30.2
Rub a plastic object on your sleeve and then hold it near (but not touching) a thin stream of water from a tap.   What happens?

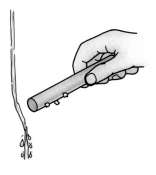

Because of friction with your sleeve, the plastic object has been **charged** with stationary or **static electricity**.

Have you ever heard a crackling noise as nylon clothes rub together when you undress?
In the dark you can see the tiny sparks of electricity (like tiny flashes of lightning).

Experiment 30.3
Rub a strip of polythene (a grey plastic) on a dry woollen cloth and hang it in a paper stirrup as shown.

We say the electricity on polythene is a **negative** charge.

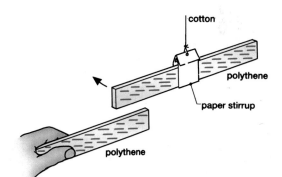

cotton

polythene

paper stirrup

polythene

Rub another strip of polythene (so it is also negatively charged) and bring it close.
What happens?

Repeat the experiment with two strips of cellulose acetate (a clear plastic). What happens?

Now repeat the experiment with one strip of charged polythene and one strip of charged cellulose acetate. What happens?

Because it behaves differently, we say that cellulose acetate is charged **positively**.

We find that: **Like electric charges repel.**
                Unlike electric charges attract.

The closer the charges, the greater the force.

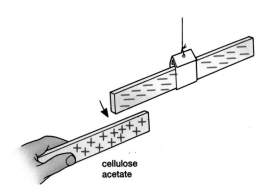

cellulose acetate

▷ The electron theory

Objects become charged because the atoms of all substances contain both negative and positive charges (see also page 344).

The positive charges (called **protons**) are in the central core or **nucleus** of the atom.
The negative charges (called **electrons**) are spread in orbits round the outer part of the atom.

In an uncharged or **neutral** atom, the number of protons (+) is equal to the number of electrons (−).

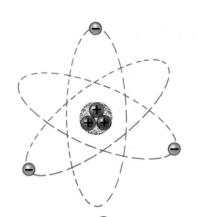

a neutral atom: 3 protons ⊕ and 3 electrons ⊖

We believe that when a polythene strip (or an ebonite rod) is rubbed on wool, some of the outer electrons are scraped off the wool and move on to the polythene. This means that the polythene has an extra number of electrons (so it is negatively charged) and the wool then has fewer electrons than protons (so it is positively charged).

Notice that only the negative electrons can move – the positive protons remain fixed.

The rubbing does not make the charges – it simply separates them.

When cellulose acetate (or glass) is rubbed, it becomes positively charged. Draw a diagram to show the movement of electrons in this case.

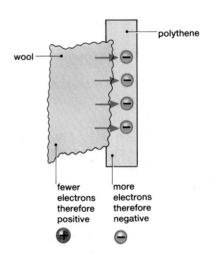

Electrostatic induction

Experiment 30.4
Rub a balloon on your clothes and then put it on the wall or ceiling so that it stays there!

This happens because the negative charge on the rubbed balloon repels some of the electrons in the ceiling away from the surface. This leaves the surface positively charged and so the negative balloon is attracted to the ceiling.

The separated charges in the ceiling are called **induced** charges.

This also explains why the paper and the water were attracted in experiments 30.1 and 30.2.

Why do your pop records tend to gather a lot of dust? Why is it difficult to clean nylon carpets?

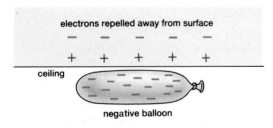

▷ The gold-leaf electroscope

This instrument can be used to tell whether an object is charged. The metal cap is connected to a metal rod and a strip of very thin metal (the gold leaf). This gold leaf is hinged at the top so that it can move.

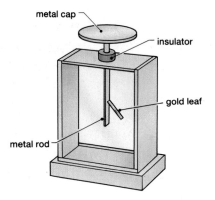

Experiment 30.5
Rub a polythene strip and bring it close to, but not touching, the cap of an electroscope.

What happens to the leaf?

The negative polythene induces charges in the electroscope by repelling negative electrons down to the metal rod and the leaf. Because like charges repel each other, the light leaf is repelled and lifted.

What happens when you remove the polythene? Why?

What happens if you bring up a positively-charged strip? Draw a diagram to show how the electrons move in this case.

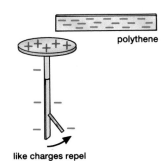

Experiment 30.6
Charge a polythene strip and then slide the strip firmly across the edge of the metal cap of the electroscope.

What happens? Does the leaf stay up even if you remove the polythene?

In this case, some of the negative electrons moved from the polythene to the electroscope, where they spread out over all the metal parts and made the leaf rise by repulsion.

What happens if you now bring up a positively-charged strip?

Experiment 30.7   Conductors and insulators
Charge an electroscope as in experiment 30.6, then touch the cap with your finger.

What happens to the leaf? Why does it fall?
What happens to the electrons on it?

Your skin is a **conductor** of electricity because it allows electrons to move through it. The moving electrons are an **electric current**.

Charge the electroscope again and then touch it with a plastic pen. Does anything happen this time?
Plastic is an **insulator** because it does not allow electrons to pass through it.
What happens with other materials?

Conductors	Insulators
Skin	Plastic
Metals	Air
Water	Rubber

▷ Experiments with a Van de Graaff generator

This is a machine used to charge objects easily.

These experiments should be done only under the supervision of a teacher – be careful you do not get an unexpected shock from the machine (it is safer to approach the dome while holding a pin in front of you – see experiment 30.12).

Experiment 30.8
Place a piece of fur on the dome and start the machine running.

What happens to the hairs on the fur?
Why is this? (Remember all the hairs have the **same** kind of charge.)

Experiment 30.9   A hair-raising experience
If someone in the class has soft dry hair, ask them to stand on a thick sheet of polythene (to insulate them from the Earth) while they touch the dome.

Experiment 30.10
Without touching the machine, blow soap bubbles near it.

What happens? Why are the bubbles attracted to the dome? How is this used in electrostatic paint spraying? (Page 244.)

There is a force on the bubbles even though they are not touching the dome. We say there is an **electric field** near the dome. We imagine that there are electric *field lines*, as shown in the diagram:

The electric field is strongest near the dome: the force on the bubble is strongest there.

Experiment 30.11   An electric wind
While standing on an insulator and touching the dome as in experiment 30.9, aim the point of a pin at a candle flame.

Which way does the flame move?

This happens because, if the point is positively charged, it attracts to it some of the electrons from atoms in the air. This leaves the atoms positively-charged.
We say the air has been *ionised*.
The positively-charged atoms (called *ions*) are repelled away from the point, by the electric field.

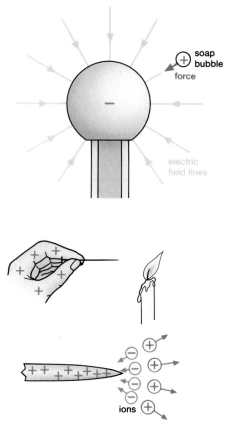

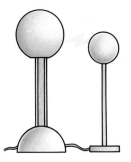

This is how a **lightning conductor** works –
if a negatively-charged cloud passes overhead, it
induces a positive charge on the point at the top of
the lightning conductor (as in experiment 30.4).
The point then repels positive ions to the cloud (as
in experiment 30.11) to neutralise it, so that it is
less likely to produce a lightning flash.

The electrons that are attracted to the conductor
travel down it to the Earth. An electric current flows
through the wire.

When is it dangerous to use an umbrella?

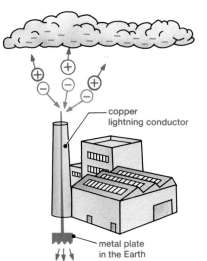

copper
lightning conductor

metal plate
in the Earth

Capacitors

Even after you switch off the machine you can get a
shock from it. It has *stored* some electric charge.
Objects that store charge are called **capacitors**.

Capacitors are used in radio and TV sets, and many
electronic circuits (see page 323). The size of their
capacitance is usually measured in micro-farads (μF).

A common capacitor is made from two long strips of
metal foil separated by a strip of insulator (waxed
paper or plastic), and then rolled up like a swiss roll:

If one metal strip is charged positively, and the other
metal strip is charged negatively, then the capacitor
stores electrical charge. It is storing energy, rather
like a small rechargeable battery.

The amount of charge that is stored is measured in
units called **coulombs**, usually shortened to C.

A coulomb is a very large amount of charge (about
6 million million million electrons), so we often use
micro-coulombs (1 μC = 10^{-6} C, one millionth C).

The dome of the Van de Graaff generator stores
about 1 μC, a lightning flash might contain 10 C.

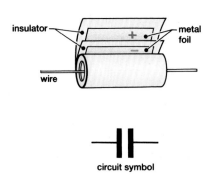

insulator

metal
foil

+

−

wire

circuit symbol

▷ Physics at work: Static electricity

An electrostatic precipitator

Power stations and factories produce huge amounts of smoke pollution. This smoke is a cloud of small dust particles or ash. It can be removed by using static electricity.

Some thin wires are stretched across the centre of the chimney: These wires are charged positively to about 50 000 V and they cause the gas around them to be charged or *ionised* (see also experiment 30.11 on page 242).
Because of this, the smoke particles become positively charged. These positive particles are then repelled by the wires, towards the earthed metal plates, where the dust sticks.

A mechanical hammer hits the plates every few minutes and the ash falls down into a bin. Later, it is used to make house-bricks.

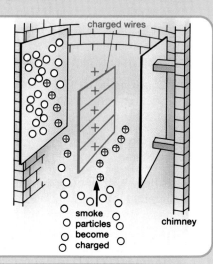

Fingerprinting

A similar method can be used to show up fingerprints on paper: The paper is placed near a charged wire like one of the plates in the chimney and a fine black powder is used instead of smoke. The powder sticks to the fingerprint but not to the clean paper.

Photocopiers are based on a similar method.

Paint spraying

Bicycles and cars are painted using an electrostatic paint spray. The nozzle is given a charge and this makes a better spray – the droplets all have the same charge and repel each other so that the paint spreads out to form a large even cloud. Less paint is needed because the charged droplets are all attracted to the object (even the back of it) because it has an opposite charge (see page 239).
The same idea is used to make crop-spraying more efficient for farmers.

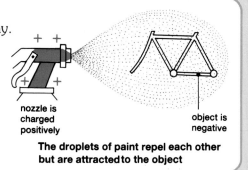

The droplets of paint repel each other but are attracted to the object

Preventing fires

A liquid flowing through a pipe can become charged in the same way as two objects rubbed together. This can be dangerous if it causes a spark and the liquid is inflammable.
For this reason, whenever an aeroplane is being re-fuelled by a tanker lorry, they are always connected together by a copper wire.

For the same reason, spare petrol for cars should always be carried in metal cans, never plastic.

Large ships have been known to explode because their tanks were being cleaned out by a high speed water-jet which became charged and caused a spark. Nowadays the tanks are filled with an inert gas before cleaning starts.

Walking on nylon carpets can give you a shock when you touch the door handle or a radiator. Modern carpets can be made conducting to prevent this.

Summary

When rubbed, polythene (or ebonite) becomes charged negatively (it has an excess number of electrons).
Cellulose acetate (or glass) becomes charged positively (it has a shortage of electrons).

Like charges repel; unlike charges attract.

A gold-leaf electroscope can be used to detect charges. Charges are measured in coulombs.

Charged objects can separate (induce) charges in other objects (by attracting or repelling electrons).

A conductor allows electrons to move through it; an insulator does not.

Sharp points lose their charge easily. This is used in a lightning conductor.
Capacitors are used to store charges.

▷ Questions

1. a) Like charges . . . . , unlike charges . . . . .
   b) When polythene is rubbed with wool, some . . . . leave the . . . . and move on to the . . . . so that the polythene becomes . . . . charged and the wool is left . . . . charged.
   c) When rubbed, cellulose acetate becomes . . . . charged because it . . . . some electrons.
   d) A negatively-charged object attracts a piece of paper because it . . . . electrons away from the surface of the paper. This leaves the surface of the paper . . . . -charged so that it is . . . . to the object. These separated charges are called . . . . charges.
   e) In a gold-leaf electroscope, the leaf rises because like charges . . . . .
   f) Conductors (unlike . . . .) allow . . . . to travel through them.
   g) Sharp points . . . . their charge easily by . . . . the air. The . . . . at the top of a lightning conductor loses . . . . to the clouds to neutralise them so that they are less likely to produce a flash of . . . . .
   h) Charges are measured in . . . . and can be stored in . . . . .

2. Explain the following:
   a) Nylon clothing crackles as you undress.
   b) In dry weather, people walking on nylon carpets may get a shock if they touch a radiator or a metal door knob.
   c) Passengers sliding off a car seat on to the ground sometimes get a shock.
   d) Petrol road-tankers usually have a length of metal chain hanging down to touch the ground.
   e) Because some anaesthetics are explosive, the floor tiles in an operating theatre are made of a conducting material.

3. Explain the following:
   a) A rubbed balloon will stick to the wall for some time.
   b) A mirror or window polished by a dry cloth on a dry day soon becomes dusty.
   c) Cassette-cases, records and other plastic objects soon become dusty.
   d) As sellotape or 'cling film' is pulled off a roll it is attracted to your hand.
   e) When spraying an object with paint, less paint is wasted if the object is charged.
   f) TV screens soon become very dusty.
   g) The amount of smoke leaving a factory can be reduced by placing a charged object inside the chimney.

4. Explain the following:
   a) It might be dangerous to raise an umbrella in a storm.
   b) The dome of a Van de Graaff machine is spherical with no sharp edges on the outside.
   c) It is safer to approach it while holding a pin in front of you.
   d) On high voltage electrical equipment, the blobs of solder should be rounded and smooth.

Further questions on page 275.

Have you started revising for your examinations? See page 366.

chapter 31

CIRCUITS

Experiment 31.1
Connect a cell or a battery to a lamp. This can be done by loose wires or on a **circuit board**:

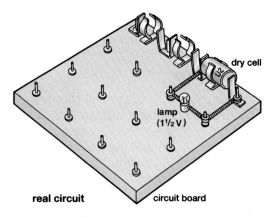

real circuit   **circuit board**

When the lamp lights, we say an electric **current** is flowing round the **circuit**.

In fact, negative electrons are being pushed out of the negative pole of the cell and are drifting slowly round the circuit, from atom to atom in the wire, to the positive pole of the cell.
A current flowing in one direction like this is called a **direct current** (d.c.).

When drawing diagrams of circuits we use symbols for each component as shown here:

circuit diagram

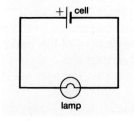

Now make a gap in your circuit.
What happens?
How can you switch the lamp on and off?
Does it matter where you make the gap?

For an electric current to flow, there must be a complete circuit, with no gaps.

The diagram shows the circuit with a switch added. Look at some switches of different kinds and include one of them in your circuit.

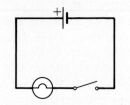

How can you use this circuit to send messages?
How was the Morse Code used in the old days?

Here are some other symbols used in electric circuit diagrams:

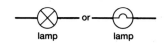

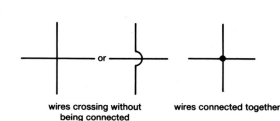

wires crossing without   wires connected together
being connected

Experiment 31.2 Conductors and insulators
Connect a circuit as shown (very like the last
experiment but with a gap in it).

Place a piece of copper across the gap.
What happens?
Place a piece of plastic across the gap.
What happens?

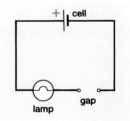

Some materials resist the flow of electrons more
than others. Repeat the experiment and draw up
a table of *conductors* and *insulators*:

Conductors	Insulators
Copper Aluminium	Plastic Rubber

All metals are good conductors.
This is because inside them they have a large
number of free electrons that can move easily from
atom to atom, so that a current flows.

Is air an insulator or a conductor?
Why are the handles of electricians' screwdrivers
made of plastic?
Why are copper wires usually covered with plastic?

Experiment 31.3
Connect a dry cell to a lamp as shown.
Use a length of thick copper wire to connect one
side of the lamp to the other side for just a second.
(!Do not do this with anything but a dry cell!)

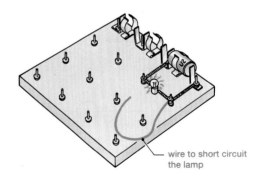

wire to short circuit
the lamp

What happens? If the current is not flowing
through the lamp, where is it going?

We say you have **shorted** the lamp or caused a
'short circuit'. Your wire has *less resistance* to the
flow of electrons than the thin wire inside the lamp.
Electricity travels by the easiest path, not
necessarily the shortest path.

Conventional current and electron flow

We know now that an electric current is really a
flow of electrons from negative to positive.
However when cells were first invented scientists
guessed, *wrongly*, that something was moving
from positive to negative and they marked arrows
this way on their diagrams.
Unfortunately we have kept to this convention
when drawing circuit diagrams, but you should
remember that in a metal the electrons are really
moving the opposite way.

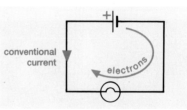

▷ Series circuits

Experiment 31.4
Connect a cell to two lamps as shown:

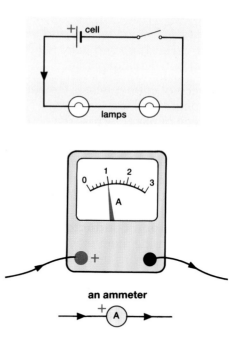

All the electrons that go through one lamp must also go through the other. We say the lamps are connected **in series**.

The electric current flowing in a circuit is measured in **amperes** or **amps** (usually shortened to A). 1 ampere is a flow of about 6 million million million electrons per second past each point!

The size of a current can be measured using an **ammeter** (see page 293).

In experiments, it is important to connect an ammeter so that its red (+) terminal is always nearer to the positive (+) pole of the cell than to the negative pole. Otherwise the pointer will be moved the wrong way.

an ammeter

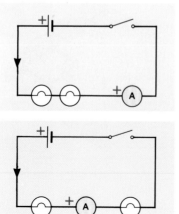

Experiment 31.5   Using an ammeter
Add a suitable ammeter to the circuit of the last experiment.
Read the scale. (On some ammeters, the numbers may have to be scaled up or down. Your teacher will explain if this is necessary.)

How much current is flowing in your circuit?

Experiment 31.6
Move the ammeter to different positions in the series circuit. What do you find?

The same current flows through each part of a series circuit.

Experiment 31.7
If you unscrew one lamp in your series circuit, what happens to the other lamp?

Christmas tree lamps are usually wired in series. What happens if one lamp breaks?

Are the lights in your house wired in series?

▷ Parallel circuits

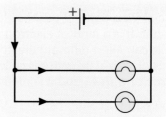

Experiment 31.8
Connect a cell (or a power-pack) to two lamps in the way shown here.

We say the two lamps are connected **in parallel**.

Experiment 31.9
Add switches to your circuit as shown (or loosen the wires instead).

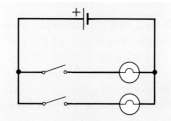

Can you switch the lamps separately?

Are the lamps in your home wired in series or in parallel?

Where in this circuit would you put a switch to control **both** lamps together? Try it.

Experiment 31.10
If you have 3 ammeters, connect the circuit shown here. (With just one ammeter, place it in each position in turn.)

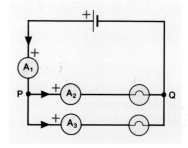

Read the ammeters. What do you notice about the reading on A_1 compared with the total of the readings on A_2 and A_3?

The current in the main circuit is the sum of the currents in the separate branches (sometimes called Kirchhoff's First Law).

As the electrons (from the negative pole of the cell) reach point Q, they divide between the two branches of the circuit until they reach P where they join together again.

Experiment 31.11   *Different parallel circuits*
There are many different ways of drawing the same parallel circuit.

Because bending the connecting wires does not change the movement of the electrons, all the circuits shown below are ***electrically the same*** as each other! (In each case the current divides to go through the lamps and then rejoins.)
Try connecting some of these circuits yourself.

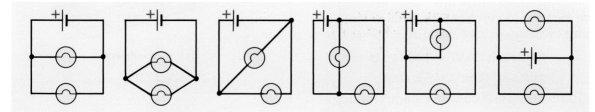

▷ E.M.F. and potential difference

A cell (or a battery) pushes electrons round a circuit. It is a kind of electron pump. Different cells can exert different electrical pressures. This electrical pressure is called *electromotive force* or **e.m.f.**, and it is measured in **volts** (V).

The e.m.f. of a dry cell is $1\frac{1}{2}$ V; the e.m.f. of a Van de Graaff generator may be 100 000 V.

When a cell is connected to some lamps, as in the diagram, the e.m.f. of the cell pushes electrons round the circuit. The energy to do this comes from the chemical energy of the cell.
The electrons give up this energy to the thin wires inside the lamps, so that they get hot and glow.

Across each lamp there is an electrical pressure difference, called a *potential difference (p.d.)* or 'voltage'. It is measured in *volts* (V) using a *voltmeter*.

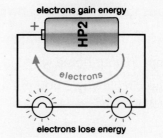

electrons gain energy

electrons

electrons lose energy

Experiment 31.12   *Using a voltmeter*
Connect a cell to two lamps in series as shown in the diagram above.
Then connect a suitable voltmeter *across* one of the lamps as shown here:

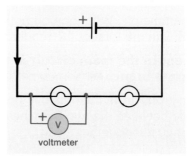

voltmeter

As with an ammeter, the voltmeter must be connected the correct way round (+ to +).
What is the reading on your voltmeter?

Reconnect the voltmeter (always in **parallel**, *across* a component) to find the potential difference across the other lamp. Then use your voltmeter to find the potential difference across the cell.
What do you notice about your three readings?

The total voltage across the lamps *equals* the voltage across the cell.

Prefixes

These are used to describe very small or large currents or voltages. (See the table on page 7.)

e.g.   1 kilovolt (1 kV) = 1000 volts
      1 milliamp (1 mA) = $\frac{1}{1000}$ amp
100 milliamp (100 mA) = $\frac{100}{1000}$ amp = $\frac{1}{10}$ amp

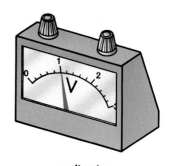

a voltmeter

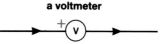

▷ Ohm's Law and resistance

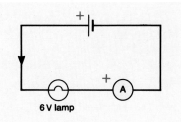

6 V lamp

What do you notice about the current as you
increase the potential difference?

In 1826, Georg Ohm discovered that:
**the current flowing through a metal wire is
proportional to the potential difference across it
(providing the temperature remains constant).**

'Proportional' means that if you double the
p.d., the current is doubled (see page 374).
Ohm's Law doesn't apply precisely to a lamp
because its temperature changes (see p. 257).

Resistance

The thin wire in a lamp tends to resist the movement of electrons in
it. We say that the wire has a certain *resistance* to the current. The
greater the resistance, the more voltage is needed to push a current
through the wire. The resistance of a wire is calculated by:

$$\text{Resistance, } R = \frac{\text{p.d. across the wire } (V)}{\text{current through the wire } (I)}$$

or, in symbols: $R = \dfrac{V}{I}$ where V = p.d. in volts (V)
I = current in amps (A)
R = resistance in a unit
called an **ohm** (Ω).

or, rearranging: $I = \dfrac{V}{R}$

or, easiest to
remember: $$V = I \times R$$

*This fact you should know evermore,
(the formula comes from Ohm's Law)
Overcome all resistance,
Learn by dogged persistence
That $V = I \times R$*

2 V accumulator

wire of resistance 20 Ω

What resistance would give you twice the current with the same accumulator?

▷ Four factors affecting resistance

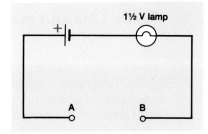

Now move the clips farther apart.
Is the lamp brighter or dimmer?
Is the current greater or less?
Has the resistance decreased or increased?

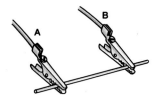

1. As the length increases, the resistance increases.
This fact is used in a rheostat (see next page).

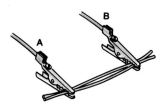

Is the lamp brighter or dimmer?
Is the current greater or less?
Has the resistance decreased or increased?

2. As cross-sectional area increases, the resistance decreases.

Which substance has the greater resistance?

3. Copper is a good conductor and is used for connecting wires.
Nichrome has more resistance and is used in the heating elements
of electric fires.

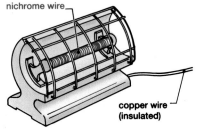

nichrome wire

copper wire
(insulated)

The resistance of a wire also changes as the temperature changes
(you can investigate this in experiment 31.23).

4. As temperature increases, the resistance of a wire increases.
This is used in a resistance thermometer (see page 33).

Which has the greater resistance – a long, thin, hot nichrome wire
or a short, thick, cool copper wire?

▷ Resistors

Resistors (sometimes made of a length of nichrome wire) can be used to reduce the current in a circuit.

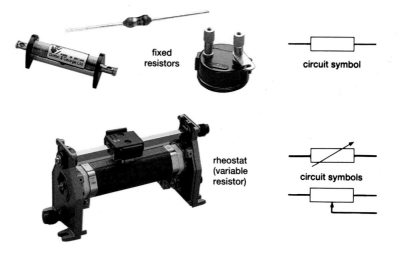

fixed resistors

circuit symbol

A variable resistor or **rheostat** is used to vary the current in a circuit. As the sliding contact moves, it varies the length of wire in the circuit.

rheostat (variable resistor)

circuit symbols

Experiment 31.17   Using a rheostat
Connect a rheostat in a circuit as shown and use it to control the brightness of the lamp.

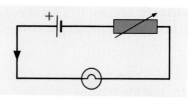

How are spotlights dimmed in a theatre?
How is a rheostat used in the speed control of a model racing car?

Measuring resistance

Experiment 31.18   Measuring resistance
To find the resistance of a resistor, use the circuit shown here. Connect all the series components first (the black parts of the diagram) and then add the voltmeter (in parallel) last of all.

Make sure that the meters are the correct way round and that you understand the marks on their scales.

Use the rheostat to adjust the current to a convenient value and note the readings of the ammeter and voltmeter.

Move the rheostat and take more readings.

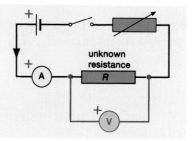

unknown resistance

Calculate the resistance from the formula (page 251)

$$\text{Resistance (in } \Omega) = \frac{\text{potential difference (in V)}}{\text{current (in A)}}$$

The average of the last column of the table should give you an accurate value of the resistance.

Some sample results:

p.d. (in V)	Current (in A)	Resistance (in Ω)
2.0	1.0	2.0
3.0	1.5	2.0
4.0	2.0	2.0
6.0	3.0	2.0

▷ Resistors in series

Experiment 31.19
Get two resistors whose values are marked (in Ω)
and connect them in series as shown here (in black).

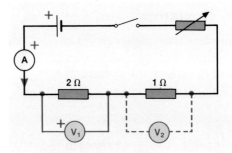

We know already (from page 248) that **the same
current flows through each part of a series circuit.**

Now investigate the potential differences in a
series circuit by connecting a suitable voltmeter:
a) in position V_1 across one of the resistors
b) in position V_2 across the other resistor
c) in position V_3 across both together.
What do you notice about the voltmeter readings?

**For resistors in series, the total p.d. is the sum of
the p.d.s across the separate resistors ($V_3 = V_1 + V_2$).**

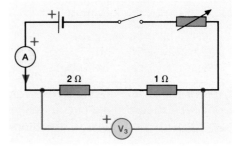

What is the combined resistance of the two resistors in series?
To find out, use the reading of V_3 and the ammeter reading
to calculate the combined resistance (using $R = \dfrac{V}{I}$).

Do you find that a 2 Ω resistor and a 1 Ω resistor in series combine
to act like a 3 Ω resistor?   Try this with other resistors.

**The total resistance (R) of a series circuit is equal to the sum
of the separate resistance (R_1, R_2):**

$$\boxed{R = R_1 + R_2}$$

Example
A p.d. of 4 V is applied to two resistors (of 6 Ω and 2 Ω)
connected in series.
Calculate a) the combined resistance,
          b) the current flowing,
          c) the p.d. across the 6Ω resistor.

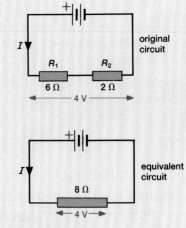

a) Formula first: Combined resistance $= R_1 + R_2$
   Then numbers: $= 6\,Ω + 2\,Ω$
                                  $= 8\,Ω$

b) Formula first: $V = I \times R$ for the combined resistance
   Then numbers: $4 = I \times 8$

   $\therefore\ I = \frac{4}{8} = \frac{1}{2}$ amp $= 0.5$ A

c) Formula first: $V = I \times R$ for the 6 Ω only
   Then numbers: $V = 0.5 \times 6$
                  $V = 3$ V   across the 6 Ω resistor.

What is the p.d. across the 2 Ω
resistor?

▷ Resistors in parallel

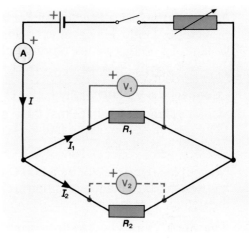

**For resistors in parallel, the potential difference
across each is the same.**

We already know (from page 249) that
**the current in the main circuit is the sum of the
currents in the parallel branches ($I = I_1 + I_2$).**

What is the combined resistance of the two resistors in parallel?
Use your voltmeter reading with the ammeter reading to find the
value of the combined resistance (using $R = \dfrac{V}{I}$).

Notice that the **combined resistance (R) is *less* than either of the
separate parallel resistances (R_1, R_2)** because the electrons find it
easier to travel when they have more than one path they can take.

The combined resistance R can be found from: $\dfrac{1}{R} = \dfrac{1}{R_1} + \dfrac{1}{R_2}$ or $R = \dfrac{R_1 \times R_2}{R_1 + R_2}$

When the two resistors are *equal*,
the combined resistance is *half*.

Example
A p.d. of 6 V is applied to two resistors (of 3 Ω and 6 Ω) connected
in parallel. Calculate a) the combined resistance,
　　　　　　　　　　　　 b) the current flowing in the main circuit,
　　　　　　　　　　　　 c) the current in the 3 Ω resistor.

a) Formula first: $\dfrac{1}{R} = \dfrac{1}{R_1} + \dfrac{1}{R_2}$

Then numbers: $\dfrac{1}{R} = \dfrac{1}{3} + \dfrac{1}{6} = \dfrac{2 + 1}{6} = \dfrac{3}{6} = \dfrac{1}{2}$

∴ Combined resistance $R = 2\,\Omega$

b) Formula first: $V = I \times R$ for the combined resistance
Then numbers: $6 = I \times 2$

∴ $I = \dfrac{6}{2} = \underline{3\,\text{A}}$ in the main circuit.

c) Formula first: $V = I_1 \times R$ for the 3 Ω resistor only.
Then numbers: $6 = I_1 \times 3$

∴ $I_1 = \dfrac{6}{3} = \underline{2\,\text{A}}$ in the 3 Ω resistor.

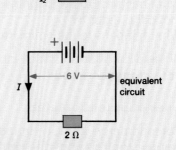

What is the current, I_2, in the
6 Ω resistor?

▷ Potential divider (voltage divider)

A battery provides a fixed e.m.f. For example, a 6 V battery provides 6 V only. However we often wish to provide a different voltage. To do this we use a **potential divider** circuit.

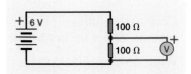

Experiment 31.21
Connect a 6 V battery to two 100 Ω resistors in series, as shown:

Use a voltmeter to check that the p.d. across the 2 resistors is 6 V. Then use the voltmeter to find the p.d. across each resistor in turn. Do you find that the 6 V is shared equally between the 2 resistors?

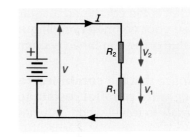

Now replace one resistor by a 200 Ω resistor. The 200 Ω resistor has two-thirds of the total resistance. Do you find two-thirds of the voltage is across it?

If we apply Ohm's Law to this circuit:

$$V_1 = IR_1 \quad \text{for resistor } R_1 \qquad \ldots (1)$$

$$V = I(R_1 + R_2) \quad \text{for the resistors in series} \ldots (2)$$

Dividing equation (1) by equation (2):

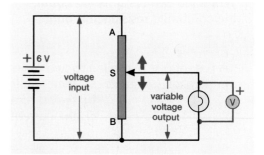

$$\boxed{\frac{V_1}{V} = \frac{R_1}{R_1 + R_2}} \quad \text{or} \quad \boxed{V_1 = \frac{R_1}{R_1 + R_2} \times V}$$

This is the potential divider equation.

Experiment 31.22
Connect a battery to a rheostat (see page 253) so that all three of its terminals are used:

The fixed voltage of the battery is applied across the full length of the wire AB.
As the slider S is moved to different positions, the voltage across BS varies. If S is at B, the output voltage is zero. If S is at A, the output voltage is the full input voltage. If S is halfway between B and A, the output voltage is half the input voltage; and so on.

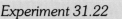

What happens to the brightness of the lamp as you move the slider S?
What happens to the voltmeter reading as you move S?

When a rheostat is used like this it is called a **potentiometer** or 'pot'.

Uses of potentiometers

Potentiometers are used in radios and record players as volume controls and tone controls. They are often circular with a rotating slide or 'wiper':

Potential dividers are often used in electronic circuits.

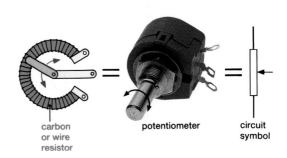

carbon or wire resistor   potentiometer   circuit symbol

▷ Current : Voltage graphs

In experiment 31.18 (on page 253), you measured several values of the **current I** through a resistor, and the **voltage V** across the resistor. If you plot a graph of these values, you get a *straight line through the origin:*

This means that the current **I** is *proportional* to the voltage **V**.
This is Ohm's Law (see page 251).
The steeper the graph, the lower the resistance.
The flatter the graph, the higher the resistance.

A substance that gives a straight graph like this is called an **ohmic conductor**. Copper wire and all other metals give this shape of graph, unless they change temperature.

Not all objects and substances give a straight graph like this. You can use the same circuit as before (experiment 31.18) to investigate what happens with other objects and materials. Alternatively, you can adapt the potentiometer circuit shown on the opposite page (experiment 31.22). Where should you include an ammeter in that circuit?

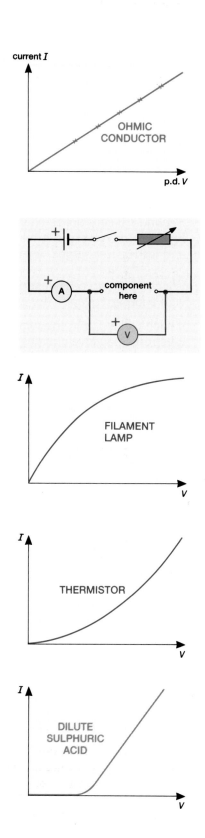

Experiment 31.23   A filament lamp
Use your circuit to investigate how the current **I** varies with the voltage **V** in a filament lamp (e.g. a torch bulb).

You can see that the lamp is a **non-ohmic** conductor because the graph is not a straight line. It does not obey Ohm's Law.

As more current flows, the metal filament gets hotter and so its resistance *in*creases. This means the graph gets flatter, as shown:

Experiment 31.24   A thermistor
A thermistor is used in electronics and is made of a **semi-conductor** substance (see page 317).

Look at the graph. Is a thermistor an ohmic conductor?
Why does the graph bend the opposite way to the lamp?

As more current flows, the thermistor gets hotter and so its resistance *de*creases (see page 317). The graph gets steeper.

Experiment 31.25   An ionic solution
An ionic solution (like dilute sulphuric acid) can carry an electric current (see page 269).

In this case, no current flows until a certain voltage is reached. This voltage depends on the chemicals that are present.

Another non-ohmic conductor is a diode (see graph, page 314).

▷ Cells

Cells and batteries are useful sources of electricity. They change chemical energy into electrical energy.

A **zinc–carbon cell** (dry Leclanché cell):
This is the common cell used in torches. It is an electron pump with an electrical pressure of 1.5 volts. The zinc is slowly eaten away as the chemical energy of the zinc is converted to electrical energy.
This is a *primary* cell – once the chemicals are used up you throw it away.

Secondary cells are re-chargeable. For example, a **lead–acid battery** in a car turns the starter motor and is then re-charged when the engine is running.

We can put two cells *in series* to make a *battery* with a larger voltage – for example, in a torch:

If each cell is 1.5 V, the total is 3 volts.
Electrons (negative charges) are repelled from the negative terminal of the battery, and flow as a current through the wire and the lamp. The more charge that goes through the wire in each second, the bigger the current.

In fact:

$$\text{Current } I \text{ (in amps)} = \frac{\text{charge } Q \text{ (in coulombs)}}{\text{time } t \text{ (in seconds)}}$$

In symbols:

$$I = \frac{Q}{t} \qquad \text{or} \qquad Q = I \times t$$

1 coulomb is defined as the charge that passes through the lamp if 1 amp flows for 1 second.

As the electrons flow round the circuit, they gain potential energy in the cell and then lose this energy in the lamp. The change in energy depends on the potential difference (p.d.) or 'voltage'.

1 volt is defined as the p.d. between two points if 1 joule of energy is needed to move 1 coulomb of charge between the two points.
If the voltage across the torch bulb is 3 volts, this means that each coulomb going through the lamp gives up 3 joules of energy (to heat the filament).

If **Q** coulombs pass through a p.d. of **V** volts, then:

$$\boxed{\textbf{Energy converted} = Q \times V \textbf{ joules}}$$

From the section above, **Q** = **I** × **t**, and so:

$$\boxed{\textbf{Energy converted} = I \times t \times V \textbf{ joules}}$$

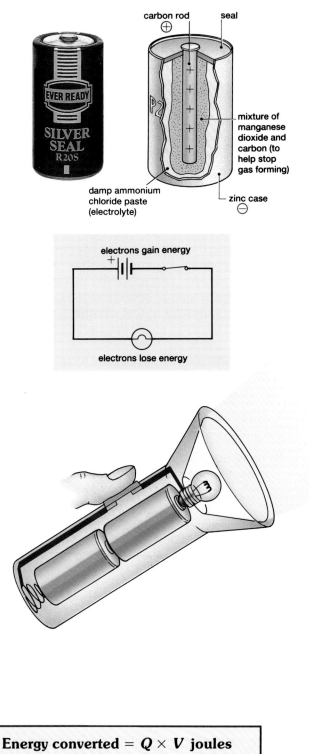

carbon rod seal

mixture of manganese dioxide and carbon (to help stop gas forming)

zinc case

damp ammonium chloride paste (electrolyte)

electrons gain energy

electrons lose energy

▷ Energy changes in a circuit

Experiment 31.26   E.m.f. and terminal p.d.
Connect a high resistance voltmeter across a dry cell on a
circuit board. Note the reading of the voltmeter.

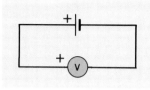

If we assume for the moment that the voltmeter is not taking any
current from the cell, then this voltage is called the *electromotive
force (e.m.f.)* of the cell. It is measured in volts and is the *total
energy per coulomb* that is available from the cell.

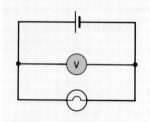

If you now connect a lamp, as shown, so that a current flows from
the cell, then you will find that the voltmeter reading is smaller. If
you take more current, the p.d. across the cell's terminals falls lower.

This terminal p.d. is always *less* than the e.m.f. of the cell because
the cell has some *internal resistance*.
Some of the cell's energy (the 'lost volts') is used up in pushing the
current through this internal resistance.

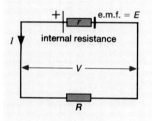

In the diagram, the e.m.f. = **E** volts, the internal resistance = **r** ohms.
Then: Energy **supplied**   energy **wasted**
    from chemical   = energy **delivered** + in internal
    energy of cell   to external circuit   resistance

∴ e.m.f., **E** = terminal p.d., **V**   + 'lost volts'

From page 251: $E = I \times R$ $+ I \times r$

Factorising: ∴ $\boxed{E = I(R + r)}$

Example

A battery of e.m.f. 6 V and internal resistance 2 Ω is
connected to a resistance of 4 Ω. What current flows?

Formula first: $E = I(R + r)$
Then numbers: $6 = I(4 + 2)$ ∴ $\underline{I = 1\text{ A}}$

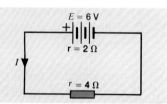

Tips for solving circuit problems

1. Simplify the circuit as much as you can.
   Sometimes it helps to re-draw it without
   any ammeters or voltmeters.

2. Use the equation:

3. **When the resistors are in series:**	4. **When the resistors are in parallel:**
a) Total resistance = $R_1 + R_2$.	
b) The current is same through each one.	a) Total resistance of two resistors = $\dfrac{R_1 \times R_2}{R_1 + R_2}$.
c) The larger resistor has the larger p.d. across it.	b) Parallel resistors have same p.d. across them.
d) The voltage across each resistor can be found by: $V = I \times R$.	c) The smaller resistor carries the larger current.
	d) The current through each resistor can be found by: $I = V/R$.
e) The voltages across the series resistors add up to the voltage across the cell.	e) The currents in the parallel branches add up to the current in the main circuit.

Summary

An electric current in a wire is a flow of electrons and can be measured in amperes (A) by a (low resistance) ammeter placed in series in a circuit.

The potential difference (p.d.) can be measured in volts (V) by a (high resistance) voltmeter placed in parallel, across a component.

The resistance of a wire increases with length and temperature; it decreases as the cross-sectional area is increased. It also depends on the substance.

Ohm's Law: the current flowing in a wire is proportional to the p.d. across it, providing the temperature is constant.

$$V = I \times R \quad \text{or} \quad I = \frac{V}{R} \quad \text{or} \quad R = \frac{V}{I}$$

▷ Questions

1. Copy out and complete:
   a) A current is a flow of . . . . . For this to happen there must be a . . . . circuit.
   b) Copper is a good . . . . .
      Plastic is an . . . . .
   c) Current is measured in . . . . using an . . . . placed in . . . . in a circuit.
   d) Potential difference (p.d.) is measured in . . . . using a . . . . placed in . . . . across a component.
   e) Ohm's Law: the . . . . flowing through a conductor is . . . . to the . . . . difference across it, providing the . . . . is constant.
   f) The formula connecting **V** (. . . . difference), **I** (. . . .) and **R** (. . . .) is
      . . . . .
   g) Resistance is measured in . . . . . The resistance of a wire increases as the length . . . . ; as the temperature . . . . ; and as the cross-sectional area . . . . .

2. *The Prof. is lost, can't see the light,*
   So help him get his circuits right:

3. Copy and complete the diagrams below:

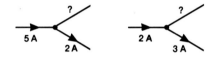

4. Draw a circuit diagram to show how two 4 V lamps can be lit brightly from two 2 V cells.

5. Calculate the combined resistance of each case:

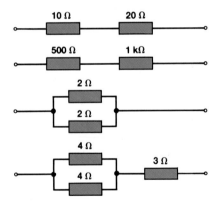

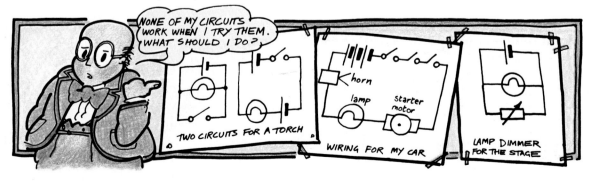

6. Explain what happens to the identical lamps X and Y in the circuit shown, when a) switch S_1 only is closed b) S_2 only is closed c) S_1 and S_2 are closed.

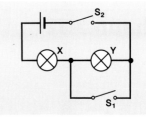

7. Draw a circuit diagram to show how 3 lamps can be lit from a battery so that 2 lamps are controlled by the same switch while the third lamp has its own switch.

8. In the diagram, the lamps are identical and ammeter A_2 reads 0.5 A.
   a) What are the readings on A_1, A_3, A_4?
   b) Redraw the diagram to include a switch to control both lamps together.
   c) Redraw the diagram to include two switches to control the lamps separately.
   d) Redraw it to include a rheostat to control lamp P only.

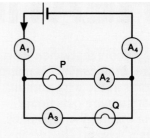

9. a) What p.d. is needed to send 2 A through a 5 Ω resistor?
   b) A p.d. of 6 V is applied across a 2 Ω resistor. What current flows?
   c) What is the value of a resistance if 20 V drives 2 A through it?

10. In the circuit shown, the p.d. across the 4 Ω resistor is 8 V.
    a) What is the current through the 4 Ω resistor?
    b) What is the current through the 3 Ω resistor?
    c) What is the p.d. across the 3 Ω resistor?
    d) What is the p.d. applied by the battery?

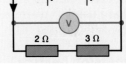

11. In the circuit shown, the voltmeter reads 10 V.
    a) What is the combined resistance? b) What current flows?
    What is the p.d. across c) the 2 Ω resistor? d) the 3 Ω resistor?

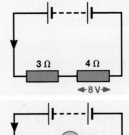

12. If the battery in question 11 has an internal resistance of 1 Ω, what is its e.m.f.?

13. In the circuit shown: a) What is the combined resistance?
    b) What is the p.d. across the combined resistance?
    c) What is the p.d. across the 3 Ω resistor?
    What is the current in d) the 3 Ω resistor? e) the 6 Ω resistor?

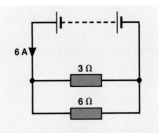

14. A flash of lightning carries 10 C of charge which flows for 0.01 s. What is the current? If the voltage is 10 MV, what is the energy?

15. *It seems the Prof's gone out of his mind. How many errors can you find?*

chapter 32

HEATING effect of a current

You know that when electrons pass through a wire, they can give some of their energy to the atoms in the wire and make them vibrate more, so that the wire gets hotter.
The greater the resistance, the hotter it becomes – this is why an electric fire glows red-hot while the connecting wires stay cool (see also page 252).

Electric fires

Radiant electric fires glow red-hot at about 900 °C. Sometimes the coil of nichrome wire is inside a silica tube. Why is this safer?
What is the concave mirror for?

Convector electric fires are cooler at about 450 °C, and are used to warm a room by convection currents (see page 50).

Immersion heaters

The diagram shows an *immersion heater* fitted into a hot-water tank. The heater contains resistance wire, like an electric fire but insulated from the water.
Where is the tank hottest?   Why?
Through which pipe is hot water delivered?
What is the other pipe for?
Why should the tank be thickly lagged?

Make a list of all the electrical heaters in your home – remember to include electric kettles and cookers, hairdryers, toasters, electric blankets, etc.

Circuit breaker

The heating effect of a current can also be used to protect circuits from dangerous over-heating. The diagram shows a 'thermal' *circuit breaker* using a bi-metallic strip (see page 25).

What happens to it if too much current flows?

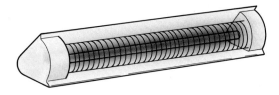

Alec Trishan once plugged in his fire,
To raise the room temperature higher,
But he soon got too hot,
And he thought he should not,
So he tied a large knot in the wire.
(Find two things wrong with this.)

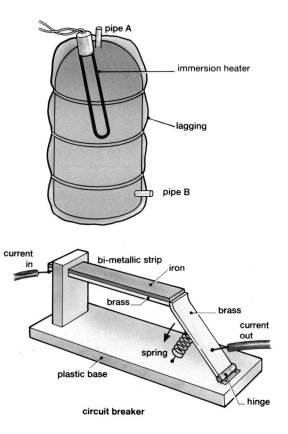

pipe A
immersion heater
lagging
pipe B

current in
bi-metallic strip
iron
brass
brass
current out
spring
plastic base
hinge

circuit breaker

Electric filament lamp

Look closely at a light bulb – can you see that the filament is a coil of wire which has been coiled again to give a hotter 'coiled coil'? The filament (heated to 2500 °C) is made of **tungsten** which has a high melting point.
An inert gas is usually included in the bulb to reduce the evaporation of the tungsten.

These lamps are not very efficient. Less than 10% of the electrical energy is converted into light energy – the rest is heat.
The yellow sodium lamps used for street lighting and the fluorescent tubes used for shop lighting are about 4 times as efficient, and so are cheaper to run.

Which kind of lamp would you use to reduce the shadows in a classroom?

How do you know that the lamps in your home are wired in parallel and not in series?

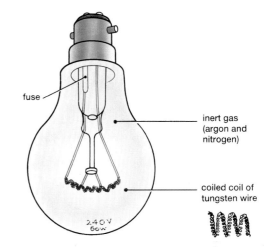

fuse

inert gas (argon and nitrogen)

coiled coil of tungsten wire

240 V 60 W

The circuit diagram shows how a lamp is wired to a ceiling rose and a switch (L and N refer to the 'live' and 'neutral' wires from the a.c. mains supply – see also page 266).

The switch **must** be in the live wire for safety when you are changing lamps.
NB *Never* inspect any mains wiring without first switching it off at the mains fuse box.

Why should wall switches not be fitted in bathrooms? Why are string pull switches safer?

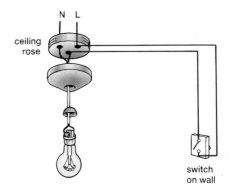

N L

ceiling rose

switch on wall

Two-way switches

With a two-way switch as shown in the diagram wire C can be connected to either A or B.

These switches can be used to switch a lamp from two positions – for example, the top and bottom of the stairs:

Can you control the lamp with either switch?
In the diagram, would the lamp be on or off?

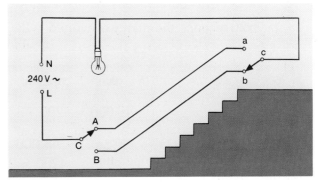

N
240 V ~
L
A
C
B
a
c
b

▷ Electrical power

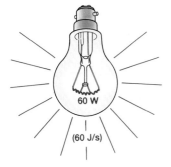

60 W

(60 J/s)

In Physics, we always measure energy in joules (see page 113). Therefore the *rate* of using energy, called *power*, is measured in joules per second, called *watts* (W), (see page 120).

If a lamp is marked 60 W, it means it is converting electricity to heat and light at the rate of 60 joules per second.

From page 258,
the change in electrical energy (in joules) = current, I × time × p.d., V

Dividing by time, to find the power:

| **Power, P = current, I × potential difference, V**
(in watts) (in amps) (in volts) | or | $P = I \times V$ |

For resistors, we can combine $P = I \times V$ with $V = I \times R$ (Ohm's Law), so that we get alternative equations:

$$\boxed{P = I \times V} \quad \text{or} \quad \boxed{P = I^2 \times R} \quad \text{or} \quad \boxed{P = \frac{V^2}{R} \text{ watts}}$$

If these equations are multiplied by the time t, then they become equations for the electrical *energy* converted to heat:

$$\boxed{\textbf{Electrical energy converted} = IVt = I^2Rt = \frac{V^2t}{R} \text{ joules}}$$

Example
An electric kettle connected to the 240 V mains supply draws a current of 10 A. Calculate:
a) the power of the kettle,
b) the energy produced in 20 seconds,
c) the rise in temperature if *all* this energy is given to 2 kg of water.
   (Specific heat capacity of water = 4200 J/kg °C, see page 43.)

a) Formula first: Power, P = current, I × p.d., V
   Then numbers: = 10 A × 240 V
                       = 2400 W (= 2.4 kW)   (where 1 kW = 1 kilowatt = 1000 W).

b) A power of 2400 W means that the energy changed is 2400 joules per second.

   ∴ In 20 seconds, the energy converted = 2400 × 20   joules
                                          = 48 000 J (= 48 kJ)

c) From page 43:

   $$\frac{\text{energy}}{\text{given}} = \frac{\text{specific heat}}{\text{capacity}} \times \frac{\text{mass of}}{\text{water}} \times \frac{\text{change in}}{\text{temperature}}$$

   Then numbers: $48\,000 = 4200 \times 32 \times$ rise in temperature

   ∴ $\dfrac{\text{rise in}}{\text{temperature}} = \dfrac{48\,000}{4200 \times 2} = 5.7\,°C$

Power of appliances in the home

This can vary widely, from 3 watts for an electric clock to 5 kilowatts for an electric cooker. The power is usually marked on a label on the back of the appliance. Look carefully at some in your home and at the values shown in the table.

Appliance	Power
Lamp	100 W = 0.1 kW
TV	200 W = 0.2 kW
Vacuum cleaner	500 W = 0.5 kW
One-bar fire	1000 W = 1.0 kW
Kettle	2500 W = 2.5 kW

The cost of electricity

If you look at your electricity meter at home you will see it is marked in **kilowatt-hours** (kW h).

The *kilowatt-hour* is a unit of *energy* (*not* power) and is calculated by:

> **Energy consumed = power × time**
> (in kW h) (in kW) (in hours)

Since 1 kW = 1000 W and 1 hour = 3600 seconds, it follows that 1 kW h = 3 600 000 joules.

We used a joulemeter on page 42.

Example 1
A 2000 W electric fire is used for 10 hours.
What is the cost, at 8p per kW h?

Formula first: Energy consumed = power × time
                    (in kW h)   (in kW)   (in hours)
Then numbers: = 2 kW × 10 hours
                                 = 20 kW h

∴ Cost at 8p per kW h = 20 × 8p
                      = 160p = £1.60

The Prof. got a bill for his heater,
The shock turned him into a cheater.
He tried to reverse
The effect on his purse
By reversing the wires to the meter.
Can you explain why this would not work?

Example 2
Here's part of a bill from an electricity company. Unfortunately it has been chewed by Professor Messer's dog. What is the total due to be paid?

∴ Number of units used = 2586 − 1486
                       = 1100 kW h

∴ Cost at 8p per unit = 1100 × 8p
                      = 8800p = £88.00

∴ Total = £88 + £11 ('standing charge' – this is a fixed rental charge)
        = £99

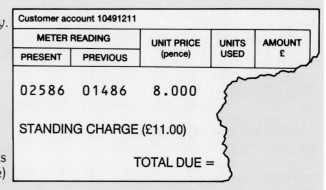

Customer account 10491211

METER READING		UNIT PRICE (pence)	UNITS USED	AMOUNT £
PRESENT	PREVIOUS			
02586	01486	8.000		

STANDING CHARGE (£11.00)

TOTAL DUE =

▷ Fuses

Experiment 32.1
Connect the circuit shown in black in the diagram, with crocodile clips at A and B.
Complete the circuit with a short length (about 3 cm) of 36-gauge tin wire, placed between A and B. Check that the lamp is on.

Now imitate a fault in the circuit by using a piece of copper wire to 'short' out the lamp (as shown in red). What happens to the thin wire?

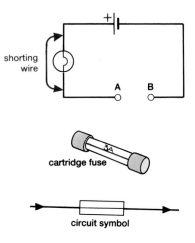

The thin tin wire melted or *fused* and so stopped the current flowing.

Fuses are included in circuits as deliberate weak links for safety, so that if a fault occurs and too much current flows, the fuse wire melts before anything else is damaged or starts a fire.

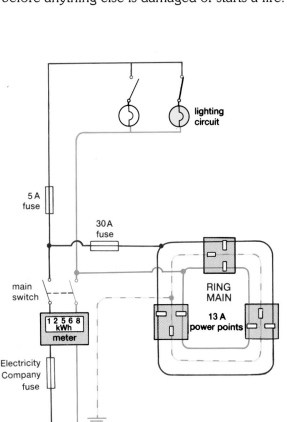

A young electrician called Hyde,
Learned nothing because of his pride.
He got his wires wrong,
And before very long,
He cried – then he sighed – then he died!

Household wiring

Here is a simplified diagram of a modern household circuit. It is in two parts: a lighting circuit protected by a 5-amp fuse and a *ring main* power circuit (for electric fires, etc.) protected by a 30-amp fuse.
Each power point is also protected by a fuse inside the plug (see opposite page).

The current supplied to your home is **a.c.**, *alternating current* (see page 299).
This means that the current flows one way and then the opposite way, again and again.
It goes each way 50 times in each second.

L and **N** in the diagram refer to the Live and Neutral wires coming from the power station. The third wire (**E**) is connected to Earth.

The *Live* wire is the most dangerous – the mains voltage may easily kill you.
The *Neutral* wire also carries the current, but because it is earthed back at the power station, its voltage is usually not as high as the live wire.
The *Earth* wire is for safety and only carries a current when there is a fault, so that a fuse melts (see opposite page).

!*Never* touch any part of a mains circuit without first switching off at the main switch near the meter!

The Earth wire

If a fault develops in an appliance and a live wire touches the metal case, a large current flows from the live wire to the earth wire and the fuse breaks. This makes the appliance safe until the fault is mended and then the fuse replaced.

Notice that the fuse (like a switch) **must** be placed in the **live** wire so that it can isolate the appliance from the live wire.

Your hairdryer probably does not have an earth wire connected to it. This is because it has an insulating plastic case. It is **double-insulated** and marked with the symbol:

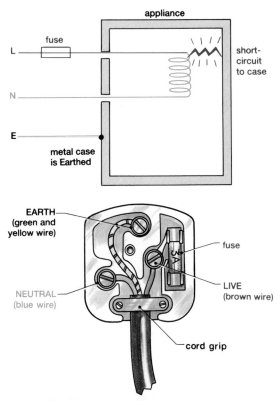

Fused plugs

It is absolutely essential that a plug is wired with the coloured wires going to the correct places:
Live – brown wire (the old colour was red),
Neutral – blue wire (old colour was black),
Earth – green and yellow (old colour was green).

Practise wiring a plug, taking care not to damage the insulation or the wires by cutting too deeply. Do not make the bare ends longer than necessary, and ensure that the cord-grip screws are firm. Make sure that you fit a fuse of the correct value.

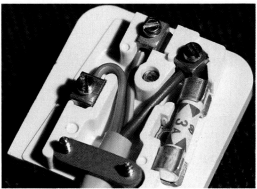

Fuse rating

The fuse rating is the maximum current that the fuse can carry without melting.
Only certain values are available: e.g. 3 A, 5 A, 13 A.

Example
A vacuum cleaner has a rating of 480 W on the 240 V mains. What fuse should be fitted in the plug?

Formula first: Power = current × p.d. (page 264)
480 W = current × 240 V

$$\therefore \text{Current} = \frac{480}{240} = 2\text{ A}$$

A 3 A fuse should be fitted. A 13 A fuse could allow a damaging current to flow before melting.

What fuse rating would you choose for:
a) a table lamp of 100 W, b) a kettle of 2400 W?

A careless young student called Hughes,
Once fitted the wrong size of fuse.
This fault caused a wire
In his house to catch fire.
– His death made the 6 o'clock news!

Summary

$$\text{Power} = IV \text{ watts} \quad (\text{or} \quad I^2R \quad \text{or} \quad \frac{V^2}{R})$$

where I is in amperes, V in volts, R in ohms.

$$\frac{\text{Energy}}{\text{(in joules)}} = \frac{\text{power}}{\text{(in W)}} \times \frac{\text{time}}{\text{(in seconds)}}$$

$$\frac{\text{Energy}}{\text{(in kW h)}} = \frac{\text{power}}{\text{(in kW)}} \times \frac{\text{time}}{\text{(in hours)}}$$

A fuse is used to protect a circuit and is always placed in the live wire.
Mains wiring colours: brown (live), blue (neutral), green and yellow (earth).

▷ Questions

1. Copy out and complete:
   a) The power (in watts) in an electric circuit is equal to the current (in . . . .) multiplied by the . . . . (in . . . .).
   b) To calculate the energy (in joules), the power (in . . . .) is multiplied by the . . . . (in . . . .).
   c) To calculate the energy (in kilowatt-hours), the . . . . (in . . . .) is multiplied by the . . . . (in . . . .).
   d) A fuse (or a light switch) should always be placed in the . . . . wire of a mains circuit.
   e) When wiring a fused plug, the brown wire should go to the . . . . terminal, the blue wire to the . . . . terminal and the green and yellow wire to the . . . . terminal.
   f) The earth wire should be connected to the . . . . of an appliance.

2. A vacuum cleaner is labelled 250 V 500 W. When connected to a 250 V supply, how much current does it take?

3. Copy out and complete the table opposite:

4. *The Prof. has got a mental block – Now tell me why he gets a shock.*

5. A 60 W lamp is connected to the 240 V mains.
   a) What current does it take?
   b) What is the resistance of the filament?

6. A 2 kW fire is switched on for 6 hours. What is the cost if one unit (kW h) costs 10p?

7. A 2 kW fire, a 200 W TV and three 100 W lamps are all switched on from 6 p.m. to 10 p.m. What is the total cost at 8p per kW h?

8. The diagram shows the inside of a 3-pin plug:
   a) What is the colour of each of the wires attached to A, B, C?
   b) Which of these is i) the earth wire ii) the live wire?
   c) What is placed in the gap CD? d) Why is this component necessary? e) How is its value chosen? f) Why is the direction of the wire round screw A not as satisfactory as those round B and C?

Appliance	Power (W)	p.d. (V)	Current (A)	Correct fuse (from 3 A, 5 A, 10 A, 13 A)
Car headlamp	48	12		
TV	240	240		
Hairdryer		240	2	
Iron	960	240		
Kettle		240	10	

Further questions on page 274.

chapter 33

CHEMICAL effect of a current

What happens if you try to pass electricity through a liquid?

Experiment 33.1
Connect the circuit, as shown, to two electrodes in a beaker.
Pour a liquid into the beaker and use the lamp to see if the liquid can conduct electricity. Try this with distilled water, salt water, vinegar, paraffin, acids and copper sulphate solution. (Remember to wash out the apparatus each time you change the liquid.)

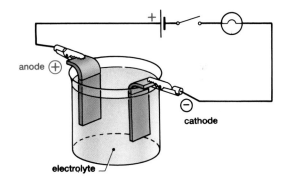

Liquids which conduct electricity are called *electrolytes*. Show your results in a table of electrolytes and non-electrolytes.

When electricity is passed through a liquid, it may change the liquid – this is called **electrolysis**. The apparatus used for electrolysis is called a *voltameter* (not a voltmeter).

Experiment 33.2   Electrolysis of water
Pass electricity through water, using a water voltameter with two platinum electrodes connected to a battery as shown. Water is a poor conductor of electricity, so some sulphuric acid must be added to make an electrolyte.

What happens?

When a current is flowing, bubbles appear at both the anode (+) and the cathode (−).
If you can collect and test these gases, you will find that they are oxygen at the anode and hydrogen at the cathode.

What do you think has happened?

The electricity has a chemical effect on the water and has split it into oxygen and hydrogen. (The volume of hydrogen is twice that of oxygen and so the formula for water is H_2O.)

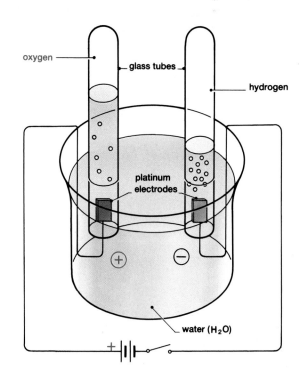

▷ Electroplating

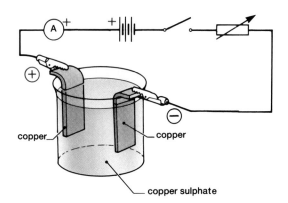

copper — copper

copper sulphate

Experiment 33.3   Electrolysis of copper sulphate
Connect the circuit shown here to a copper voltameter consisting of a beaker of copper sulphate
solution and two clean copper electrodes.
Label and find the mass of each electrode (to
0.1 gram) before putting them in the electrolyte.

Pass a current of about $\frac{1}{2}$ A for about $\frac{1}{2}$ hour.

Then carefully dry the electrodes and find the
new mass of each one. What do you find?

The cathode (−) has gained mass and is covered
with a bright new coating of copper.
We say it has been **electroplated** with copper. The
anode (+) has lost an equal mass of copper.

Experiments (first done by Michael Faraday) show:

> the **mass of metal** deposited in electrolysis is
> **proportional** to the **current** and **proportional**
> to the **time** for which it flows.

Experiments also show that:
> **the metal (or hydrogen if it is present) is
> always deposited at the cathode.**

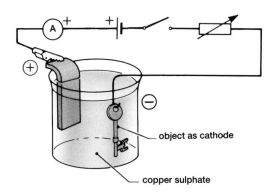

object as cathode

copper sulphate

Experiment 33.4   Copper-plating an object
Replace the copper **cathode** of experiment 33.3
by a clean metal object (e.g. a key or a badge).

Pass a very small current for several minutes
until your object becomes plated with copper.

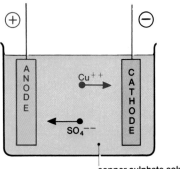

copper sulphate solution

The ionic theory

A molecule of copper sulphate has the chemical
formula $CuSO_4$ (= 1 atom of copper Cu, 1 atom
of sulphur S, 4 atoms of oxygen O_4).
When copper sulphate is dissolved in water, we
believe that it splits into two charged parts (Cu^{++}
and SO_4^{--}). These are called **ions** (see page 242).

The copper atom Cu loses 2 electrons to become
a positive ion Cu^{++}.
As unlike charges attract, this Cu^{++} ion drifts
toward the negative cathode where it gains 2
electrons to become an atom of copper-plating on
the cathode. The SO_4^{--} ion drifts to the positive
anode.

▷ Uses of electrolysis

1. Electroplating
In a similar way to your copper-plating in experiment 33.4, spoons can be silver-plated if they are used as the cathode of a voltameter containing silver. Chromium plating for cars and bicycles is achieved by a similar method.

2. Refining copper
In experiment 33.3, the copper transferred to the cathode was very pure even though the anode may have been made of impure copper. This process is used to produce pure copper for electric cables.

3. Manufacture of sodium and aluminium
These metals are obtained by electrolysis (of common salt for sodium, and of aluminium oxide for aluminium).

▷ Questions

1. Copy out and complete:
   a) A liquid which conducts electricity is called an . . . . .
   b) The positive electrode of a voltameter is called the . . . . and the negative electrode is called the . . . . .
   c) In the electrolysis of water, hydrogen forms at the . . . . , and . . . . forms at the . . . . .
   d) In the electrolysis of copper sulphate, ... is deposited on the . . . . .
   e) A metal or hydrogen is always deposited at the . . . . .
   f) Michael Faraday discovered that the . . . . of metal deposited is proportional to the . . . . and proportional to the . . . . .
   g) The current in the electrolyte is carried by charged atoms called . . . . . The positive . . . . move towards the . . . . . In copper sulphate solution, the atoms of copper have lost . . . . and so have a . . . . charge which causes them to move towards the . . . . .

2. Explain why Professor Messer had difficulty in electroplating a plastic toy.

3. The diagram shows some apparatus used to copper-plate an object at B.
   a) What is the substance at A?
   b) What is the liquid C?
   c) What would be connected to the positive of the battery?
   d) What factors determine the mass of copper deposited?

4. In a particular voltameter, 1 gram of copper is deposited. The experiment is repeated at three times the current for twice the time. What mass of copper is deposited?

5. The mass of copper deposited on a cathode when 1 amp flows for 1 second is 0.000 33 g. How much copper is deposited
   a) if 2 A flow for 500 s
   b) if 2000 coulombs are passed?

More questions on page 275.

How much revision have you done?
See page 366.

▷ Circuits

1. A pupil made an electric circuit like the one below in order to measure the current through two lamps.

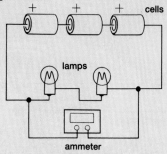

a) Are the lamps in *series* or *parallel*? [1]
b) The pupil has made a mistake in this circuit. What is the mistake? [2]
c) Draw a diagram showing the **correct** way to connect this circuit. Use the proper circuit symbols in your diagram. [4] (NEA)

2. Jim is a do-it-yourself enthusiast. He fitted a fog-lamp to his car and wired it into the supply to the headlamp. He wired it incorrectly. The circuit he used for the fog-lamp is shown:

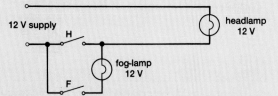

H = headlamp switch, F = fog-lamp switch.
a) Jim saw that the fog-lamp was lit when he pressed its switch, F, but it was not at full brightness. Use the circuit to explain why. What would Jim notice about the brightness of the headlamp? [3]
b) Jim saw that the fog-lamp did not work when he turned on the headlamp switch as well. Why? [2]
   What would Jim notice about the brightness of the headlamp now? [1]
c) Jim tried to find out what was wrong. He removed the headlamp bulb.
   What happened when he then turned on the switches? [1]
d) Sketch a diagram of the circuit Jim should have wired to make both the headlamp and the fog-lamp work properly. [2] (MEG)

3.

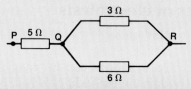

The p.d. across PR is 14 volts. Calculate
a) the equivalent resistance of network QR,
b) the current flowing through PQ,
c) the potential difference across PQ,
d) the current through the 3 ohm resistor.

4. The table shows corresponding values of potential difference across a torch bulb and the current passing through it.

Potential difference (V)	0	0.02	0.1	0.5	1.0	1.65	2.3	3.1	4.0
Current (A)	0	0.04	0.08	0.12	0.16	0.20	0.24	0.28	0.32

a) Draw a diagram of a circuit which could have been used to obtain these data. [5]
b) On graph paper plot a graph of current on the *y*-axis against p.d. on the *x*-axis. [10]
c) i) Use the graph to find the potential difference across the bulb when the current through it was 0.25 A.
   ii) Calculate the resistance of the bulb filament when the current through it was 0.25 A. [5]
d)

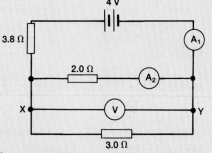

In the circuit shown, assume that the battery and ammeters have negligible resistance and that the voltmeter draws a negligible current. Calculate
   i) the resistance of the two resistors in parallel,
   ii) the total resistance of the circuit,
   iii) the reading on A_1,
   iv) the reading on V,
   v) the reading on A_2,
   vi) the power dissipated in the 2 ohm resistor. [15] (NEA)

5. a) For a particular specimen of wire, a series of readings of the current through the wire for different potential differences across it is taken and plotted as shown.
Describe how the resistance is changing. Suggest the most likely explanation. [4]

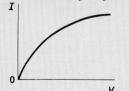

b) How would the resistance of a piece of wire change if
  i) the length were doubled? [2]
  ii) the diameter were doubled? [2]

c) The growth of cracks in a brick wall can be detected by measuring the change in resistance of a fine piece of wire. The wire is stretched tightly and fixed firmly to the wall on each side of the crack:

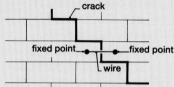

If the crack opens slightly, what happens to i) the length of wire? [1]
  ii) the thickness of wire? [1]
  iii) the resistance of the wire? [2] (NEA)

6.

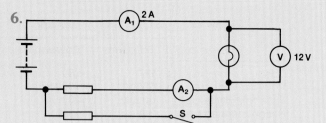

In this circuit, ammeter A_1 reads 2 A and the voltmeter reads 12 V. Switch S is open.
a) What is the reading of ammeter A_2? [1]
b) Calculate the resistance of the lamp. [3]
c) Calculate the power of the lamp.
   (Use $P = VI$) [2]
Switch S is now closed.
d) What happens to the readings of:
   i) Ammeter A_1? [1] ii) Ammeter A_2? [1]
e) Explain your answers to d). [2] (MEG)

7. The following diagram shows a circuit containing a lamp L, a voltmeter and ammeter.

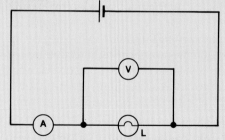

The voltmeter reading is 3 V and the ammeter reading is 0.5 A.
a) What is the resistance of the lamp?
   A 1.5 Ω   **B** 2.5 Ω
   C 4.5 Ω   **D** 6.0 Ω
b) What is the power of the lamp?
   A 1.5 W   **B** 2.5 W
   C 4.5 W   **D** 6.0 W   (LEAG)

8. An experiment was set up to investigate the heating effect of an electric current.
A constant current I was passed through a coil of constant resistance immersed in water. T, the temperature rise per minute, was measured for various values of current I. The values obtained are given in the table.

I (in A)	0	1.00	2.00	2.50	3.00	3.50	4.00
T (in K/min)	0	0.06	0.26	0.40	0.57	0.74	0.93

a) Draw up a table of values I^2 and T. [4]
b) Draw a graph of T (y-axis) against I^2 (x-axis). [6]
c) Determine the slope of the straight line portion of the graph. [4]
d) Determine the slope of the graph at a rate of temperature rise of 0.8 K/min. [4]
e) Suggest a reason for the change in the slope of the graph. [2]
   (SEG)

Further reading

Physics Plus: *Electrostatics outside the laboratory* (CRAC)
Over 1000 Physics formulae – Moore (Hamlyn)
Mr Tompkins in Wonderland – Gamow (CUP)

▷ Electric power

9. This question is about some of the many problems with which an electrician might be faced at work. Imagine that you are an electrician and that the following problems and questions are brought to you by some of your customers. In each case describe how you would answer them. Remember, your business depends on satisfied customers.

In each of the first four cases, explain a possible cause for your customer's trouble and how it might be solved.

Customer 1 'A lamp in my house has gone out.'   [2]

Customer 2 'The lights downstairs have all gone out.'   [2]

Customer 3 'All the lights in the house have gone out.'   [2]

Customer 4 'All the lights in my street have gone out.'   [2]

The next customer would like a clear wiring diagram.
Customer 5 'I would like two lamps in the dining room both worked from the same switch.' Draw a suitable wiring diagram.   [4]

The next customer is quite bewildered by the problem and you must give this customer a clear, full explanation of how you make the calculation.
Customer 6 'How much a week will it cost me to run my 1 kW electric fire and 2 kW electric radiator if they are both used for 4 hours each evening?' (1 kW h costs 5p)   [4]

The last customer does not even know there is a problem!
Customer 7 'I should like you to come and fit a mains socket in the bathroom so that I can have the electric fire on when I have a bath.' Explain what the problem is.   [3]
(NEA)

10. For each of the following situations,
i) explain the danger and,
ii) give a way of putting it right.
a) A wall switch for the bathroom.
b) Two flexes connected to the same 13 A plug.
c) A metal table lamp connected with a two-core mains flex.
(LEAG)

11. The diagram shows a 240 V electric hair-dryer with a plastic case.

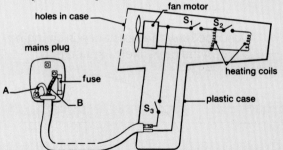

a) i) Write down the colours the cables in the plug should have.
   ii) One pin in the plug is not being used. Does this make the dryer dangerous?
b) Which switches need to be closed to put on: i) the fan alone?
   ii) the fan and one heating coil?
c) When the hairdryer is working at full power, the voltage is 240 V and the current in each coil is 2 A.
   i) What is the resistance of one coil?
   ii) The fan motor takes a current of 0.5 A. What is the total current from the supply when both coils are in use?
   iii) What fuse is needed in the plug? Choose from: 1 A   3 A   7 A   10 A
   iv) Suppose you find a 13 A fuse in the hairdryer! Explain why this might be dangerous.   (SEG)

12. Some information about two electric lawnmowers – 'Supatrim' and 'Powermo' – is shown:

	Supatrim	Powermo
Cost	£60	£80
Power rating	500 W	1000 W
Time to mow lawn	30 min	20 min
Recommended fuse	3 A	13 A
Width of cut	30 cm	50 cm

The cost of electrical energy is 6p/kW h.
a) Explain the meaning of 'kW h'.   [1]
b) Calculate the cost of the electrical energy used when mowing the lawn with
   i) the Supatrim,   ii) the Powermo.   [2]
c) Explain why a different fuse is recommended for each of the lawnmowers.   [1]
d) Give **two** reasons why a person might choose to buy the Powermo rather than the Supatrim.   [2]   (LEAG)

13. When a current of 4.00 A passes through a certain resistor for 10 minutes, 2.88×10^4 J of heat are produced. Calculate:
a) the power of the heater,
b) the voltage across the resistor. (WJEC)

14. Shown below is a diagram of a domestic 'instant' warm water system. This contains an electrical immersion heater. On the outer casing are two gauges showing
    the current through the heater, and
    the temperature of the water delivered.
The power is from a 240 V a.c. supply.

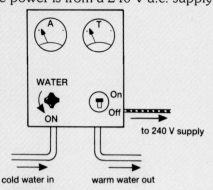

a) When the heater was turned on the reading on the ammeter was 20.0 A. Calculate
  i) the power of the heater, [2]
  ii) the energy supplied to the heater in 60 seconds. [2]

The temperature of the cold water was taken. The heater was turned on, and the water supply adjusted to deliver 3 kg of warm water every 60 seconds. The temperature of the water produced was measured when it was steady.

Temperature of cold water = 16 °C
Temperature of warm water = 36 °C
Mass of water delivered = 3 kg
Time = 60 s

b) Calculate the energy gained by the water. (The specific heat capacity of water = 4.2 kJ/kg °C) [3]
c) Calculate the percentage efficiency of the heater. [2]
d) Why is the energy gained by the water less than that supplied by the heater? [1]
e) Why is it essential that the water heater is fitted with a thermostat? [1] (NI)

▷ Electrostatics

15. When Darren takes his jumper off, it becomes negatively charged as it rubs against his shirt. This is because the jumper has
A lost electrons **B** gained electrons
C lost protons **D** gained protons (LEAG)

16. An aircraft flies just below a negatively-charged thunder cloud as shown. Electrostatic charges are induced on the aircraft.

a) Copy it and show the positions and signs of the induced charges on the aircraft.
b) Explain, in terms of the movement of electrons, the distribution of the charges you have shown.
c) What would happen to the induced charges when the aircraft flies away from the cloud? (LEAG)

▷ Chemical effect

17. You are asked to copper-plate some brass.
a) What electrolyte would you use?
b) What would you use for the anode?
c) Why would the brass have to be clean?
d) What type of supply of electricity would you use?
e) State **two** ways in which the mass of copper deposited could be halved.
f) Name **two** other metals that could be deposited by electrolysis.

18. Chromium is deposited on steel by electrolysis when making car hub caps. The reaction at the cathode is:
$$Cr^{3+}(aq) + 3e^- \rightarrow Cr(s)$$

Coulombs = amperes × seconds
Density of chromium = 6.4 g/cm^3
52 g Cr is deposited by 289 500 coulombs

a) Calculate the mass of chromium deposited when a current of 150 A is used for **ten** hours.
b) Calculate the average thickness of chromium deposited if the hub cap has a surface area of 2000 cm^2. (LEAG)

chapter 34

MAGNETISM

What is the quickest way of picking up some pins that have been spilt on the floor?

If a magnet is used, pins stick to the ends or **poles** of the magnet. Can you think of any other uses for magnets?

Experiment 34.1
Place a bar magnet in a paper stirrup as shown and hang it from a wooden table or a wooden retort stand.

Leave it hanging until it stops moving. From a map or from the position of the Sun, find out which direction is North.

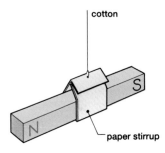

cotton

paper stirrup

Which direction does the magnet point to?

You have made a **compass**. A freely moving magnet comes to rest pointing roughly North–South.
The end pointing North is called the **North-seeking pole** or **North pole (N-pole)** of the magnet. The other end is the **South-seeking pole (S-pole)**.

Experiment 34.2
Mark the N-pole of the magnet in experiment 34.1. Then take that magnet well away and repeat the experiment with another magnet.

Now bring the N-pole of the first magnet close to the N-pole of the hanging magnet. What happens?

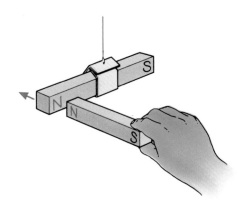

What happens if you bring the S-poles close together?
What happens if you bring a N-pole near a S-pole?

We find that: **Like poles repel each other.**
Unlike poles attract each other.

All physics students should learn well
This scientific fact:
Like magnetic poles repel
And unlike poles attract.

▷ Making a magnet

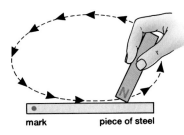

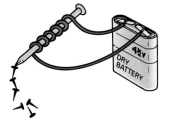

mark   piece of steel

Destroying magnetism

~a.c.

A. Stroking method

Experiment 34.3
Mark one end of a piece of unmagnetised steel. Check that it is not magnetised by dipping it into a dish of small iron nails.

Stroke the piece of steel with the N-pole of a magnet. Start at the marked end of the steel and lift the magnet clear at the end of each stroke. Do this about 10 times in the same direction.

How many nails can the steel pick up now?
Do you find the marked end is now a N-pole?
(Use a small compass to test this, remembering that like poles repel.)

B. Electrical method

Experiment 34.4   An electromagnet
Wind about 1 metre of insulated copper wire round and round a large nail. Then connect the ends of the wire to a battery, as shown.

How many nails can it pick up?
This is an **electromagnet** (see also page 286).

Disconnect the battery and test the large nail again.
Is it magnetised less strongly than before?

Why are cranes in steelworks often fitted with electromagnets?

A. Hammering

Experiment 34.5
Magnetise the piece of steel strongly (as in experiment 34.3) and then hammer it. Test it with nails.
Is it magnetised as strongly as before?

B. Heating

Experiment 34.6
Magnetise the piece of steel as before and then heat it in a Bunsen flame until it is red hot.
When it cools down, test it. Is it demagnetised?

C. Alternating current method

Experiment 34.7
This can be done only if your teacher can provide you with a safe supply of alternating current.
DO NOT use mains voltage – it is dangerous!
Pass alternating current (see page 299) through the coil you used in experiment 34.4 and then slowly reduce the amount of current in the coil (or gradually pull the nail out of the coil).
Then test the nail. Is it completely demagnetised?

Which of these three methods would you use to demagnetise a watch?

▷ Theory of magnetism

Experiment 34.8
Magnetise a thin piece of steel, as in experiment 34.3.
Then break it in half and use a compass to test each piece.
Is each half a magnet?

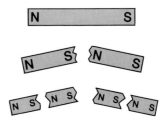

What happens if you break each piece again?

If we repeated this again and again, then we can imagine
ending up with a very very small 'atomic magnet'.
In iron and steel, these atomic magnets line up with each
other in groups, called **domains**.

It seems that all pieces of iron and steel are made of
millions of these tiny magnetic domains.
In the following diagrams, the tiny magnetic domains are
shown by arrows, with the arrowhead as the N-pole.

Magnetised and unmagnetised objects

In an unmagnetised piece of iron, the magnetic domains
are pointing in all directions and so cancel out each other:

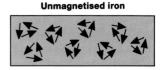

Unmagnetised iron

If the magnetic domains can be turned round to point the
same way, then the piece of iron becomes magnetised.
This is because all the tiny N-poles add up at one end and
all the S-poles add up at the other end:

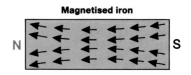

Magnetised iron

Experiment 34.9
Fill a small test-tube with iron filings and test it to check
that it is unmagnetised.

Stroke it with the N-pole of a magnet, while looking
carefully inside the test-tube. Can you see the iron
filings being pulled round by the magnet?

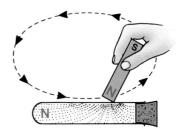

Bring a small compass near each end of the test-tube.
Is the test-tube now a magnet?
Which end is its N-pole? Can you explain this (remem-
bering that like poles repel)?

Shake the test-tube (as though you are hammering it)
and test it again.
Is it demagnetised? Why?

When an object is heated, do the molecules vibrate slower
or faster?
When an object is heated, why is it demagnetised?

▷ Magnetic fields

The region round a magnet where it has a magnetic effect is called its *magnetic field*.

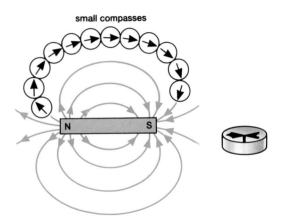

Experiment 34.10
Place some small plotting compasses near a weak bar magnet (preferably on an overhead projector):

The compasses point along curved lines as shown in the diagram. These lines are called *lines of flux* (or lines of force) and they show the direction of the magnetic field at each point.

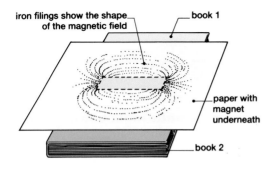

iron filings show the shape of the magnetic field — book 1

paper with magnet underneath

book 2

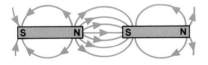

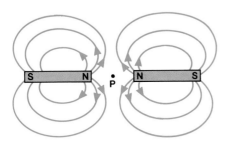

Experiment 34.11
Place a sheet of paper over a bar magnet.

Sprinkle a thin layer of iron filings over the paper and then tap the paper gently. The iron filings act like thousands of tiny compasses and point themselves along the lines of flux.

Look carefully at the pattern that appears. Can you see that it is the same shape as in the diagram at the top of the page?

Experiment 34.12
Use the iron filings method to find the shape of the magnetic field round two magnets placed so that the N-pole of one is near the S-pole of the other.

Where do the iron filings gather most thickly? Where is the field strongest?

Experiment 34.13
Use the same method to find the shape of the field when a N-pole is placed near another N-pole.

At point P in the diagram, there is no magnetic field. P is called a *neutral point*.

The magnetic field of the Earth

How does a compass show you that the Earth has a magnetic field?

In fact, the Earth acts as if there is a bar magnet inside it (although such a magnet cannot really exist because the centre of the Earth is too hot).

Notice in the diagram that the *S-pole* of the imaginary magnet is in the northern hemisphere so as to attract the N-pole of a compass.

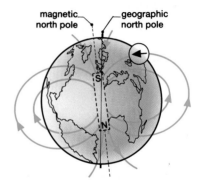

magnetic north pole — geographic north pole

▷ Magnetic induction

Experiment 34.14
Hold an unmagnetised piece of iron near some nails. If it is unmagnetised, it will not attract the nails.

Now bring near a strong magnet as shown in the diagram.

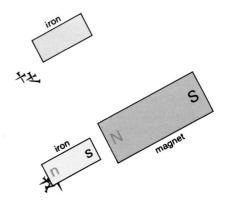

What happens to the nails?
This means that without being touched or stroked, the piece of iron has magnetism *induced* into it.

What do you think has happened to the magnetic domains in the piece of iron?

Which end of the iron becomes a S-pole? Why?
Why is the iron now attracted towards the magnet?

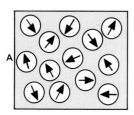

Experiment 34.15
To understand **magnetic induction** more clearly, place some small 'plotting' compasses on a table (or on an overhead projector if the compasses are clear-sided).

Count how many compasses (or 'magnetic domains') are pointing in each direction.
Does this represent a slightly magnetised or a strongly magnetised specimen?

Now bring a magnet nearby. What happens?

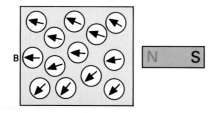

At the side of the specimen nearest the magnet's N-pole, do you get a N-pole or a S-pole?
Why is there attraction now between the end of the specimen and the magnet?

Because this magnetic induction causes attraction, the only sure way to test if an object is already magnetised, is to see if it *repels* a compass.

Objects like steel tools and girders in bridges are often found to be slightly magnetised by magnetic induction from the Earth's magnetism (particularly if the object is hammered or shaken by vibrations).

In a hospital, a strong magnet may be used to induce magnetism in a steel splinter and attract it from someone's eye.

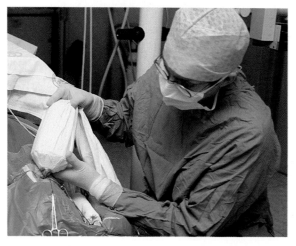

▷ Magnetic materials

Experiment 34.16
Get some specimens of many different substances e.g. iron, steel, nickel, plastic, wood, copper, etc.
Use a magnet to test which ones are attracted (by magnetic induction).

Draw up lists of magnetic and non-magnetic substances.

Magnetic substances	Non-magnetic substances
Iron Nickel	

Iron and steel are both magnetic but do not behave in exactly the same way.

Experiment 34.17
a) Hold an *iron* bar above some iron nails (or iron filings), with a magnet over the iron bar to magnetise it (by induction).

   How many nails will it hold up?
   Now take away the magnet. Do the nails fall off?

b) Now repeat the experiment with a *steel* bar the same size. Do some nails stay on when you remove the magnet?

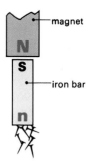

Iron is called a 'soft' magnetic material because it is easy to magnetise and also loses its magnetism easily. Iron is used in electromagnets (page 286) and transformers (page 302). Iron is also used as the magnetic material in a reed switch (see page 318).

The diagram shows another kind of magnetic switch:

What happens if the magnet moves closer?
How could you use this switch, with a battery and an electric bell, to make a burglar alarm for your door?

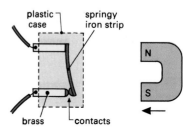

Steel is called a 'hard' magnetic material because it is harder to magnetise and also does not lose its magnetism so easily. It is used to make permanent magnets.

Permanent magnets are used in compasses and as magnetic catches to keep cupboard doors shut.
Fridges and freezers have magnetic strips all round the door to keep the door completely closed and so keep the cold air inside.

A permanent magnet is placed in the oil sump tank of a car engine so that it collects any bits of steel that are worn off the engine:

Iron oxide is another 'hard' magnetic material. It is used to make computer discs, videotapes, and the recording tape for tape-recorders (see page 307).

A magnetic drain plug removes bits of steel worn off a car engine

Paint thickness gauge

If paint is to protect steel from rusting – for example, on ships or cars – it has to cover all the surface. The paint is usually about $\frac{1}{5}$ mm thick and this has to be checked with a paint thickness gauge:

The gauge uses a magnet and a spring balance.

Experiment
Hang a bar magnet from a spring balance (using sellotape). Touch the magnet to a piece of steel and measure the force (in newtons) needed to pull it away. Then measure the force with one or more pieces of paper ('the paint') between the magnet and the steel. What do you find?

This is how a paint thickness gauge works.
(It can also tell you if the steel has rusted away or been replaced by fibreglass.)

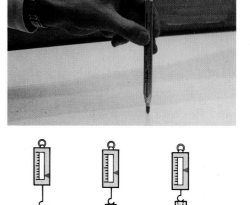

Magnetic inks and paints

It is possible to make magnetic inks by mixing very small particles of a magnetic substance with a liquid. The mixture (a dark brown colour) can be used as a paint to make magnetic tape for tape-recorders (see page 307), and floppy discs for computers.

Banks use magnetic ink on cheques so that the cheques can be sorted automatically by a machine which detects the magnetic field round each number:

Automatic cash-cards also use magnetic ink to store information on the card.

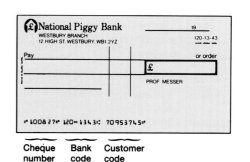

Cheque number   Bank code   Customer code

Magnetic liquids

Magnetic ink can be made with oil so that it does not dry out. This magnetic liquid can be used to detect very small cracks in the surface of steel pipelines. These cracks can be very dangerous if left untreated.

The pipe is magnetised by a simple coil and painted with magnetic liquid:

If there is a tiny crack, some of the magnetic lines of force leak out of the pipe at the crack. Then the magnetic liquid shows up the shape of the crack.

Magnetic oil is also used to lubricate the shafts of motors and the coils of loudspeakers. A magnet is used to keep the magnetic oil in the right place.

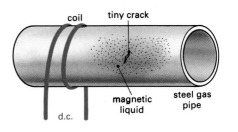

Summary

Like magnetic poles repel. Unlike poles attract.

Strong magnets can be demagnetised by
a) hammering   b) heating
c) using a coil with alternating current.
Iron (unlike steel) is easily magnetised and easily demagnetised.

Magnets can be made by
a) the stroking method, or
b) the electrical method
   (more details on page 285).

Lines of flux show the direction of the field round a magnet.

▷ Questions

1. Copy out and complete:
   a) Like poles . . . . ; unlike poles . . . . .
   b) If a N-pole is used in the stroking method of magnetisation, the end where the stroking begins is a . . . . pole.
   c) Strong magnets can be demagnetised by i) a coil carrying . . . . current, ii) . . . . or iii) . . . . .
   d) If the N-pole of a magnet is brought near an unmagnetised nail, then magnetism is . . . . into the nail, with a . . . . pole at the end of the nail nearest the N-pole of the magnet. The nail is then . . . . to the magnet.
   e) The Earth's magnetic field is rather like that of a . . . . magnet with its . . . . pole in the northern hemisphere.

2. Explain the following:
   a) Magnets are often fitted to the doors of refrigerators and cupboards.
   b) A crane in a steelworks is fitted with a large electromagnet (see page 238).
   c) Flour is usually passed near a magnet before being packed.
   d) A steel ship becomes magnetised as it is constructed.
   e) The sump plug on a car is usually magnetised.
   f) A magnetised screwdriver can help to place screws in holes that are hard to reach.

3. You are given a small plotting compass and three grey metal bars. One is aluminium, one is iron, one is a magnet. How can you identify each one?

4. Use the idea of magnetic domains to explain:
   a) If a magnet is broken in half, each half is a magnet.
   b) Heating or hammering destroys magnetism.
   c) Magnetism is felt strongly at the poles but not at the centre of a magnet.
   d) There is a limit to how strongly an iron bar can be magnetised (**saturation**).
   e) Magnetism is induced into an iron bar placed near a magnet.

5. Choose a material, giving a reason, for each of the following: a) the core of an electromagnet, b) a compass needle, c) the case of a compass.

6. Draw diagrams showing the magnetic fields round a) a single bar magnet, b) two magnets with their unlike poles close together, c) two magnets with their N-poles close together, d) the Earth.

7. Professor Messer says that 'like poles must attract' because the North pole of the Earth attracts the N-pole of a compass. Explain why he is wrong.

8. *Look, here's a new compass the Prof. has designed. But how many silly mistakes can you find? (8?)*

Professor Messer's new compass

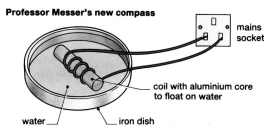

mains socket

coil with aluminium core to float on water

water

iron dish

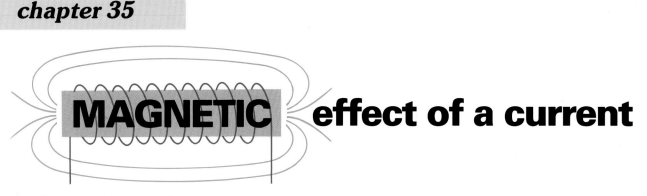

chapter 35

MAGNETIC effect of a current

In 1819, Hans Christian Oersted discovered that electricity has a *magnetic* effect.

Experiment 35.1   Oersted's experiment
Let a compass settle in a North–South direction.
Then, with the circuit shown, hold the wire over the compass and press the switch (for a few seconds only).

What happens to the compass needle when the current flows?
What happens to the deflection of the compass if you reverse the direction of the current?

We find that **a wire carrying a current has a magnetic field round it.**

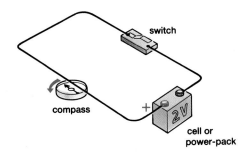

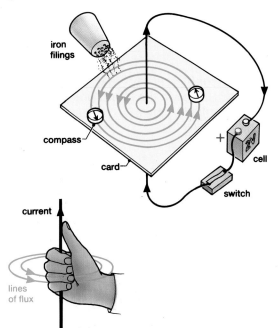

Experiment 35.2 Finding the shape of the field
Using the apparatus shown, sprinkle some iron filings on to the horizontal card.
Pass a large current through the vertical wire (briefly), while tapping the card so that the iron filings show the shape of the magnetic field.

Can you see that **the magnetic field round the wire is in the shape of circles?**

Place a small compass on the card to find the direction of the magnetic field lines.

Does your result agree with the diagram?
What happens if you *reverse* the current?

The diagram shows how to remember this result using the **Right-hand Grip Rule**: imagine gripping the wire, with your right thumb pointing the same way as the current . . . then your fingers are curling the same way as the magnetic field lines.

▷ The magnetic field near a coil

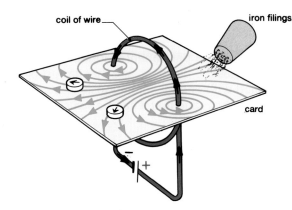

coil of wire — iron filings

card

Experiment 35.3   A short coil
Sprinkle iron filings round a coil as in the
diagram. Pass a large current and tap the card.

Study the shape of the magnetic field:
Use a small compass to find the direction of the
field.
Does the direction of the field round each wire
agree with the result of experiment 35.2?

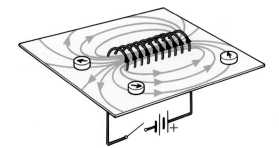

Experiment 35.4   A long coil (or 'solenoid')
Repeat the experiment but with a long coil or
solenoid.

Study the shape of the field. Does it look
familiar? (Look at page 279 if you are not sure.)

**The magnetic field round a solenoid has the
same shape as the field round a bar magnet.**

Notice that the field *inside* the solenoid is very
strong and uniform. It can be used to magnetise
objects (see also page 277).

Use a small compass to find which end of the
solenoid behaves like a N-pole.
Now look at this N-pole end of the solenoid.
Which way is the current flowing – clockwise or
anticlockwise?

Now do the same for the S-pole end of the
solenoid. What do you find?

A rule for the polarity of a coil
*The diagram shows an easy way to remember this
result. Use this rule to check the diagrams above.*

Experiment 35.5   Adding an iron core
Using the same apparatus, move a small
compass away from the coil until its direction
shows that it is only just affected by the
magnetic field. Now slide an iron bar into the
solenoid. What is the effect on the compass?

Why does an iron core make a stronger field?

The strength of the magnetic field can be
*in*creased by:
● using a larger current,
● using more turns of wire on the coil,
● using a 'soft' iron core,
● bringing the poles together, as shown:

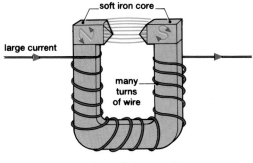

soft iron core

large current

many
turns
of wire

a strong electromagnet

▷ Uses of electromagnets

Compared to a permanent magnet, an electro-magnet has the disadvantage that it must be supplied with energy continuously. However, it has the advantage that it can be switched on and off.

1. Electromagnets are used on cranes in steel-works and scrapyards (see also page 238).

2. In hospitals dealing with eye injuries, electro-magnets are used to remove splinters of iron or steel (see page 280).

3. The electric bell

Study the diagram.

a) When the bell-push completes the circuit, a current flows through the electromagnet.

b) The soft-iron armature is attracted toward the electromagnet and the hammer hits the gong.

c) This movement breaks the circuit at point X, so that the current stops flowing and switches off the electromagnet.

d) The spring pulls the armature back so that contact is made and the sequence begins again.

The 'make-and-break' contact screw at X is adjustable to give the best sound. If the gong is removed, the bell becomes a buzzer.

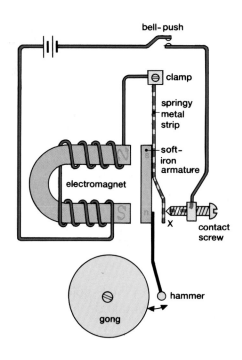

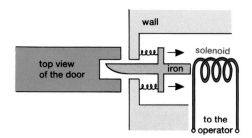

4. An electromagnetic door lock

The diagram shows a top view of a security door, perhaps in a bank. There is an iron bar keeping the door locked.

If the security officer agrees to let you in, he presses a switch, so that a current flows through the solenoid, which becomes magnetised.

What happens to the iron bar?
What are the springs for?

This idea is often used in factories to operate valves and other machinery by remote control.

5. The electromagnetic relay

A relay is a magnetic switch, used so that a *small* current can switch on a large current.

For example, in a car, the starter motor needs a very large current (100 A or more), which should be carried in thick wires kept as short as possible. A relay is used so that this large current is switched on by a smaller current in the dashboard switch.

In the diagram, a small current in the long thin wires to the key switch on the dashboard is used to energise the solenoid.
The solenoid attracts a soft-iron armature, which moves into the coil and so completes the motor circuit. What is the spring for?

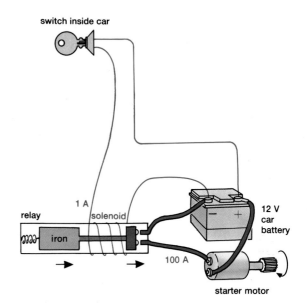

Relays are often used in electronic circuits, to control a large current (see page 319).

In car factories, car bodies are welded together by robot machines, controlled by a computer. The computer uses a small current and a relay, to switch on the large currents for the robot.

6. The telephone earpiece (or radio earphone)

This converts varying electric currents into sound waves. The varying electric currents pass round the coils of an electromagnet, which attracts an iron disc:

As the currents vary, the movement of the disc varies. This makes a sound wave in the air (see also page 220).

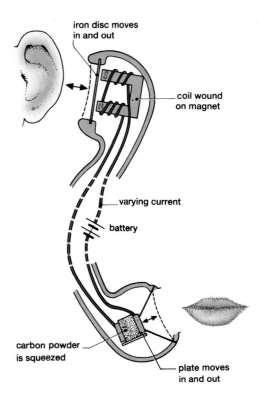

The varying electric currents can be produced by a **carbon microphone** in the mouthpiece of another telephone.
The sound waves force a plate in and out, so that it squeezes some carbon powder. When the carbon powder is squeezed, its resistance decreases and so the current from the battery varies in the same way that the sound wave varies.

▷ The motor effect

Experiment 35.6
Hang a flexible wire between the poles of a
powerful magnet.
Connect the wire to a cell and a switch as shown:

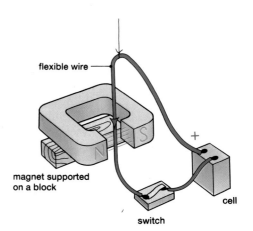

flexible wire

magnet supported
on a block

switch

cell

What happens when you press the switch briefly?

Does the movement reverse if you reverse the
direction of the current?
What happens if you turn over the magnet?

This movement is put to practical use in electric
motors.

The force on the wire can be increased by:
● using a larger current,
● using a stronger magnetic field,
● using a greater length of wire in the field.

The wire moves because the magnetic field of the
permanent magnet reacts with the magnetic field
of the current in the wire.

The diagram shows the combined field of the
magnet and wire. The wire tends to be
'catapulted' outwards.

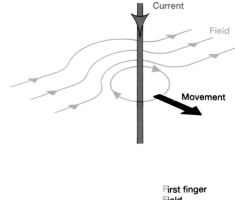

Current

Field

Movement

To remember the direction of movement we
use **Fleming's Left-hand Rule**:
Hold your *left* hand in a fist and then spread out
the thumb, first finger and second finger so that
they are at right-angles to each other.

– Point your **F**irst finger in the direction of the
   magnetic **F**ield (from N to S).

– Rotate your hand about that finger until your
   se**C**ond finger points in the direction of the
   Current (conventional current, from + to −).

– Then your thu**M**b points in the direction of the
   Movement of the wire.

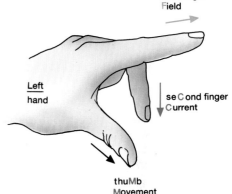

First finger
Field

Left
hand

seCond finger
Current

thuMb
Movement

Check this rule with the movement in your
experiment.

The moving-coil loudspeaker

A loudspeaker is used in your radio, TV, and in hi-fi systems. It converts electrical energy to sound energy. It changes electrical vibrations into vibrations of the air molecules.

It contains a movable coil attached to a large cone. The coil fits loosely over the centre of a cylindrical permanent magnet so that the coil is in a strong magnetic field.

If a current flows in the direction shown, the coil will move to the right (use Fleming's Left-hand Rule to check this). If the current reverses, the coil moves the opposite way.

As the current (from an amplifier) varies rapidly, the coil and the cone vibrate rapidly.
As the cone vibrates out and in, it produces the compressions and rarefactions of a sound wave (see page 221).
If the current varies at a frequency of 1000 hertz, then you will hear a note of frequency 1000 hertz.

The same apparatus can be used in reverse as a moving-coil microphone (see page 297).

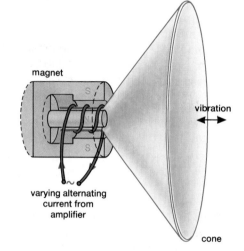

magnet

vibration

varying alternating
current from
amplifier

cone

A coil pivoted in a magnetic field

Experiment 35.7
Wind several turns of wire round a cork and hang it, from a clamp, between the poles of a magnet as shown. Pass a current briefly through the coil.

What happens? Does the coil turn the opposite way if you reverse the current?

This twisting movement is used in electric **motors** and in moving-coil galvanometers.

There are two ways to find the direction of rotation in the diagram:
a) Use Fleming's Left-hand Rule for each side of the coil.
b) Find which end of the coil becomes a N-pole. This end will be attracted to the S-pole of the magnet.

Apply both these methods to the diagram and check that they give the same result.

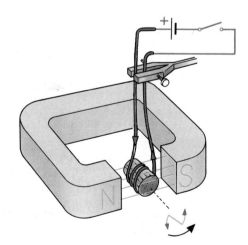

▷ The simple electric motor

The coil in the last experiment turns so that its
N-pole moves to the S-pole of the magnet but
then stops.

To keep the coil turning round toward the other
pole as in a motor, we have to reverse the current
in the coil at just the right moment.
In a direct current motor, the current is reversed
every half turn by a ***commutator*** which looks
like a copper ring cut into two halves.

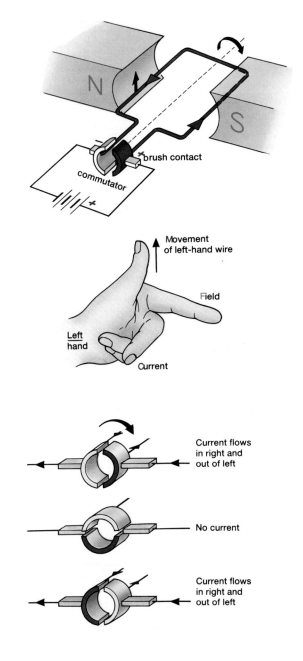

The diagram shows the current from the battery
travelling through a brush contact to one half of
the commutator. After travelling round the coil,
the current passes through the other half of the
commutator and back to the battery.

Use Fleming's Left-hand Rule for the wire that is
near the N-pole of the magnet.
Do you find that this wire moves upwards, so that
the coil turns clockwise?

When the coil turns through 90° (so that the coil
is vertical), the current stops flowing because the
gaps in the commutator break the circuit. However,
the coil keeps turning because of its own
momentum.
When the brushes make contact again, both the
commutator and the coil have turned over so that
the wire which is ***now*** nearest the N-pole has
current coming out towards us.
This means that the force on it is upwards and so
the coil still turns clockwise.

That is, the commutator makes sure that which-
ever wire in the diagram is nearest the N-pole, it
always has current moving out toward us and so
the wire keeps turning clockwise.

What would happen if you reversed the battery?

Of course, in practice the coil would have more
than one turn of wire.

The brushes are usually made of carbon and
held against the commutator by springs.

A designer of lifts, L. E. Vator,
Built a motor with no commutator,
It turned, with a wheeze,
Through just 90 degrees,
And trapped, between floors, its creator.

Experiment 35.8   Building an electric motor

a) Push a short knitting needle through a large cork.

b) Push 2 pins in one end of the cork as shown, to make a simple commutator.

c) Now wind round the cork about 30 turns of thin insulated copper wire, starting at one pin and finishing at the other.
It helps if you can cut channels in the cork to take the wire.

d) Scrape the insulation off the ends of the wire (using emery paper or a knife).
Wrap each end of the wire round a pin, making good electrical contact.

e) Push 2 pairs of long pins into a baseboard to support the axle at each end.

f) Strip the insulation from the ends of 2 wires and use drawing pins to hold them in position so that they just touch the commutator.

g) Use plasticine to support magnets on each side of the coil (with opposite poles facing).

h) Connect the wires to a 2 V or 4 V supply and give the coil a flick to start it.

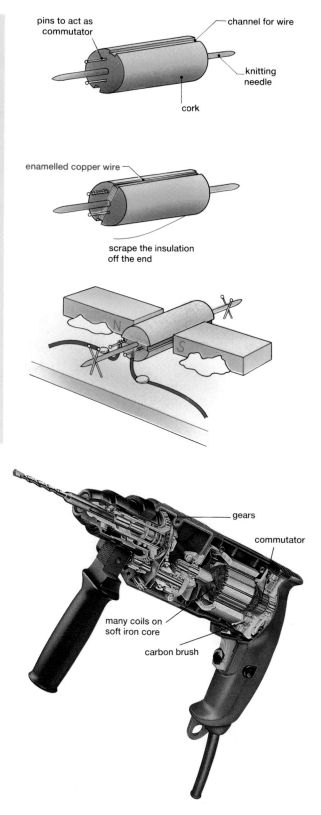

How could you improve your motor?

Practical electric motors

These can be made more powerful in 4 ways:

1. A large number of turns are wound on the coil.

2. A soft-iron core is used (instead of cork) so that the magnetic field is stronger.

3. The permanent magnets can be replaced by electromagnets which give a stronger field. (This also allows the motor to work from an a.c. supply as well as a d.c. supply, because when the current reverses, **both** the magnet's field and the coil current reverse at the same time.)

4. The motor will be more powerful (and run more smoothly) if extra coils are added round the core. This means that the commutator must be split into more than two parts: 4 coils would need a commutator with 8 segments.

▷ Galvanometers

Galvanometers are used to detect electric currents. The simplest galvanometer is a wire and a compass, as in Oersted's experiment (page 284).

The moving-coil galvanometer

This is built like a motor but without a commutator.

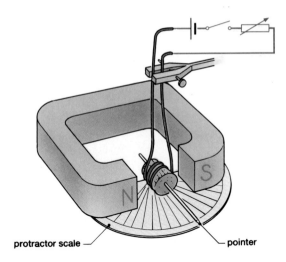

Experiment 35.9
Use the apparatus of experiment 35.7, with a pointer and a scale added.

What happens if you use a rheostat to vary the current?
What happens if you reverse the current?

protractor scale — — pointer

Experiment 35.10
Connect a manufactured moving-coil voltmeter in series with a cell and a rheostat.
Look closely at the coil's movement as you vary the rheostat.

A manufactured moving-coil meter has:
a) Two coil springs to return the pointer to zero when the current stops.
   These springs also conduct the current in and out of the coil.
b) The magnet has curved pole faces and there is a soft-iron core fixed at the centre of the moving coil.
   This means that the coil is always in line with a strong uniform magnetic field. This ensures that the meter has the advantage of a *linear*, evenly-divided scale.

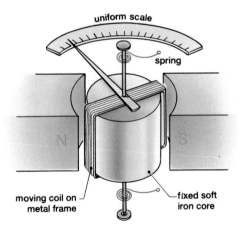

uniform scale

spring

moving coil on metal frame — — fixed soft iron core

Some moving-coil galvanometers do not have a pointer: a mirror is fixed to the coil and this reflects a beam of light on to a scale.

Moving-coil meters can be made so sensitive that they can measure a micro-amp or even less.

Disadvantages of this meter: a moving-coil meter must be connected the correct way round in a circuit (so that the pointer moves the right way). It cannot be used with alternating current (unless a *rectifier* is used – see page 315).

Meters in the control room in a power station

Uses of galvanometers

Galvanometers can measure small currents and they can show us how the current is changing. This makes them very useful.
For example, the diagram shows a galvanometer in a **car fuel gauge**:

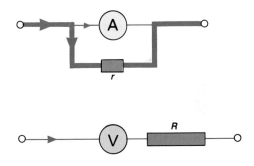

As the fuel in the tank is used up, the float moves lower down and the sliding contact on the rheostat changes the amount of resistance in the circuit. The meter on the dashboard shows a different current for a different level of fuel.

In a similar way, a galvanometer can be used to show the **temperature**. This is used in a *thermocouple* thermometer (page 33) and in a *thermistor* thermometer (page 317).

An **ammeter** can be used to measure a large current, but this would damage a galvanometer. To prevent this, a 'shunt' resistor with a very *low* resistance is placed in **parallel** across the galvanometer. This takes most of the current, rather like a motorway by-pass will take most of the traffic round a village instead of through it.

A **voltmeter** also uses a galvanometer, but this time a *high* resistance is placed in **series** with the meter, so that only a small current can flow through it. The larger the voltage, the larger the current and the more the pointer moves.

Summary

An electric current always produces a magnetic field.

For a straight wire, the lines of force are circular.
Right-hand Grip Rule:
Right thumb pointing with the current,
your fingers curl with the magnetic field lines.

For a coil or a solenoid, the lines of force are like those of a bar magnet. The poles can be found by the N–S rule.

Electromagnets are used in electric bells, relays and telephone receivers.

If a wire carrying a current is placed across a magnetic field, there is a force on the wire.
Fleming's Left-hand Rule:
First finger for **F**ield (N to S)
se**C**ond finger for **C**urrent (+ to −)
thu**M**b then gives **M**ovement.

A coil carrying a current in a magnetic field tends to twist. This is used in an electric motor where a commutator reverses the current every half revolution so that the coil keeps turning.

A moving-coil galvanometer is similar but with springs instead of a commutator.

▷ Questions

1. a) The lines of . . . . round a straight current-carrying conductor are in the shape of . . . . .

   b) The Right-hand Grip Rule: if you grip the . . . . with your right . . . . pointing with the . . . . , your fingers curl with the . . . . field lines.

   c) For a solenoid, the magnetic field is like that of a . . . . .

   d) If a coil is viewed from one end and the current flows in an anticlockwise direction, then this end is a . . . . pole.

   e) The magnetic effect of a coil can be increased by increasing the number of . . . . , increasing the . . . . , or inserting a . . . . core.

2. a) Fleming's Rule for the motor effect uses the . . . . hand.

   b) A motor contains a kind of switch called a . . . . which reverses the current every half . . . . .

   c) A moving-coil galvanometer has the advantage that it has a . . . . scale and can be made very . . . . to small currents. Its disadvantage is that it must be connected the . . . . way round in a circuit and it cannot measure . . . . current.

3. For the coil in the diagram, when the switch is pressed:
   a) What is the polarity of end A?
   b) Which way will the compass point then?

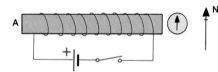

4. a) Draw a large circuit diagram of an electric bell and explain, step by step, how it works.

   b) Draw a wiring diagram to show how one electric bell can be rung by switches at both the front and back doors, using only one battery.

5. Draw a circuit diagram to show how a small current through a relay can be used to switch a large current. How is this used in cars?

6. The diagram shows a relay circuit for a burglar alarm. Explain, step by step, how it works. In what way is it 'fail-safe'?

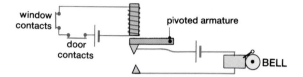

7. The diagram shows a design for an electrically operated model railway signal.
   a) Explain, step by step, how it works.
   b) Explain how, by adding a scale and a spring, you could use it as an ammeter.

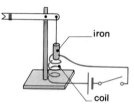

8. Look at the **ticker-timer** on page 133. The electromagnet is supplied with alternating current, of frequency 50 Hz. Describe carefully how this machine works.

9. Which way does the wire in the diagram tend to move?

10. a) Explain, with the aid of a diagram, how an electric motor works.
    b) Show on your diagram which way your motor would rotate.
    c) In what ways may a motor be made more powerful?

11. Draw a large diagram of a moving-coil galvanometer and explain how it works. What are its advantages and disadvantages? In what ways can it be made more sensitive?

12. *A motor has two parts, you'll note,*
    A rotor and a stator.
    So now please state (and learn by rote)
    Which has the commutator.

13. Two students investigated how the strength of an **electromagnet** depended on the current:

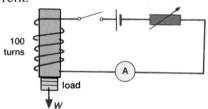

Their results are shown in the table:

Current (A)	0	0.5	1.0	1.5	2.0	2.5	3.0	3.5	4.0	4.5	5.0
Load (N)	0	0.2	0.8	1.6	3.0	5.4	10.0	13.6	14.6	14.9	15.0

a) Use the data to plot a graph of load (**y**-axis) against current (**x**-axis).

b) What load do you think could be supported if the current was i) 2.75 A ii) 6.0 A?

c) Sketch the graph you would expect to get if the coil had only 50 turns.

d) What is happening to the domains in the iron (see chapter 34), and why does the graph level off at the top?

14. A motor draws a current of 2 A at 10 V and lifts a load of 10 N through a distance of 3 m in 2 seconds.

a) What is the work done on the load? (See page 113.)

b) What is the rate of working on the load (power output)?

c) What is the electrical power input?

d) What is the efficiency?

16. The diagram shows an electromagnetic **circuit-breaker**. Study the diagram:

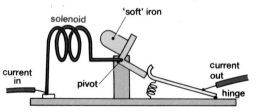

a) Describe, step by step, what happens if too much current flows.

b) Why does this make the circuit safer?

c) How could you reset the circuit-breaker?

d) How could this be combined with a 'thermal' circuit-breaker (see page 262)?

17. Comment on each of these statements made by Professor Messer:

a) The N-pole of an electromagnet is where the current leaves.

b) The N-pole of a coil is the end where the current flows clockwise.

c) The coil in an electric bell should have a core of hard-steel.

d) The armature of a relay should be a permanent magnet.

e) A motor with 5 coils needs a carbon commutator with 5 segments.

f) An ammeter can be made more sensitive by removing the springs.

Further questions on page 334.

How much revision have you done? See page 366.

15. *Professor Messer's not too bright, So help him get his circuits right:*

chapter 36

ELECTRO-MAGNETIC INDUCTION

As soon as Oersted discovered that electricity produced magnetism, many people began to look for the reverse effect.

Eventually, in 1831, Michael Faraday discovered how to make electricity using magnetism.

Experiment 36.1
Connect a straight piece of wire to a sensitive galvanometer. Then move the wire across a strong magnetic field.

What does the galvanometer show you?

We call this current an *induced* current.

Faraday found that **if a wire is moved to cut across lines of flux, then a current is induced in the wire** (if there is a complete circuit).
This is called **electromagnetic induction**.

The electrons in the wire have been given a push as the wire moves across the magnetic field.
If the wire moves along the magnetic field lines, there is no current. It has to cut *across* the field.

The current can be increased by:
● using a stronger magnetic field,
● moving the wire faster.
If the wire stops moving there is no current.

To remember the direction of the induced current, we use **Fleming's Right-hand Rule:**
Hold your *right* hand in a fist and then spread out the thumb, first finger and second finger so that they are at right-angles to each other.
– Point your **F**irst finger in the direction of the magnetic **F**ield (from N to S).
– Rotate your hand about that finger until your thu**M**b points in the direction of the **M**ovement of the wire.
– Then your se**C**ond finger points in the direction of the **C**urrent.

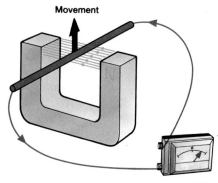

sensitive galvanometer

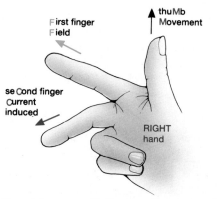

Does this agree with the experiment above?

Connect a coil with a large number of turns to a centre-zero galvanometer.

a) Then move a bar magnet quickly into the coil. What happens?
    This is shown on the diagram:

b) Hold the magnet still, inside the coil. Does any current flow?

c) Then pull the bar magnet out of the coil. Which way does the current flow?

d) Move the magnet in and out repeatedly. What do we call this kind of current?

e) Move the magnet very slowly and then quickly. Which way gives more current?

f) Move in a weak magnet and then a strong magnet at the same speed. Which is better?

g) Try the experiment with another coil having only a few turns. Do you get less current?

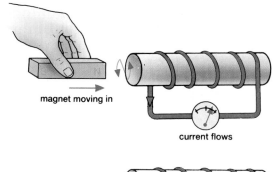

magnet moving in

current flows

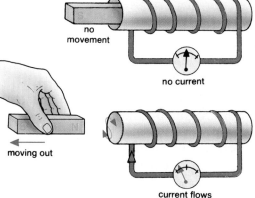

no movement

no current

moving out

current flows opposite way

Faraday's Law: He found that the induced voltage (and the current) can be increased by:
- using a stronger magnet,
- moving the magnet faster,
- increasing the number of turns on the coil.

The **direction** of the current in the coil can be found by **Lenz's Law**:

the direction of the induced current is such that it <u>opposes</u> the change producing it.

This is shown in the diagrams above. When the N-pole of the bar magnet is moving *in*, the current flows so as to produce a N-pole at that end of the coil. This means that the current is opposing the movement of the magnet (by trying to repel it). When the N-pole moves *out*, the current flows the other way to make a S-pole which again *opposes* the movement (by trying to attract the magnet).

These induced currents are used in record-player pick-ups (page 306), in tape-recorders (page 307), and in a moving-coil microphone:

This is built like a small loudspeaker (page 289). It converts sound energy into electrical energy (see the sound experiments on page 226).

Singers rely on induced currents in a microphone

▷ Current generators (dynamos)

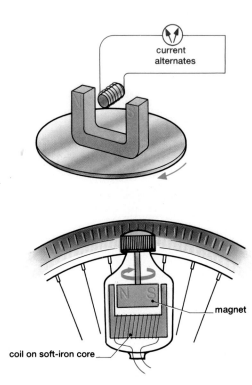

current
alternates

What happens to the pointer of the galvanometer
when the magnet rotates?
What happens if you speed up the turntable?

A current that reverses to and fro like this is
called *alternating current (a.c.)*.

Bicycle dynamo

A *bicycle dynamo* works in the same way. A
magnet rotates near a coil of wire so that the lines
of flux are cut by the wire.
The coil is wound on a soft-iron core so that the
magnetic field is stronger.

magnet

coil on soft-iron core

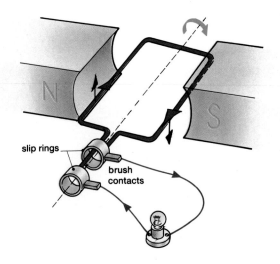

slip rings

brush
contacts

The simple a.c. generator (an alternator)

A generator can be built rather like a motor, with
a rotating coil and a fixed magnet. As the coil
rotates, it cuts lines of force and a current is
induced.

Use Fleming's **Right**-hand Rule to find the direction
of the current at the instant shown in the diagram
(if the coil is rotating clockwise). Do you agree
with the arrows on the diagram?

As the coil rotates, the current is conducted in and
out by way of *slip-rings* and carbon brushes.

Sketch a diagram of this alternator when the coil
has turned through half a revolution and mark in
the direction of the current.
Which way does the current flow through the
lamp now? What does 'alternating' mean?

Can you find three ways in which this generator
could be improved to increase the current?

▷ Alternating current (a.c.)

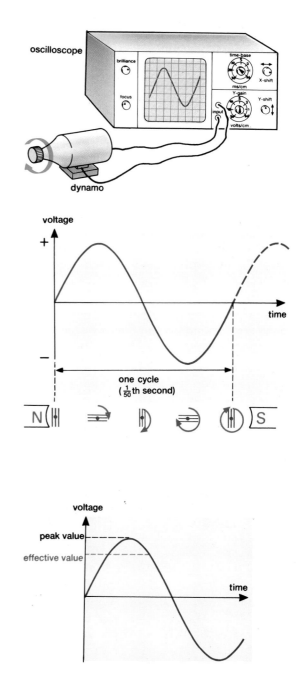

Look at the shape of the graph on the oscilloscope (this shape is called a **sine curve**).

Each complete cycle of the graph corresponds to one complete revolution of the generator.
For the mains supply (in Britain), each cycle takes $\frac{1}{50}$th second. That is, the frequency of the mains supply is 50 cycles per second (50 Hz).

The diagram shows how the varying voltage corresponds to the different positions of the coil:

The voltage is largest when the coil is horizontal.
The voltage is zero when the coil is vertical.

Peak value and effective value

The **peak** voltage occurs only for a moment.
The **effective value** is the value which is equivalent to a steady direct current and gives the same heating effect in an electric fire.
It is found that:

> **Effective value = 0.7 × peak value**
> or
> **Peak value = 1.4 × effective value**

Example
The effective mains value in Britain is 240 V.
∴ Peak value = 1.4 × 240
= 340 V

This means that mains wiring must be insulated to at least 340 V. The **Live** wire in a mains circuit varies between +340 V and −340 V during each a.c. cycle.

▷ The simple d.c. dynamo

This is built in just the same way as a d.c. motor.
It has a **commutator** so that the current in the
brush contacts **always flows the same way**:

Use Fleming's Right-hand Rule to find the
direction of the current in the diagram.

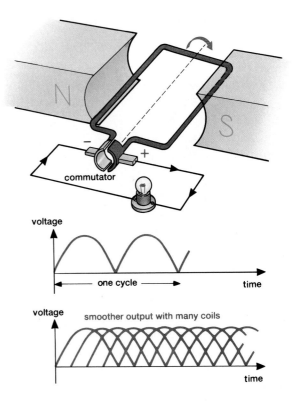

Experiment 36.6
Connect a d.c. motor to a lamp and spin the
motor rapidly. What happens?

Experiment 36.7
Connect an oscilloscope to this d.c. dynamo.
Sketch the graph of the output voltage:

There is a smoother output if the dynamo has
many coils and the commutator is split into
many segments:

▷ Practical generators

The voltage (and the current) of a generator can
be increased by:
- using a coil with more turns,
- using a stronger magnet, or using a powerful
  electromagnet to make the field stronger,
- winding the coil round a soft-iron core so that
  the magnetic field is stronger,
- rotating the coil faster.

The photograph shows a very large alternator in a
power station. It makes electricity for your home.

In large generators like these, it is usually the
magnet that rotates, as in a bicycle dynamo.

In power stations, the alternator is usually driven by
steam power (see the experiments on page 117).
The heat energy comes from burning oil or coal,
or from nuclear energy (see page 350). The steam
turns a turbine fan. This turns the generator and the
kinetic energy is converted into electrical energy.

A car speedometer

Experiment 36.8
Suspend a thick aluminium disc or dish over a strong magnet which can rotate on a record-player turntable. Use a straw as a pointer.

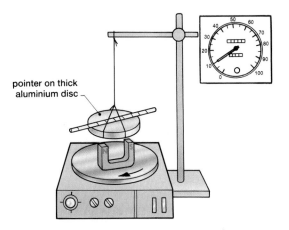

pointer on thick
aluminium disc

What happens when the magnet rotates?
Does the pointer turn through a larger angle if you speed up the turntable?

The movement of the magnet induces currents (called **eddy currents**) in the disc.
By Lenz's Law, these currents try to **oppose** the relative movement between the two objects, and the disc tries to follow the magnet.

How can this be converted into a practical design for a speedometer?

Two coils linked by magnetism

Experiment 36.9   Mutual induction
Place two coils side by side as shown.
Connect coil **A** to a battery and a switch.
Connect coil **B** to a centre-zero galvanometer.

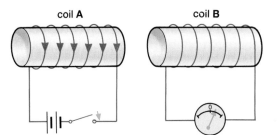

coil **A**   coil **B**

What happens as you switch on the current?
Does anything happen if you keep it on?
What happens if you switch off coil **A**?

A current is induced here, even though the coils are not moving. When you switch **on** coil **A** it becomes an electromagnet and it puts a field round coil **B**. The effect is just the same as pushing a magnet quickly into coil **B**. An induced current flows for a moment.

When the magnetic field is steady, the current stops. When you switch **off** coil **A**, the field stops quickly. This is the same as pulling a magnet out of coil **B**, and an induced current flows the opposite way. The induced current is caused by a **changing** field.

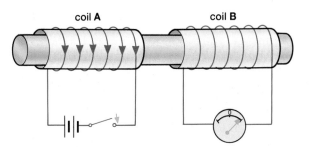

coil **A**   coil **B**

Now try it with a soft-iron core through **A** and **B**: Is the induced current larger now? Why is this?

If the current is switched on and off repeatedly, what happens in the second coil?

If the first coil carried alternating current, what would you expect to find in the second coil? This is a simple **transformer**.

▷ The transformer

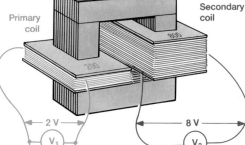

laminated soft-iron core

Demonstration experiment 36.10
(!Danger – high voltages – demonstration only!)
Two coils are placed on a soft-iron core as shown:
One coil (the **primary** coil) is connected to an a.c.
voltmeter and a source of alternating current. The
secondary coil is connected to an a.c. voltmeter.

The two coils are connected to each other by a
magnetic field (as in the last experiment).
This means that there is an induced alternating
current in the secondary coil.

If you measure the voltages you will find that
there is a connection between these voltages
and the number of turns on each coil. In fact:

$$\frac{\text{Secondary voltage}}{\text{Primary voltage}} = \frac{\text{Number of turns on secondary coil}}{\text{Number of turns on primary coil}} \quad \text{or} \quad \frac{V_2}{V_1} = \frac{N_2}{N_1}$$

If the secondary coil has more turns than the primary
coil, then it is a **step-up** transformer, because the
secondary voltage is bigger than the primary voltage.

Step-up (ratio 1 : 4)

If the secondary coil has fewer turns than the primary,
it is a **step-down** transformer:

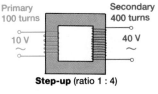

Step-down (ratio 2 : 1)

Transformers are very efficient machines.
If a transformer was 100% efficient, then:

$$\frac{\text{Power supplied}}{\text{to primary coil}} = \frac{\text{Power delivered}}{\text{in secondary coil}}$$

or $\quad I_1 \times V_1 = I_2 \times V_2$

(from page 264)

This means that if the voltage is stepped **up** by a
certain ratio, then the current in the secondary is
stepped **down** by the same ratio.

In practice, the efficiency is never 100%, although it
may be 99%. Energy is lost if some of the primary
magnetic field does not pass through the secondary.
In practice, the coils are wound on top of each other:

Energy is also lost because some **eddy currents** are
induced in the soft-iron core as well as in the coil.
These eddy currents can be reduced by making the
core laminated from thin metal strips, or 'lamina'.

A transformer inside a record-player

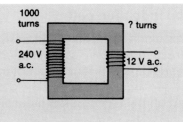
Uses of transformers

1. Transformers are used to get low voltages from the 240 V a.c.
mains, for door bells, radios, computers, hi-fi equipment, etc.
In your TV set, a step-up transformer produces a very high voltage.

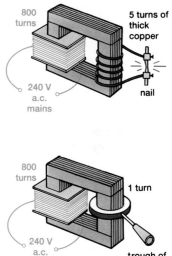

circuit symbol

2. *Demonstration 36.11   Resistance welding*
If the primary coil has 800 turns and the secondary only 5 turns,
it is a step-down transformer with a ratio of 800 : 5, which is 160 : 1.
In a perfect transformer, the current is stepped up in the same
ratio, so that a current of 1 A in the primary might give a
secondary current of almost 160 A.
This large current heats the nail until it melts. The current can
also be used to weld together two nails. This is electric welding.

3. *Demonstration 36.12   An induction furnace*
If the secondary coil has just one turn, then the secondary
current is even greater. This large current will quickly boil water
or melt solder. 'Induction heating' is sometimes used in industry.

4. The national Grid system
To transfer energy from the power station to your home we
could use ***either*** a) a low voltage and a high current
           or b) a high voltage and a low current.

Method a) is wasteful as the high current heats the power lines.

Method b) is used, with transformers at each end of the Grid
system to change the voltage. This is the main reason for using
a.c. in the Grid (because transformers do not work with d.c.).

A Grid engineer called Sid,
Thought he'd change what he
usually did,
He very soon found,
With transformers changed round,
That he melted the national Grid.

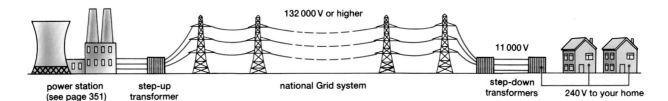

The induction coil

This works like a step-up transformer but with a *direct* current. We saw in experiment 36.9 that a steady direct current does not induce a current in the secondary coil, but if the d.c. is switched on and off then a current is induced.

The petrol inside a car engine is ignited by the spark from an induction coil:

A rotating cam switches on and off the primary circuit. This induces a large voltage across the many turns of the secondary coil. A rotating distributor applies this high voltage to the correct sparking plug.

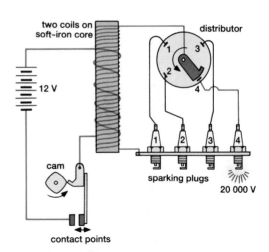

Summary

If there is relative movement between a magnetic field and a wire, then a current may be induced in the wire.

Fleming's **Right**-hand Rule:
First finger for **F**ield (N to S),
thu**M**b for **M**ovement,
se**C**ond finger then gives direction of **C**urrent.

Lenz's Law: the direction of the induced current is such that it **opposes** the change producing it.

A simple a.c. generator has a rotating coil with slip-rings. A d.c. generator has a commutator.

Effective value of a.c. = 0.7 × peak value.

Transformers change alternating voltages: $\dfrac{V_2}{V_1} = \dfrac{N_2}{N_1}$

If the voltage is stepped-down, the current is stepped-up in almost the same ratio.
Energy losses are reduced by a laminated core.

The Grid system: electricity is distributed at high voltage and therefore low current.

▷ Questions

1. Copy out and complete:
   a) When there is relative movement between a wire and a magnetic . . . . then . . . . may be . . . . in the wire.
   b) Fleming's Rule for the dynamo effect uses the . . . . hand.
   c) Lenz's Law: the . . . . of the . . . . current is such that it . . . . the change producing it.
   d) The strength of the induced current depends on the . . . . of the movement, the . . . . of the magnet and the number of . . . . on the coil.
   e) An a.c. generator is fitted with . . . . whereas a d.c. generator is fitted with a . . . . .

2. Copy out and complete:
   a) A transformer will work only with . . . . current.
   b) The formula for a transformer is . . . . .
   c) When the voltage is stepped-down, the current is stepped- . . . . in the . . . . ratio (if the transformer is . . % efficient).
   d) The core of a transformer is . . . . to reduce . . . . currents.
   e) In the national Grid system, the electricity is distributed round the country at . . . . voltage and . . . . current.
   f) In the Grid, ... current is used because transformers will not work on . . . . current.

3. A coil is connected to a galvanometer. When the N-pole of a magnet is pushed in the coil, the galvanometer is deflected to the right. What deflection, if any, is observed when a) the N-pole is removed, b) the S-pole is inserted, c) the magnet is at rest in the coil? State 3 ways of increasing the deflection on the galvanometer.

4. a) Explain, with the aid of a diagram, the action of a simple a.c. generator.
   b) Sketch a graph to show how the voltage varies during one revolution of the coil.
   c) Mark on the graph where the coil is
      i) horizontal ii) vertical.
   d) On the same graph, sketch and label the graphs you would get if the coil rotated
      i) at twice the speed ii) at the same speed but in the opposite direction.

5. Prof. Messer says that he is careful with the 240 V a.c. mains because he does not want to be killed by a 340 V shock. Explain this.

6. An electric guitar has steel strings vibrating next to a small magnet and a 'pick-up' coil connected to an amplifier:

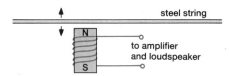

   If you pluck the string so that it vibrates at a frequency of 500 Hz,
   a) what happens to the magnetic field near the coil?
   b) what happens in the coil?
   c) what do you hear?
   Electric guitars can't use nylon strings. Why?

7. Complete the following table of transformers:

8. Explain what is wrong in each diagram:

9. A 24 V lamp, needing 2 A, is supplied from a transformer connected to the 240 V mains.
   a) What is the turns ratio?
   b) How much power is supplied to the lamp?
   c) How much power is supplied by the mains? What have you assumed here?
   d) How much current is taken from the mains?

10. a) 240 000 W of power might be delivered through the national Grid at i) 240 V or ii) 240 000 V. Calculate the size of the current in each case.
    b) Why would 240 000 V be used, despite the dangers? c) Why is a.c. used?
    d) What are the energy changes in the process which begins with coal arriving at a power station and ends in a lamp in your home?

Primary p.d.	Secondary p.d.	Primary turns	Secondary turns	Step-up or step-down
100 V a.c.		10	100	
100 V a.c.		100	10	
240 V a.c.	12 V a.c.	200		
11 000 V a.c.	132 000 V a.c.		12 000	

Further questions on page 336.

The best way to revise? See page 366.

▷ Physics at work: In your home

Records

Music is recorded on an LP record disc by making a narrow groove in the plastic (see also page 220). The vibrations of the sound are copied into the wobbles of the groove. The groove is an **analogue** of the sound (see page 324). A louder sound gives a larger wobble (with a large amplitude). A higher note gives a more rapid wobble (with a higher frequency).

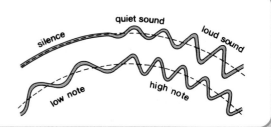

Magnetic pick-up

The wobbles of the groove are converted to electricity by means of a needle, or **stylus**, attached to a small magnet. The magnet is near a small fixed coil of wire. As the needle wobbles with the groove, the magnet also wobbles. This makes the magnetic field in the coil vary and so the coil has a varying current in it (by **electromagnetic induction**, see page 296). The varying current in the coil is passed to an amplifier (page 324) and then to a loudspeaker (page 289).

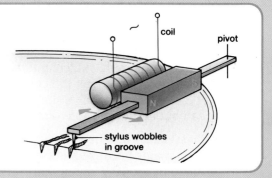

Stereo records

Old records had only one soundtrack and were played back through only one loudspeaker. Stereo records give a more realistic sound. They are recorded by two microphones (one to the left of the singer and one to the right) and played back through two loudspeakers. Both sounds are combined into one groove, with the left-hand side of the groove carrying the wobbles for the left-hand sound, and the right-hand side of the groove carrying the right-hand signal.

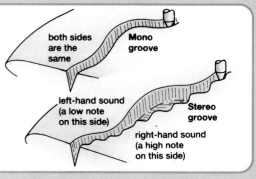

Compact discs

Compact discs are different: they are **digital** discs (see also page 324). Instead of the wobbling grooves there is a pattern of 'pits' cut into the shiny disc. There are billions of pits, with a varying length and spacing, depending on the sound.
The pattern of pits is read by using a weak laser beam (see page 195). The laser beam is focussed on one track and moves across the disc like a stylus. The pits and flats reflect the beam differently, producing a beam which is off or on. This shines on to a light-detector and de-coder which converts it to a varying electric current for an amplifier and loudspeaker.

Video discs work in a similar way and can store 50 000 pictures on one disc.

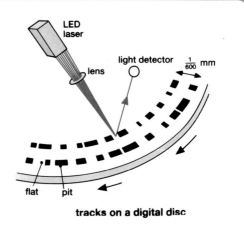

tracks on a digital disc

▷ Physics at work: In your home

Tape-recorder

A tape-recorder uses a 'tape head' to magnetise the iron oxide on the tape. The tape head is an electro-magnet with a very small gap (about $\frac{1}{1000}$ mm).

The strong magnetic field near the gap magnetises the tape (with the same frequency as the sound). The electromagnet changes the direction of the tiny magnetic domains (see page 278):

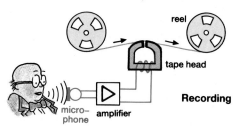

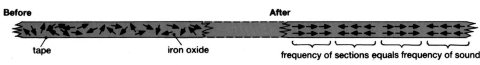

On playback, the magnetised tape goes past a tape head again. The magnetic field of the tape induces a small current in the coil (by *electromagnetic induction*, see page 296). This current is amplified and passed to the loudspeaker.

A recording can be erased by a similar 'erase head'. This carries an alternating current at a very high frequency (60 kHz). It demagnetises the tape (so the tiny magnets point in many directions).

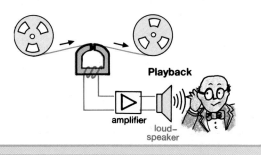

Television

A television tube is built like a CRO (page 310). However, in a TV, the electron beam is deflected by currents flowing in the *deflection coils*:

The magnetic field of a coil bends the electron beam to different parts of the screen (see page 309). The beam moves to build up a new picture every $\frac{1}{25}$ second (see page 311).

In a colour TV set there are *three* electron guns. One gun varies in brightness according to the *red* signal from the TV station. The other two guns vary with the *green* and *blue* signals. Because these are the three primary colours of light (see page 218) a complete range of colours can be produced.

If you look closely at a TV screen you will see it is covered with groups of red, green and blue dots. These dots glow when the electrons hit them.

A *shadow mask* is fitted just behind the screen. This has a hole in it for each group of coloured dots. It ensures that the beam from the 'red' gun only hits the red dots; and similarly for the green and blue.

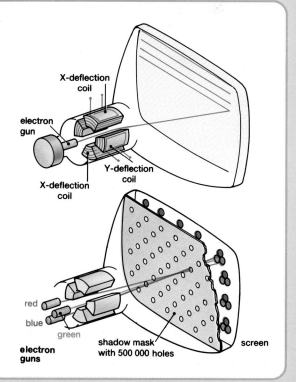

chapter 37

ELECTRON BEAMS

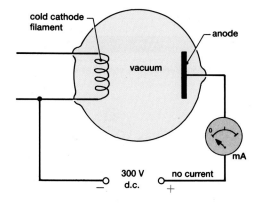

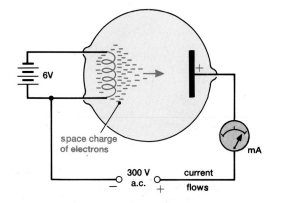

Experiment 37.1   Diode valve
(!Danger – high voltage – demonstration only!)
The apparatus, called a **diode**, consists of a glass
bulb containing two metal electrodes in a vacuum.
One electrode, a metal plate, is called the anode.
The other electrode (the cathode) is a filament
which can be heated as in an electric lamp.

Connect a sensitive ammeter and a high-voltage
battery (or power-pack) across the diode as
shown in the diagram.

Does any current flow? Why not?
Reverse the battery connections. Does any current
flow through a vacuum?

Now connect a 6 V battery across the cathode
filament so that it is heated and glows.

Does a current flow through the ammeter now?
Reverse the connections to the high-voltage
battery, as before. Does a current flow through
the ammeter now?

A current can flow only when the cathode is hot.
We believe that **electrons** are 'boiled' off the
hot cathode because at this high temperature they
have a lot of energy.
This process is called **thermionic emission**.

The negative electrons form a **space charge** near
the cathode.
If the anode is **positive**, these negative electrons
are attracted across the gap and a current flows.
If the anode is **negative**, the negative electrons are
repelled and no current flows.
This shows that **electrons are negative**.

Because the current can flow only one way
through the apparatus, it is called a diode **valve**.

▷ Cathode-ray tubes

This glass bulb has a hot cathode with an anode close to it. The anode has a hole in it so that when it is positive and attracts electrons, some electrons pass through the hole and shoot across the vacuum. This arrangement is called an *electron gun*.

In the middle of the bulb is a second anode in the shape of a Maltese cross. The end of the tube is a *fluorescent screen* as in a television set.

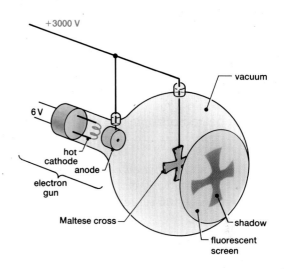

Experiment 37.2   Maltese cross tube
Switch on the cathode so that it glows white hot. Now switch on the 3000 V supply to shoot electrons across the vacuum.

What do you see on the screen?

The streams of electrons leaving the cathode and shooting across the vacuum are called *cathode-rays*. The edges of the cathode-ray shadow on the screen are sharp. This is because the electrons are travelling in straight lines.

Mrs Messer: How do you make a Maltese cross?
Professor Messer: Pour water down his trousers.

Bring a strong magnet near the tube.
What happens to the shadow? Why is it distorted?
What happens if you reverse the magnet?

Experiment 37.3   Cathode-ray deflection tube
In this tube an electron gun sends a beam of electrons (cathode-rays) through the vacuum so that the beam cuts across a fluorescent screen. The screen is placed between two horizontal metal plates (labelled Y_1 and Y_2).

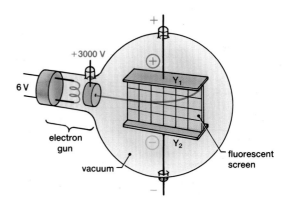

The electron beam can be deflected in two ways:

a) By **magnetic deflection**, using bar magnets as in the last experiment.
Alternatively, electromagnets (coils) can be used (as in a TV set, see page 307).

b) By connecting plate Y_1 to the positive (+) terminal of the power supply and plate Y_2 to the negative (−) terminal, as shown.
Why do the electrons deflect away from the negative plate and move nearer the positive plate?
What happens if you reverse the connections?
This **electrostatic deflection** is used in a cathode-ray oscilloscope.

Televisions and computers use cathode-ray tubes

▷ The cathode-ray oscilloscope (CRO)

This uses a cathode-ray tube very like the one in the last experiment: an electron gun sends electrons through the vacuum to a screen:

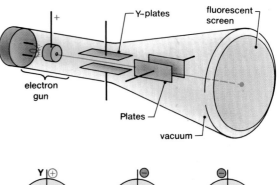

Two Y-plates can deflect the beam vertically (as in the last experiment). Two X-plates can deflect the beam horizontally.

The diagrams show how charging the plates positively and negatively can deflect the beam to any position on the screen:

How could the spot be deflected to the bottom left-hand corner?

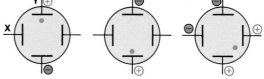

The controls

The complete oscilloscope has several controls to adjust, usually in the following order:

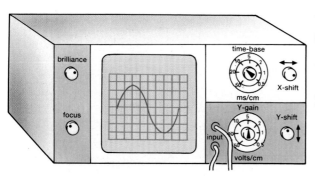

From the control settings shown, can you see that this waveform has:
peak voltage = 15 volts (3 cm on the screen)
time-period = 80 milliseconds (8 cm on the screen)

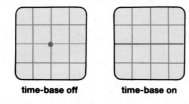

time-base off   **time-base on**

1. The **brightness** and **focus** controls.

2. A voltage *signal* (perhaps from a microphone) is applied to the input terminals. This signal is amplified before it is applied to the Y-plates. The **Y-gain control** varies the amount of amplification. A small signal needs more gain.

   In the diagram the Y-gain is set to 5 V/cm. This means that the spot moves 1 cm up (or down) for *every* 5 V across the input terminals.

3. When we use an oscilloscope (CRO) to draw graphs, we must use the **time-base control**.

   When this control is switched on, the spot moves at a constant speed across the screen and then jumps back very quickly to start again. The time-base control is used to vary the speed of the spot. At high speeds it appears as a line, due to your persistence of vision (page 203).

   In the diagram the time-base is set to 10 ms/cm. This means that it takes the spot 10 milliseconds to travel 1 cm across the screen.

▷ Uses of an oscilloscope

1. Measuring voltages

The deflection of the spot depends on the *voltage* applied to the deflection plates. A CRO is a good voltmeter, taking almost no current.

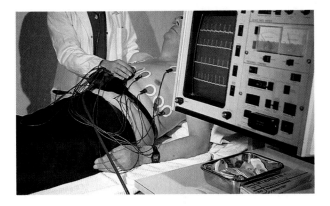

Experiment 37.4
With the time-base switched *off*, connect one 1.5 V dry cell to the input terminals.
Measure the size of the deflection.

Now connect 2 dry cells in series to the input terminals. What happens to the deflection?

What happens with three cells?
What happens if you reverse the battery?

What happens if you connect a low-voltage *alternating* current supply?

no input   d.c. input   a.c. input

2. Studying waveforms

With an alternating input, the time-base should be used to move the spot horizontally at the same time, so that we can see the shape of the voltage waveform:
A CRO with a microphone can be used to show the waveforms of sounds (see page 226).
Hospitals use CROs to monitor heart-beats.

a.c. input only   time-base only   both voltages

3. Television

A TV set is a CRO with two time-bases – one moves the spot across the screen, the other moves it vertically (like the way your eyes move when you are reading).
The spot marks out 625 lines on the screen and it does this 25 times in each second.
The signal from the aerial varies the brightness of the spot so that a picture is built up of bright and dark spots (as in a newspaper photograph).

A VDU (Visual Display Unit) for a computer is the same.

In a colour TV set there are *three* electron guns (see the diagram, page 307). These build up a full-colour picture using just the 3 primary colours: red, green and blue.

In a radar set, a CRO displays a picture of the radio echoes received by a rotating aerial (see page 215).

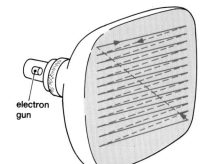

electron gun

LONDON'S RIVER

The River Thames from Gallions Reach to Southend Pier
A composite picture from the five Radars of the Thames Navigation Service

▷ X-rays

X-rays were discovered by W. Röntgen in 1895.

An X-ray tube is really a high-voltage diode valve. A hot cathode emits electrons which are accelerated to high speed by a high voltage (perhaps 100 000 V).

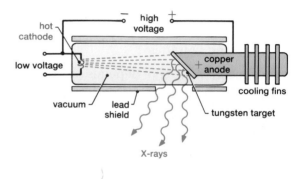

When the fast-moving electrons hit the tungsten target, some of their kinetic energy is converted to X-radiation (most of their energy is converted to heat which is conducted away to the cooling fins).

Controls for the X-ray tube

a) To give a more intense beam of X-rays, the cathode is made hotter, to give more electrons and so give more X-rays.
b) To give 'harder' X-rays (more penetrating, with shorter wavelength), the voltage across the tube is increased to give the electrons more kinetic energy.
   'Softer' X-rays (less penetrating, longer wavelength) are given by lower voltages.
   TV sets emit a small amount of soft X-radiation.

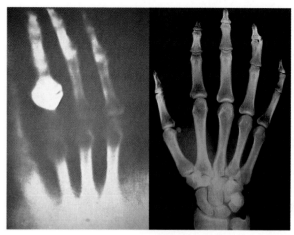

The very first X-ray (taken by Wilhelm Röntgen of his wife's hand and ring) and a modern X-ray

The snake that swallowed two light bulbs:

Properties of X-rays

1. !X-rays are dangerous! They can cause cancer.
2. X-rays pass through substances but are absorbed more by dense solids (e.g. bones or lead).
3. X-rays affect photographic film.
4. X-rays ionise gases, so that the gases become conducting (and so discharge electroscopes).
5. X-rays are not deflected by magnetic or electric fields.
6. X-rays can be diffracted (so they are **waves**).
7. X-rays are electromagnetic waves of very short wavelength (about 10^{-10} metre). See page 212.

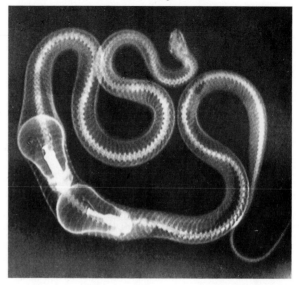

Uses of X-rays

1. In **medicine**: to inspect teeth, broken bones, etc. With care, X-rays can be used to kill cancer cells.

2. In **industry**: to inspect metal castings and welded joints for hidden faults.

3. In **science**: photographs of X-rays diffracted by crystals give information about the arrangements of the atoms in different substances.

Summary

A hot cathode produces a space charge of negative electrons by thermionic emission. The electrons are attracted by a positive anode.

This is used in an oscilloscope.

An electron beam ('cathode rays') can be deflected by a magnet or by charged plates (as in a CRO).

When fast electrons are stopped by a target, X-rays may be produced.

▷ **Questions**

1. Copy out and complete:
   a) When a wire is heated in a vacuum, it emits . . . . . This process is called . . . . emission.
   b) A current flows in a diode valve only when i) the cathode is . . . . and ii) the anode is . . . . to attract the . . . . electrons.
   c) A beam of electrons ('. . . . rays') can be deflected by i) a . . . . or ii) . . . . . The . . . . electrons are deflected away from a . . . . -charged plate.
   d) In a cathode-ray oscilloscope (CRO) the vertical deflection of the spot is proportional to the . . . . applied across the Y-plates. When studying waveforms, the spot is moved horizontally by a . . . . circuit.
   e) When fast electrons are stopped, ... of very . . . . wavelength may be produced.

2. Draw a simple labelled diagram of a cathode-ray tube for a CRO. Name and explain the purpose of the controls on a CRO.

3. *Professor Messer's full of woe –*
   He can't control his CRO.

4. A time-base is adjusted so that it draws a horizontal line, as shown, 50 times per second. Draw diagrams of what is seen when the following are connected:
   a) a battery which makes the upper Y-plate positive b) an a.c. supply of 50 Hz
   c) an a.c. supply of 100 Hz d) an a.c. supply of 100 Hz in series with a diode.

5. Here is a waveform on a CRO screen:

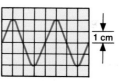

The Y-gain control is set to 2 V/cm and the time-base is set to 5 ms/cm. Calculate:
   a) the peak voltage,
   b) the time-period of the wave,
   c) the frequency of the wave.
With the same input, the controls are set to 1 V/cm and 10 ms/cm. Sketch the screen.

Further questions on page 337.

chapter 38

ELECTRONICS

Silicon diodes

Some substances (like germanium and silicon) are neither good conductors nor insulators. These substances are called **semi-conductors**.

Semi-conductors can be made **n**-type or **p**-type, by adding tiny amounts of other substances (**n** and **p** stand for **n**egative and **p**ositive charges). If a p-type material and an n-type material are joined together, we get a very useful device, called a p–n junction **diode**.

Experiment 38.1 Silicon diode
Connect a battery, a lamp and a diode in series. Connect the marked or narrow end of the diode (the 'cathode') nearest to the negative terminal of the battery. What happens to the lamp?

Now turn the diode the opposite way round in the circuit. Does the lamp still light?

A diode allows a current to flow *only one way*.
It is a *rectifier*.
The arrow on the diode symbol shows the direction in which conventional current can flow. The diode is then said to be *forward-biassed*.

If the diode is *reverse-biassed*, the current is nearly zero.

Experiment 38.2 I : V graph
Repeat experiment 31.23 (page 257) measuring the current *I* and the voltage *V* for a junction diode.

As the graph shows, almost no current flows if the voltage is applied in the reverse direction.
A current flows in the forward direction whenever the voltage is more than about 0.6 volts.

If the reverse voltage or the forward current is increased too much, then the diode will be damaged. See p. 257 for graphs of other **non-ohmic** conductors.

Diodes have many uses in radios, TV and computers.

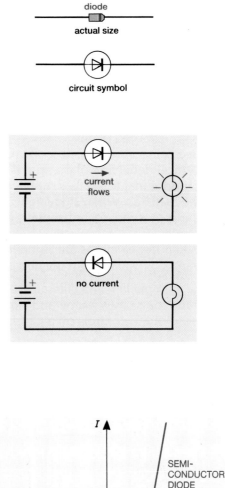

diode
actual size

circuit symbol

current
flows

no current

I

SEMI-
CONDUCTOR
DIODE

reverse direction

0.6 V

V

▷ Uses of diodes

1 Protecting equipment

Diodes are used to protect radios and computers which would be damaged if the battery or power supply was connected the wrong way round. If the battery in the diagram was connected the other way round, then no current would flow and no damage could be done.

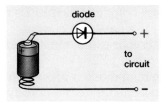

2 Switching in stand-by batteries

Here is a circuit that could be used on your bicycle:

If you are riding the bike, the dynamo will produce a higher voltage than the battery, and light the lamp using the *left*-hand circuit only, saving your battery. But if you stop at traffic lights, the dynamo will stop, and then the *right*-hand circuit will light the lamp. What does each diode do?
A similar circuit could keep a computer running even if there was a mains power cut.

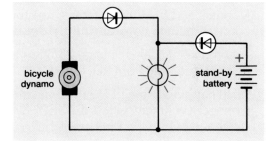

3 Rectification

The mains supply provides alternating current (a.c.), but often we want direct current (d.c.) as in a battery. How can we use a diode to do this?

> *Experiment 38.3   Half-wave rectifier*
> Connect a diode in series with a 1000 Ω resistor and a low voltage a.c. supply.
> Now use an oscilloscope (a CRO, see page 310) to look at the waveform: first connect the CRO across the a.c. supply (the 'input') and sketch the a.c. waveform (see also page 311).
> Then connect the CRO across the load resistor as shown in the diagram. Sketch the new waveform:

Why is only half the a.c. waveform still present?
The diode conducts only for half the a.c. cycle (only when it is forward-biassed).
The alternating current has been **rectified** to an uneven direct current. You are using this circuit whenever you plug your radio into the a.c. mains.

> *Experiment 38.4   Smoothing*
> Now connect a 10 µF capacitor (see page 243) in parallel with the resistor (across the input terminals).

Look at the waveform. It has been **smoothed**.
The capacitor stores some energy when the voltage is high, and then releases it when the voltage falls.

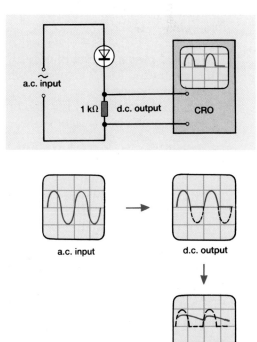

315

▷ The Light-Emitting Diode (LED)

Some special diodes called LEDs will give out light when they are passing a current.
The usual colour is red, but it is possible to get yellow or green LEDs.
A light-emitting diode is an example of a **transducer** – it changes energy from one form to another.
What are the two forms of energy in a LED?

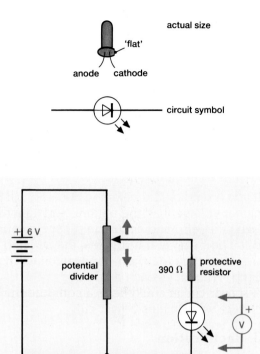

actual size

'flat'

anode   cathode

circuit symbol

Experiment 38.5
Connect an LED to a protective resistor. Then connect them to a **potential divider** circuit:

A potential divider (see page 256) lets us vary a voltage smoothly from low to high.

What happens to the LED as you vary the voltage? Connect a voltmeter across the LED. What voltage is needed before the LED will light?

Does the diode light if it is reversed in the circuit?

6 V

potential divider

390 Ω

protective resistor

The LED will not light until the applied voltage is greater than about 2 V. However if the full 6 V was applied, the LED would be damaged. A LED must always have a **protective resistor** in series with it.

Example
A conducting LED has a voltage drop across it of 2 V and should pass a current of only 10 mA (= 0.01 A).
If it is connected to a 9 V supply, what protective resistance is needed?
If there is 2 V across the LED, the other 7 V must be across the protective resistor R.

Formula first: $V = I \times R$ (see page 251)
Then numbers: $7 = 0.01 \times R$

$$\therefore R = \frac{7}{0.01} = \underline{700\ \Omega}$$

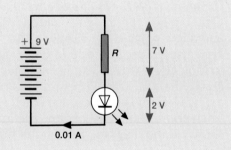

9 V

R

7 V

2 V

0.01 A

Uses:
LEDs are often used as indicator lamps. LEDs are small, reliable, need only a small current and last much longer than filament lamps.
LEDs are sometimes used in **seven-segment displays** in digital clocks, cash registers and calculators:

a seven-segment display

Professor Messer was using his calculator, sitting opposite Mrs Messer.
Why did she laugh when he typed in 0.7734 and then 5508?
Can you make other words? Try: 0.37108, 5663. Or 35006, 5637. Are they 3781637?

▷ The Light-Dependent Resistor (LDR)

When a semi-conductor has light shining on it, it can conduct electricity more easily.
Its resistance depends on the brightness of the light, so it is called a **Light-Dependent Resistor** (or photo-conductive cell).

The LDR shown in the photo is made from a semi-conductor called cadmium sulphide (CdS). In the dark its resistance is more than 10 MΩ. In sunlight its resistance is only about 100 Ω.

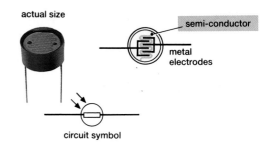

circuit symbol

Experiment 38.6   A light-meter
Connect an LDR in series with a cell, a milli-ammeter and a variable resistor. Adjust the variable resistor to give a deflection on the ammeter:

What happens to the current as you cover and uncover the LDR?

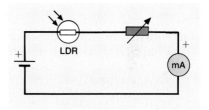

Uses:
This circuit could be used as a photographer's light-meter (if it was calibrated). The same circuit can be used to alter the aperture in automatic cameras. More uses are shown later (page 322).

▷ The thermistor

When a semi-conductor is heated, it can conduct electricity more easily. This is because the rise in temperature makes more free electrons available to carry the current.
The higher the temperature, the **less** the resistance.

circuit symbol

Experiment 38.7   An electronic thermometer
Repeat the last experiment, but with a thermistor in place of the LDR. Heat the thermistor with your hand or a hairdryer:

What happens to the current as the thermistor gets hotter?

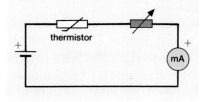

Uses:
This circuit could be used as an electronic thermometer if the ammeter was calibrated in °C (see page 33). It would have a **non-linear scale**, with **un**equal divisions.
Thermistors are used to protect the filaments of projector lamps and TV tubes from a current surge as they are switched on. The cold thermistor keeps the current low at first; as the filament heats up, so does the thermistor and it gradually lets more current pass. This extends the life of the filament.
More uses are shown later (page 321).

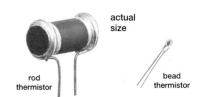

rod thermistor   bead thermistor

Experiment 38.8   I : V graph
Repeat experiment 31.23 (p. 257) measuring current **I** and voltage **V** for a thermistor. The graph is shown on p. 257.

▷ Reed switch

A reed switch can be used in electronic circuits. It consists of a glass tube with two iron reeds sealed inside it:

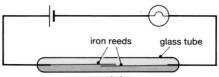

If the iron is unmagnetised, there is a gap between the reeds, and no current can flow. The switch is said to be **normally-open** (NO).

If a magnet is brought near the reed switch, it magnetises each iron reed as shown:
(This is 'magnetic induction', see page 280.)
The result is that the two reeds are attracted towards each other, and they bend to touch each other. The lamp comes on.
What happens when the magnet moves away?

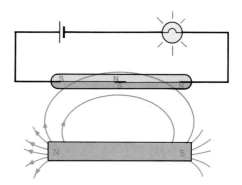

Experiment 38.9
Use a magnifying glass to look closely at the reeds while you move the magnet nearby.

Normally-closed reed switch

If an extra non-magnetic contact is added to the reed switch as shown, it can be used as a **normally-closed** (NC) switch. The current flows from B to C **until** a magnet is brought near. What happens then?
How can it be used as a change-over (CO) switch?

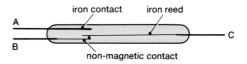

Uses:
A **normally-closed** reed switch can be used in burglar alarms. The reed switch is fixed in the door frame and a magnet is fixed to the door:

What happens if a burglar opens the door?
What are the disadvantages of this simple alarm?

A **normally-open** reed switch can be used as a safety switch on the door of a microwave oven. Draw a diagram to show how this would work.

Reed relay

If a reed switch is put inside a coil of wire (an electromagnet, see page 285), then it can be operated by a current in the coil.
This is a **reed relay**.
A small current in the coil can switch on a larger current in the reed switch. (It is a simple kind of amplifier.)

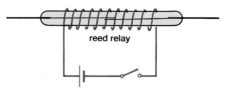

▷ Electromagnetic relay

The diagram shows another kind of relay, which can switch on a larger current than a reed relay.
If enough current flows in the coil it will attract the soft iron armature (see also page 286). The armature is pivoted and so pushes the springy contacts together, so a current can flow through them. What happens if the coil current is switched off?

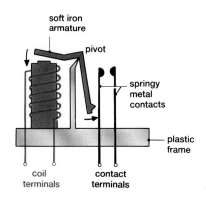

The diagram shows one pair of normally-open contacts. Others can be added side by side so that several circuits can be switched. Normally-closed and change-over contacts can also be used.

Experiment 38.10   A burglar alarm
Connect a relay and a light-dependent resistor (LDR) to a battery and a bell, as shown:
Shine a light on the LDR, so that its resistance falls.

What happens to the current in the LDR?
What happens to the relay? What happens to the bell?
If you put this circuit in a drawer (where it is dark), what would happen if a burglar opened the drawer?

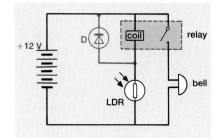

It is important to include a *diode* as shown at **D** in the diagram. This is because the relay coil acts like an induction coil (see page 304) and can give a large reverse voltage. The diode protects the circuit by shorting out this large voltage.

▷ Transducers and Systems

The LED, LDR, thermistor and reed switch are all examples of *transducers* – they convert energy from one form into another. The rest of this chapter is about electronic **systems**. All the systems have three main parts as the diagram shows:

For example, a record player has an input transducer (the pick-up), a processor (the amplifier) and an output transducer (loudspeaker).

Here is a summary of the transducers you have met:

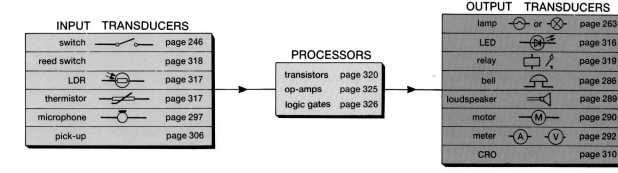

▷ The transistor

This is a device for amplifying small electric currents. It is made of three layers of n, p, n semi-conductor material. The three layers are called the **collector, base** and **emitter**.

The diagram shows one kind of transistor, but there are many different sizes and shapes.

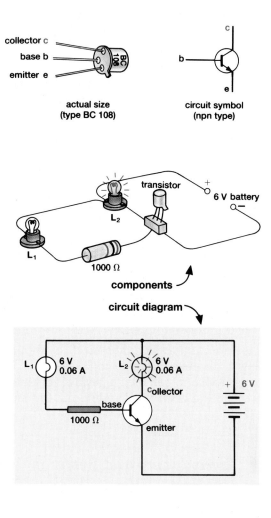

actual size
(type BC 108)

circuit symbol
(npn type)

Experiment 38.11   Current amplifier
To investigate a transistor, set up the circuit shown:

Do you find that lamp L_2 lights but lamp L_1 does not?

Now unscrew lamp L_1 so that there is a break in the circuit to the base b.
Do you find that lamp L_2 (in the collector circuit) goes out?

Now replace lamp L_1. Does L_2 light again?

This shows that:
a small current in the base circuit causes a larger current to flow in the collector circuit.

In your circuit, a current of about 1 mA in the base circuit has switched on a current of about 60 mA in the collector circuit.
A transistor acts as a **current amplifier**.

A transistorised radio contains several transistors to amplify small currents until they are big enough to drive a loudspeaker.

Experiment 38.12   Moisture detector
Change your circuit so that instead of lamp L_1, there are two probes as shown. The probes are two pieces of copper wire placed close together but not touching.

What happens to lamp L_2 if you connect the probes together with a conductor? Why is this?

What happens if you place a drop of water touching both probes? Why does the lamp light?

How could you place the probes in a water tank to light the lamp when the tank fills to a certain depth? How could you use this circuit to light a warning lamp when it is raining outside?

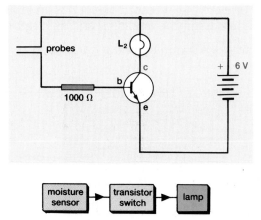

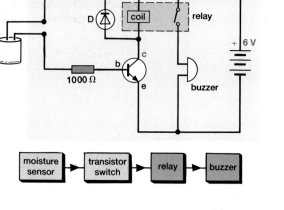

What happens when you fill the beaker with water?

Uses:
How could this be used to help a blind person know
when to stop pouring from a tea-pot?
How could it warn when a baby's nappy is wet?
Or switch on windscreen-wipers automatically?

Adding a thermistor

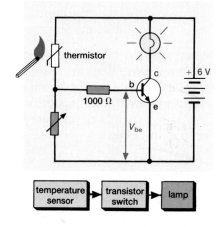

What happens when you heat up the thermistor? (See p. 317.)

The thermistor and the variable resistor are the two parts of a
voltage divider or 'potential divider' (see page 256).
As the thermistor gets hotter and its resistance *de*creases, its
'share' of the 6 V *de*creases and so the voltage across the
variable resistor *in*creases. This increases V_{be} (see the diagram)
and switches *on* the transistor.
If V_{be} is less than 0.6 V the transistor switch is off.
If V_{be} rises above 0.6 V the transistor starts to conduct.
If V_{be} is above 1.5 V the transistor is fully on.

Why is the variable resistor called a 'sensitivity control'?

Uses:
Draw a diagram showing how you could add a relay so that a
bell would ring as a fire warning.
How could it be used to warn if a chip pan was getting too hot?

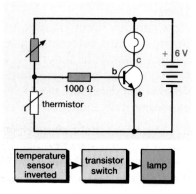

Does the lamp light now when the thermistor is hot or when
it is cold? Why?

Uses:
How could this be used as a frost alarm to warn a fruit farmer?
How could this be adapted (using a relay) to keep a tropical
fish tank or a greenhouse from becoming too cold?

▷ More uses of electronic circuits

Adding a light-dependent resistor

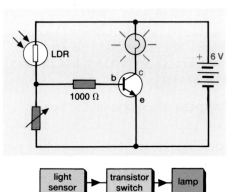

Experiment 38.16   A light-sensor circuit
Connect an LDR to a transistor switch as shown:
Cover up the LDR so it is dark. Is the lamp off?
Shine light on the LDR. What happens?

Which variable component can you adjust to
make your circuit switch on:
– even in a dim light?
– only in a bright light? Try it.

How could you use this as a burglar alarm for a
drawer? (It is usually dark inside a drawer.)

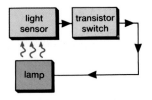

Experiment 38.17   An electronic candle
Place the LDR near the lamp and facing it, in a dark
room. Adjust it so the lamp is just off.

What happens when you bring a lighted match near
the lamp? Why does the 'candle' light?
What happens if you snuff out the 'candle' by
covering it with your fingers? Why?
This is an example of ***positive feedback*** from the
lamp to the LDR.

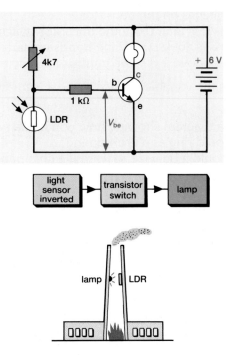

Experiment 38.18   A burglar alarm (dark-operated)
Invert the light-sensor part of your circuit as shown:
Adjust the sensitivity control until the lamp is
off in daylight.
Now cover up the LDR. What happens to the lamp?

What happens to the resistance of the LDR when it
is covered (see page 317)? Does this increase or
decrease the voltage V_{be}? Why does the lamp light?

Uses:
How can this circuit be used to switch on car lamps
or street-lamps automatically at dusk?
Draw a diagram to show how you could add a relay
to ring a bell if a burglar interrupts a beam of light.
What is the disadvantage of this circuit (see
experiment 38.29)?

It is important that factory chimneys do not pollute
the air with too much smoke. Invent a device to
warn a factory manager if the chimney is producing
too much smoke.

Causing a delay

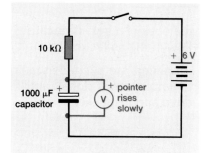

The battery is charging up the capacitor through the resistance.
Because it is a large capacitance and a large resistance it takes a
long time. The larger the resistance and the larger the
capacitance, the longer the **time constant** of the circuit.

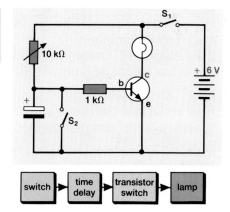

Why is there a delay before the transistor is switched on?
The transistor does not come on until V_{be} rises to 0.6 V.

Press switch S_2 to discharge the capacitor and so re-set the
circuit. How can you vary the time delay? Try it.

Uses:
How could this circuit be used to make an egg-timer for a deaf
person? Draw the circuit you would use if the person was blind.

Use: How could you use this circuit to switch off your radio
after you have gone to sleep?

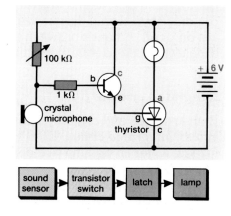

This circuit also contains a **thyristor** (silicon controlled rectifier).
This is a special diode with a third connection: the gate (g).
A thyristor does not conduct **until a current enters the gate**;
then it continues conducting **even if the gate current stops**.

Tapping the microphone gives a voltage for a short time (see
page 226). Why does the lamp light and stay lighted?
This is a 'latching' circuit (see also page 330). To turn off the
lamp you must disconnect the battery.

Uses:
How could this be used to sound an alarm if a baby cries?
How could this be used as a burglar alarm?

▷ Amplifiers

In experiment 38.11 we saw that a small current in the base of a transistor causes a larger current to flow in the collector circuit. It is a **current amplifier**.

Amplifiers are needed in radios and record-players to amplify small signals until they are big enough to drive a loudspeaker.

Experiment 38.23 One-stage amplifier (an intercom)
The diagram shows a simple one-stage amplifier:

The microphone gives a varying **input** voltage across a–b (see also page 226). This is amplified to give a larger **output** voltage across A–B. It is a **voltage amplifier**.
If the output is 100 times the input voltage, the circuit has a **voltage gain** of 100.

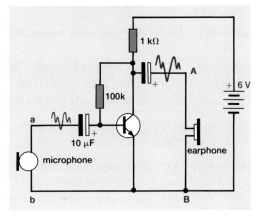

Use:
How could you use this as an intercom to connect your bedroom with another room?

In practice, amplifiers for radios or record-players usually have two or more stages to give greater amplification. This is done by connecting the output A–B of the first stage to the input a–b of the second stage.

Analogue or digital?

The last circuit is an example of an **analogue** system: the voltage can vary smoothly and have any voltage between the lowest and the highest.

Another kind of system is **digital**. Digital systems can only have certain definite values, usually just **on** or **off**. A transistor switch (page 320) and a logic gate (see page 326) are digital systems.

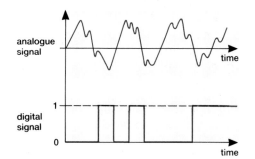

Integrated circuits (IC)

Both analogue and digital systems can be built into an **integrated circuit** (IC). The photo shows a full-size IC for an op-amp (see opposite page). The actual circuit or 'chip' is very small but can easily contain the 20 transistors, 11 resistors and 1 capacitor needed to make the op-amp circuit.

Nowadays some ICs have more than a million components on one chip. This is **micro**electronics.

actual size
of IC

actual
size
of
'chip'
inside

magnified photo
of the circuit

▷ Operational amplifier (op-amp)

One very useful IC is called an *operational amplifier* or
'*op-amp*' (see photo on opposite page).
It is an amplifier with a very high gain. It has two inputs and
two main uses:

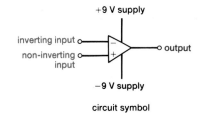

+9 V supply

inverting input o

non-inverting
input

output

−9 V supply

circuit symbol

1 Inverting amplifier

For this use, the op-amp is connected as shown in the diagram
(the power supplies are not shown).
This circuit includes a *feedback resistor*, R_F. This resistor feeds
back some of the output into the *inverting* input. This is called
negative feedback.
Although this reduces the gain of the amplifier, it also reduces
distortion and makes sure the output is an accurate copy of the
input.
The output is larger than the input and *inverted*.
The formula for the *voltage gain* (see opposite page) is:

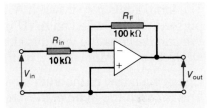

R_F
100 kΩ

R_{in}
10 kΩ

V_{in}

V_{out}

$$\textbf{Voltage gain} \; = \; \frac{V_{out}}{V_{in}} \; = \; \frac{R_F}{R_{in}}$$

With the resistance values shown in the diagram, the gain is 10.
By changing the resistors, an op-amp can give a gain up to 1000.

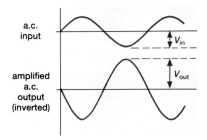

a.c.
input

amplified
a.c.
output
(inverted)

V_{in}

V_{out}

2 Comparator (to compare two voltages)

If the op-amp is used without any feedback, it has a
very high gain but is unstable. Its output swings quickly
between a high voltage and a low voltage.
It is now a *digital* circuit.
The op-amp *compares* the 2 inputs, and switches the
output (to high or to low).
If V_1 is greater than V_2 then the output is low.
If V_1 is less than V_2 then the output is high.

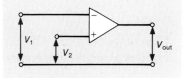

V_1

V_2

V_{out}

This op-amp is a very sensitive and rapid switch. It can
be used to make the sensor circuits on pages 320–323
much more sensitive. For example, here is a better
light-sensitive circuit adapted from page 322:

A fixed potential divider (page 256) provides a steady
voltage V_1 to one input. The second potential divider
includes an LDR, so the voltage V_2 varies with the light
shining.
If the LDR is dark, V_2 is larger than V_1 and the output is
high, so it switches on the transistor and the ammeter.

This circuit is very sensitive to changes. How could you
use it to compare the whiteness of two sheets of paper?

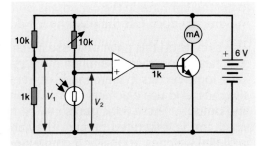

10k

10k

mA

1k

+ 6 V

1k

V_1

V_2

▷ Logic circuits

Connect the circuit shown in the diagram:
Press down the switches one at a time and then both together.

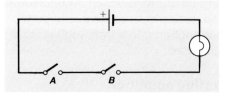

What do you have to do to get the lamp to light?

The lamp lights only if switch **A AND** switch **B** are pressed. This is called an **AND gate**.
We can describe the way this circuit or 'gate' works in a diagram called a **truth table**:

Check each row of the truth table carefully.
This is called the **logic** of the circuit.

This kind of logic is used in a washing machine where there is a main 'on' switch and another switch operated by the door of the machine. For safety, the washing machine will only work if the main switch is on **AND** the door is closed.

switch A	switch B	lamp
open	open	off
closed	open	off
open	closed	off
closed	closed	ON
inputs		output

Truth tables are usually shown with numbers rather than words. Because each switch has only two states (open, closed) we use only two numbers: **0** and **1**.
If the switch is open and so electricity cannot pass, it is given the value **0**.
If the switch is closed so electricity can go through, it is called **1**.
In the same way, a lamp which is off is **0**, and a lamp which is on is **1**.

AND

switch A	switch B	lamp
0	0	0
1	0	0
0	1	0
1	1	1
inputs		output

Here is a circuit with a different logic:

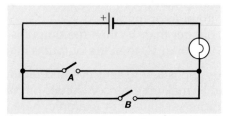

The lamp comes on if switch **A OR** switch **B** is pressed. This is called an **OR gate**.
Check its truth table carefully:

The doors of cars have switches connected with OR logic: the light inside the car comes on if either the driver's door **OR** the passenger's door is opened.

The circuits on this page only use mechanical switches. In practice, it is much more useful to have logic circuits that are made of transistor switches, and use electrical signals for inputs and outputs. These logic **gates** can be made very small, inside integrated circuits. They are used in digital watches, robots and computers.

OR

switch A	switch B	lamp
0	0	0
1	0	1
0	1	1
1	1	1
inputs		output

AND gate

A tiny integrated circuit can easily contain several AND gates, built from transistor switches (see page 320).

Each transistor switch operates on the *voltage* applied to it. The voltage can be 0 (called logical 0) or it can be higher (called logical 1).

The circuit symbol for an AND gate with two inputs is shown here: (Remember its shape: it looks like the D of AND.)

The truth table is the same as the AND gate on the opposite page. There is an output only if input **A** **AND** input **B** are at a logical 1. Remember: 0 is zero volts and 1 is a higher voltage (usually 5 V).

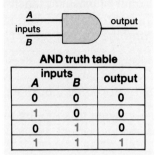

AND truth table

inputs		output
A	B	
0	0	0
1	0	0
0	1	0
1	1	1

OR gate

The circuit symbol for an OR gate with two inputs is shown here: Notice the different shape from an AND gate.

The truth table is the same as the OR gate on the opposite page. There is an output if *either* input **A** **OR** input **B** are at a higher voltage.

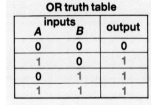

OR truth table

inputs		output
A	B	
0	0	0
1	0	1
0	1	1
1	1	1

NOT gate

The third kind of logic gate is the NOT gate or inverter. Its circuit symbol is shown here:

Its truth table is very simple: the output is always the *opposite* of the input. It is an *inverter*.

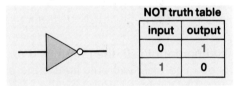

NOT truth table

input	output
0	1
1	0

Sometimes these gates are combined. For example,
AND with NOT = NAND gate.
OR with NOT = NOR gate.
Compare these symbols and truth tables with the AND and OR gates. What do you notice?

Logic gates,
Have just two states:
Just off or on,
With 0 or 1.

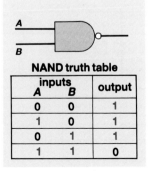

NAND truth table

inputs		output
A	B	
0	0	1
1	0	1
0	1	1
1	1	0

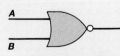

NOR truth table

inputs		output
A	B	
0	0	1
1	0	0
0	1	0
1	1	0

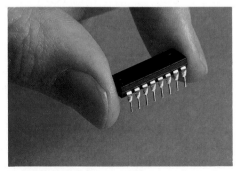

This chip contains four AND gates

▷ Using logic gates

A designer of washing machines wants to ensure that the motor will not start until the on/off switch is on **and** the door is safely closed:

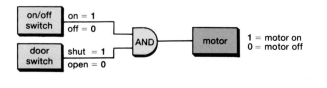

From the truth table for an AND gate (page 327), you can see that the output will be a **1** (to switch on the motor) only if **both** inputs are **1**.

If we also want to keep the motor off until the water level is correct, we add another AND gate:

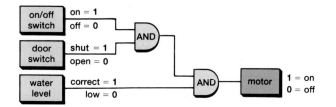

To operate these logic gates we need input circuits that give either 0 volts (logical **0**) or give about 5 volts (logical **1**). This is how we do it:

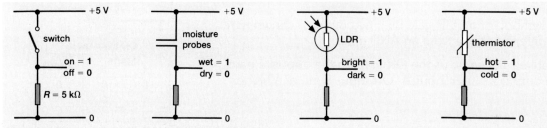

*These potential dividers give a logical **0** or **1**. If the sensor and resistor are interchanged, the logic is reversed.*

Experiment 38.26 Wet AND light sensor
Connect up this system:

If either input is **0**, the output is **0** and the LED is off. The LED will only light if the probes are wet **AND** light is shining on the LDR.

Use: To warn you (with a bell) if it was raining AND it was day-time – so you could bring in the washing from the line or the baby from the pram (but it wouldn't bother you if it rained at night!).

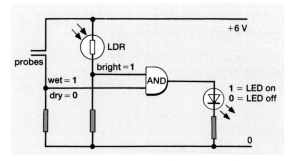

Experiment 38.27 Wet AND cold sensor
Replace the LDR with a thermistor, and add a NOT gate.

The NOT gate inverts the logic, so that it gives a logical **1** when the thermistor is *cold*.
If the output of the AND gate is **1**, it switches on the transistor and the relay circuit.

Use: How could this be used as a safety circuit in a kettle?
How can you change the circuit so that it rings a bell if it is either wet OR cold?

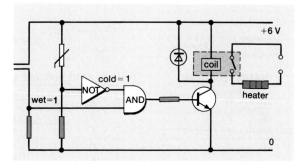

Experiment 38.28   A seat-belt alarm
Connect two switches to the logic
circuit shown:
Use LEDs to show the outputs at **P**
and **Q**. Try pressing the switches in
different combinations.

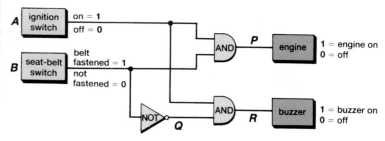

This is a kind of combination lock.
It can be used as a seat-belt alarm.

To turn on the engine, point **P** must be at logical **1** and this
can only be done by turning on both switches **A** and **B**.
However, if **A** (the driver's ignition key) is turned on before
B (the seat-belt switch) then the warning buzzer sounds.

To analyse this circuit we need to draw up a truth table to
cover all the combinations of **A** and **B**:

The third column shows the result at **P** (of **A AND B**), while
column **Q** shows the result at **Q** (of **NOT B**). You should work
through this table carefully, moving down each column in turn.

A	B	P	Q	R
inputs		engine A AND B	NOT B	buzzer A AND Q
0	0	0	1	0
0	1	0	0	0
1	0	0	1	1 = on
1	1	1 = on	0	0

Controlling an automatic washing-machine

An automatic washer needs logic to ensure that:
- it only operates if the door is closed, then
- it must be filled with enough water, then
- the water must be heated if it is cold, then
- the motor must turn the drum.

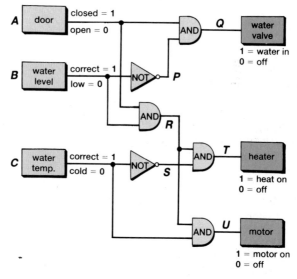

Follow through the logic of this circuit:
If the door is closed (**A** = 1) but the water level
is low (**B** = 0, so **P** = 1) then the first AND gate
switches on the water supply (**Q** = 1 because
A = 1 **AND P** = 1).
When the water level is correct (**B** = 1, so **P** = 0)
then **Q** = 0 so the water is turned off.
Now that **A** = 1 (door shut) **AND B** = 1 (water
level correct) then **R** = 1. If the water is cold
(**C** = 0, so **S** = 1) then the heater is switched
on (**T** = 1 because **S** = 1 **AND R** = 1).
The heater stays on until the water is hot
(when **C** = 1 so **S** = 0, and so **T** = 0).

Now that **A**, **B** and **C** are all correct (all at 1),
then **U** = 1 and the motor is turned on.
What happens if the door is opened (**A** = 0)?

Check through the full truth table and see
that you agree with it.

What sensors would you use for **A**, **B** and **C**?

A	B	C	P	Q	R	S	T	U
door	water	temp.	NOT B	valve A AND P	A AND B	NOT C	heater R AND S	motor R AND C
0	0	0	1	0	0	1	0	0
0	0	1	1	0	0	0	0	0
0	1	0	0	0	0	1	0	0
0	1	1	0	0	0	0	0	0
1	0	0	1	1 = on	0	1	0	0
1	0	1	1	1 = on	0	0	0	0
1	1	0	0	0	1	1	1 = on	0
1	1	1	0	0	1	0	0	1 = on

▷ The bi-stable

Some circuits can store information – they have a memory. Computers need memory circuits to store data.

The simplest memory circuit uses two NOR gates which are cross-connected.

This is called a **bi-stable** or a **flip-flop**, because it has **two** stable states. Study the diagrams:

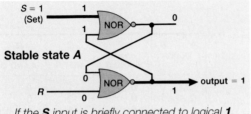

*If the **S** input is briefly connected to logical **1**, it sets the bi-stable to state **A***

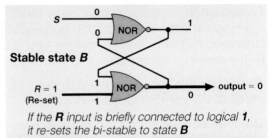

*If the **R** input is briefly connected to logical **1**, it re-sets the bi-stable to state **B***

The circuit flips between state **A** and state **B**. You should check the logic of each NOR gate (see the truth table on page 327).

In effect the circuit remembers or 'latches' the last thing that happened to it.

A computer contains millions of these bi-stables in its RAM (Random Access Memory). Each bi-stable can store one 'bit' of information.

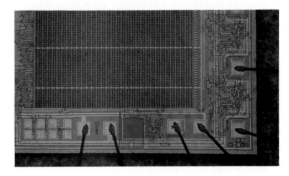

A magnified 'chip' – it contains 65 000 bi-stables!

Experiment 38.29   A burglar alarm with a latch
If a burglar triggers a sensor, the alarm bell should continue to ring even after he has moved past the sensor. Try this system:

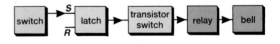

The switch might be a pressure switch under your door mat. What happens when you press the switch for just a moment?

The alarm stays on because the bi-stable latch 'remembers' that the burglar was there.
How can you re-set the circuit?

Summary

A diode allows a current to flow one way only. This can be used to rectify a.c. to d.c.
The resistance of an LDR decreases if it is illuminated. The resistance of a thermistor decreases when it is heated. An LDR, a thermistor, and a reed switch are input transducers.
A relay and a LED are output transducers.
A transistor amplifies small currents. A transistor (or an op-amp) can be used as an electronic switch.
There are several kinds of logic gates: AND, OR, NOT, NAND, NOR.

▷ Questions

1. Copy out and complete:
   a) Germanium and silicon are called . . . . .
      A diode will pass a current only when it is . . . . -biassed.
   b) A diode is useful as a half-wave . . . . to convert . . . . current to . . . . current. The output can be smoothed by a . . . . .
   c) A LED (. . . .) is a transducer; it changes . . . . energy to . . . . energy.
   d) When an LDR (. . . .) is illuminated, its resistance . . . . . When a thermistor is heated, its resistance . . . . .
   e) A . . . . (NO) reed switch does not pass a current until a . . . . is brought near it.
   f) A . . . . current in the coil of a relay can switch on or off a . . . . current through the contacts.
   g) A transistor is used to . . . . small currents. A small current in the . . . . circuit causes a . . . . current to flow in the . . . . circuit.

2. Draw the symbol and truth table for a) AND b) OR c) NOT d) NAND e) NOR gates.

3. The diagram shows a diode circuit used to switch in a battery automatically if the main power supply fails.

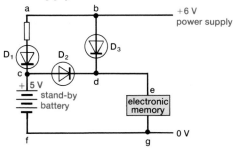

   a) Round which route does the current flow when the power is on? Why?
   b) Which route is used to 'trickle charge' the battery?
   c) Which route when the power fails? Why?

4. Draw block 'systems' diagrams:
   a) to light a lamp if it rains,
   b) to ring a bell if it rains,
   c) to switch on a fan motor if a greenhouse gets too hot,
   d) to sound an alarm if a thief opens a light-tight drawer.

5. Draw transistor circuits for the systems in question 4.

6. Draw block 'systems' diagrams:
   a) to switch on a water pump if a house plant is too dry,
   b) to switch on a heater if a fish tank gets too cold,
   c) for an anti-burglar device to switch on the house lights at night when you are away on holiday.

7. Draw transistor circuits for question 6.

8. Draw block 'systems' diagrams:
   a) to switch on a water pump if a ditch fills with water,
   b) to switch on a roadworks light at night,
   c) to switch on a fan in a storage heater if it heats up,
   d) to ring a bell if a baby cries,
   e) for a fire alarm to detect smoke ('smoke is more dangerous than fire'),
   f) to switch on a pump to fill a water tank if the level is too low,
   g) to ring a bell when an oven is up to the correct temperature,
   h) for an oven timer.

9. Draw a truth table for this system:

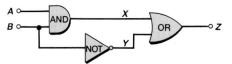

10. Draw block 'systems' diagrams:
    a) to sound an alarm only if the baby cries *and* its nappy is wet,
    b) to open a greenhouse window if it is warm in the day-time,
    c) to tell you if it is a good day for gardening because it is warm, dry and light,
    d) to sound an alarm if the tropical fish tank is dark or cold or the water is low,
    e) to allow a drilling machine to run only if the operator presses a foot switch and has not got her hands breaking a beam of light which is shining near the drill,
    f) to show a deaf person if the doorbell has rung since last they looked. (Add a re-set.)

Further questions on pages 337–339.

▷ Physics at work: Communications

A **communication system** consists of:

① *A method of transferring energy.*
A radio uses electromagnetic waves (page 213).
A telephone (page 287) uses a current in a wire.

② *Information* (for example, someone talking into a microphone).

③ *A way of en-coding the information*, combining it with the energy to be transferred. This is transmitted to the receiver, where it is de-coded.

Radio

The diagram shows how a radio wave (called the *carrier* wave) can have its amplitude changed or *modulated* by a sound wave.
This is **A**mplitude **M**odulation or **AM**:

This wave is transmitted from the radio station.

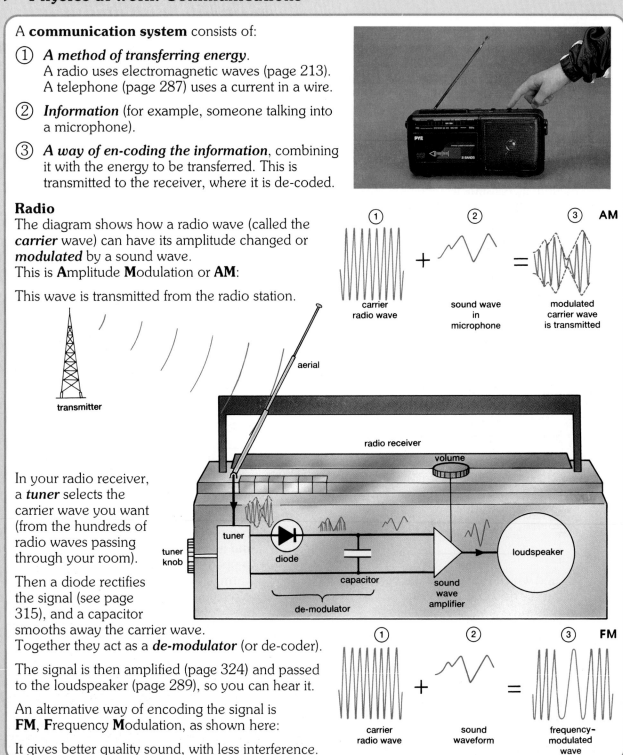

① carrier radio wave + ② sound wave in microphone = ③ **AM** modulated carrier wave is transmitted

transmitter

aerial

radio receiver

volume

tuner knob

tuner

diode

capacitor

de-modulator

sound wave amplifier

loudspeaker

In your radio receiver, a *tuner* selects the carrier wave you want (from the hundreds of radio waves passing through your room).

Then a diode rectifies the signal (see page 315), and a capacitor smooths away the carrier wave.
Together they act as a *de-modulator* (or de-coder).

The signal is then amplified (page 324) and passed to the loudspeaker (page 289), so you can hear it.

An alternative way of encoding the signal is **FM**, **F**requency **M**odulation, as shown here:

It gives better quality sound, with less interference.

① carrier radio wave + ② sound waveform = ③ **FM** frequency-modulated wave

332

▷ Physics at work: Satellites

Satellites are objects that orbit round the Earth. Like the Moon, they are held in orbit by the gravitational pull of the Earth (see page 162).

Communications satellites like 'Intelsat' are needed because microwaves (for telephone networks) cannot bend round the curve of the Earth (see the diagram on page 215).

In the same way, TV satellites can broadcast programmes to the whole of Europe:
You need a 'dish' aerial on your house to pick up the weak signal (see page 184).

Although a satellite is moving very fast, it can appear to be stationary! If it is placed in the right orbit, at a distance of 36 000 km above the Earth, it takes 24 hours to go round its orbit. This is the same period as the Earth, so the satellite appears to hover over one place. This is a *geo-stationary* or *synchronous* orbit.

Weather satellites like 'Meteosat' are often put in geo-stationary orbits (see below).

Earth observation satellites like 'Landsat' can give us a detailed picture of the Earth's surface. They can identify diseased crops and locate pollution due to oil leaks (see below).

Military satellites are used for spying.
Navigation satellites are used by ships and planes to locate their position very accurately.
Astronomical satellites can take sharper photographs outside the Earth's atmosphere.

An 'Intelsat' communications satellite

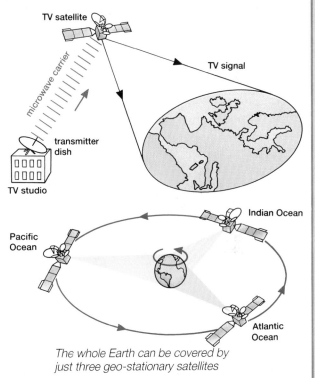
The whole Earth can be covered by just three geo-stationary satellites

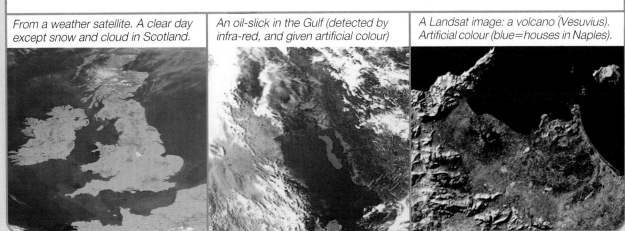

From a weather satellite. A clear day except snow and cloud in Scotland.

An oil-slick in the Gulf (detected by infra-red, and given artificial colour)

A Landsat image: a volcano (Vesuvius). Artificial colour (blue=houses in Naples).

▷ **Magnetism**

1. A small plotting compass can be used to plot a magnetic field because
  A the compass needle is a freely suspended magnet.
  B the compass needle is easily de-magnetised.
  C the Earth's magnetic field has no effect on the needle.
  D there is a vacuum inside the case of the compass.   (LEAG)

2. You are given three metal rods, which have been painted so that they all look the same, and a length of thread.
One of the rods is made of magnetised steel, another of unmagnetised iron and the third is made of copper. No extra apparatus is available.
  a) Describe how you could find out which rod was made of which material.
  b) If you knew which rod was made of magnetised steel, how would you find out which end was its North Pole?   (WJEC)

3. a) The region around a magnet over which it can exert a force is called a magnetic field. The diagram shows a small compass placed near to the North Pole of a magnet.

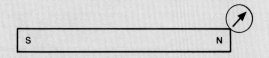

    Describe how you would plot the magnetic field pattern for this magnet.   [5 marks]
  b) You are now given: a long piece of insulated, copper wire; a rod of soft iron; a battery; a switch.
Draw a diagram to show how you would arrange these items of apparatus to give a magnetic field like the one of the bar magnet in (a).   [4]   (SEG)

4. Which of the following components must be made from a material which retains magnetism?
  A The commutator for a d.c. motor.
  B The magnet in a moving-coil meter.
  C The core for a transformer.
  D The core of an electromagnet.
  E The slip-rings of an a.c. generator. (LEAG)

5. A coil of wire is connected in series with a battery, a rheostat and a switch:

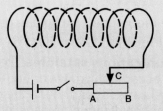

  a) Draw, on a diagram, the shape of the magnetic field **inside and outside** the coil when the switch is closed.
  b) A stronger magnetic field could be produced by winding the coil round a metal bar. Name a suitable metal.
  c) If the slider, C, on the rheostat is moved towards B, what is the effect on
    i) the resistance of the circuit?
    ii) the current through the coil?
    iii) the magnetic field in the coil?   (WJEC)

6.

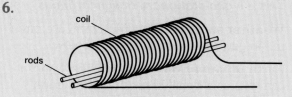

Two metal rods are placed in a long coil as shown. When a direct current flows through the coil, the rods move apart. When the current is switched off, the rods return to their original positions.
  a) Why did the rods move apart?   [2]
  b) From what metal are the rods likely to be made? Give a reason for your answer. [2]
  c) If alternating current from a mains transformer is passed through the coil, what effect, if any, will this have on the rods? Explain your answer.   [4]   (MEG)

7. The diagram on the next page represents an incomplete two-pole, single coil, direct current **electric motor** with 50 turns of wire on the armature A.
  a) Copy and complete the diagram, showing how the ends B and C of the wire wound on A would be connected through a split ring commutator to a d.c. power supply so that the armature would rotate continuously.   [6]

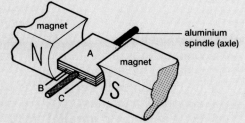

b) State what would happen to the direction of rotation of the armature if
   i) the polarity of the power supply were to be reversed,
   ii) the direction of the current in the armature were to be reversed,
   iii) the polarity of the magnetic field were to be reversed,
   iv) both the changes described in ii) and iii) were made at the same time.   [4]

c) State what would happen to the speed of rotation of the motor if
   i) the number of turns on A were to be increased, the current through A remaining unchanged,
   ii) stronger magnets were to be used with the 50-turn coil and the same power supply.   [2]

d) Explain how the commutator makes the armature rotate continuously when a power supply is connected.   [6]   (NEA)

8. A student was given a small electric **motor** and asked to measure its efficiency when lifting a load of 200 N.
The student connected the motor to an electrical circuit and read the voltmeter and the ammeter while the motor raised the load slowly and steadily. Here are the results:

Load lifted	= 200 N
Distance load was lifted	= 0.4 m
Time taken to lift load	= 5.0 s
Voltage across motor	= 10.0 V
Current through motor	= 4.0 A

a) Calculate the work done on the load.   [2]
b) Calculate the power **output** of motor.   [2]
c) Calculate the power **input** to the motor.   [2]
d) Calculate the efficiency of the motor.   [3]
e) Motors are always less than 100% efficient because of energy losses inside the motor. Give **two** energy losses saying where in the motor they occur.   [4]   (SEG)

▷ Electromagnetic induction

9. If a bar magnet is pushed into a coil of wire, a voltage will be induced across the ends of the coil. This voltage can be made larger by
A using a bar of iron instead of a magnet
B using a coil of low-resistance wire
C moving the magnet more quickly
D having a voltmeter across the coil (SEG)

10. a) Here is a drawing of a simple **alternator**. A single coil is positioned between the poles of a magnet.

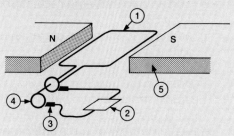

   Certain parts of the alternator have been numbered. Write down the correct name for each part 1 to 5.   [5]
b) The coil is rotated at a steady speed of 60 rotations per minute.
   i) An oscilloscope is connected across the load and adjusted to get on its screen a trace for **one** rotation of the coil. Sketch the trace for one rotation starting and finishing in the position of the coil shown in the diagram.   [3]
   ii) State **two** ways of **increasing** the voltage produced by the alternator.   [2]
c) An example of a practical alternator is a bicycle dynamo. Here, the magnet is rotated with the wheel and the coil is kept still.
   i) Give an advantage of rotating the magnet and not the coil.   [1]
   ii) Give a disadvantage of using a dynamo to power bicycle lights.   [1]   (SEG)

11. A transformer has 200 turns on the primary coil and 20 turns on the secondary. What is the output voltage if the input to the primary coil is 240 V?
A 12 V   **B** 24 V   **C** 120 V
D 240 V   **E** 2400 V   (MEG)

▷ Electromagnetic induction (continued)

12.

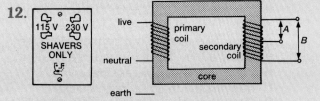

A **bathroom razor socket** contains a transformer. This is to isolate the user from the mains supply.
a) Copy the transformer diagram, and:
   i) add an S to show the correct position for a Switch.
   ii) add an F to show the correct position for a Fuse.
   iii) connect the earth lead to the correct point on the transformer. [4]
b) The transformer has two outputs, labelled A and B. Complete the table below to show the number of turns and output voltage for each coil. [5] (LEAG)

Input voltage	Primary turns	Secondary turns	A or B	Output voltage
230 V	5000	2500		
230 V	5000	5000		

13. a) i) Describe, briefly, the energy changes involved in the generation of electrical energy at a **power station.** [2]
   ii) State and explain **two** advantages of a hydroelectric power station compared with other types. [2]
b) What are the advantages of transmitting power at   i) very high voltage?
   ii) alternating voltage? [3]

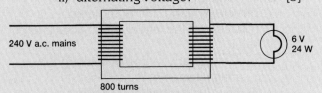

A 6 V, 24 W lamp shines at full brightness when it is connected to the output of a mains transformer. Assuming the transformer is 100% efficient, calculate
   i) the number of turns in the secondary coil if the lamp is to work at its normal brightness.
   ii) the current which flows in the **mains** cables. [2]

d) Explain whether, and how, the number of secondary turns of the transformer shown in c) should be altered if
   i) two 6 V lamps in **series** are to work at normal brightness.
   ii) two 6 V lamps in **parallel** are to work at normal brightness. [3] (WJEC)

14. The diagram shows overhead power lines transmitting electricity at 400 kV.

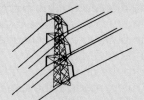

a) How is the electricity prevented from flowing to earth through the metal pylon?
b) Some people think that underground cables would be better. What are the advantages and disadvantages of underground and overhead cables?
c) The transmission lines carry 200 MW.
   i) Calculate the current in the lines.
   ii) If the total resistance of the transmission lines is 10 Ω, calculate the voltage drop when the lines carry 200 MW.
   iii) Calculate the power loss in the lines.
   iv) What happens to this 'lost' power?
d) Explain why electricity is transmitted at high voltages like 400 kV rather than low voltages like 240 V. (SEG)

15. a) A 12 V, 60 W heater can be operated by a 240 V mains transformer which has 10 000 primary turns.
   i) If the transformer is 100% efficient, calculate: the number of secondary turns, and the current flowing in the **mains** leads. [2]
   ii) Explain how an alternating voltage applied to the primary can generate an alternating voltage in the secondary. [2]
b) If the only supply available were 240 V **d.c.**, without calculation, draw a suitable circuit which could enable you to apply 12 V to the heater. [2]
(WJEC)

▷ **Electronics**

16. The diagram shows the screen, Y-gain and time-base controls from a typical oscilloscope displaying a waveform.

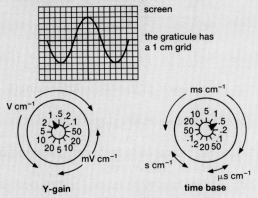

screen

the graticule has a 1 cm grid

Y-gain

time base

a) What is the setting of the Y-gain control? [2]

b) What is the peak voltage of the waveform? [2]

c) What is the time-base setting? [2]

d) What is the period of the trace? [3]

e) What is the frequency of the waveform? [4]

f) If the time-base setting is altered to 1 ms cm^{-1} and the Y-gain to 2 V cm^{-1}, sketch the resultant trace. [5]

(NEA)

17. The diagram shows a 'diode bridge' circuit which is a 'full-wave' rectifier used in power supplies. Alternating current is applied to it (from a transformer).

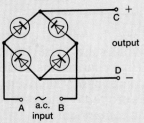

output

A a.c. B
input

a) Copy the diagram and show the path of current if A is more positive than B.

b) On another diagram show the path if B is more positive than A.

c) Why does it give a d.c. output?

d) Draw the whole rectifier circuit by including a transformer and a smoothing capacitor.

18. Electronic engineers use devices often referred to as LDRs and LEDs.

a) What are the properties of an LDR? [2]

b) What are the properties of an LED? [2]

c) i) Draw a circuit diagram, using the correct symbols, with an LDR, LED and a cell in series so that a current will flow through the circuit. [3]

ii) What will happen when a bright light is shone on the LDR? [1]

iii) What difference will it make if the cell is reversed? [1]

iv) An LED gives out only a little light. Why would a more powerful lamp not work if it were to replace the LED? [2]

d) Describe how a relay works, illustrating your answer with a suitable diagram. [4]

e) i) Design a circuit using an LDR and a relay so that a mains lamp will light when light shines on the LDR. [2]

ii) A circuit which puts a light on when daylight arrives is not much use in practice. Design a simple circuit which puts a light on when it gets dark. [3]

(MEG)

19. If a variable resistor is connected as shown, it can be used to alter the voltage smoothly.

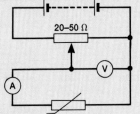

20–50 Ω

With the temperature of the thermistor at 20 °C, the circuit was adjusted until the current was 20 mA and then the voltage noted. This procedure was repeated with the thermistor at different temperatures.

Temperature (in °C)	20	40	60	80	100
Voltage (in V)	12.0	4.2	2.1	1.2	0.8

a) Plot the readings on graph paper. [5]

b) Draw a smooth curve through the points. [1]

c) Use information from your graph to help you work out: the resistance of the thermistor at:

i) 30 °C   ii) 70 °C   [5]   (SEG)

▷ Electronics (continued)

20.

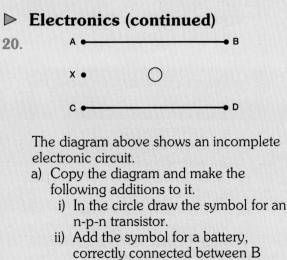

The diagram above shows an incomplete electronic circuit.
a) Copy the diagram and make the following additions to it.
   i) In the circle draw the symbol for an n-p-n transistor.
   ii) Add the symbol for a battery, correctly connected between B and D.
   iii) Draw the symbol for a thermistor between points A and X.
   iv) Draw the symbol for a variable resistor between points X and C.
   v) Draw the symbol for a lamp in the collector circuit.
   vi) Connect the base and the emitter of the transistor to the correct places in the circuit. [6]
b) As the circuit is arranged, the lamp will not light. If there is a rise in temperature the lamp will light. Explain this. [2]
c) A simple re-arrangement of this circuit would make the lamp light when the temperature falls. Explain what changes should be made, and how they will affect the working of the circuit. [2]
d) In the original circuit, instead of a variable resistor, a fixed resistor could have been used between points X and C. Explain why it might have been considered desirable to use a variable resistor for these two circuits. [2]

Further reading

Inside the Chip – Davies (Usborne)
Robotics – Potter (Usborne)
Mastering Electronics – Watson (Macmillan)
Everyday Electronics Magazine (IPC)

Have you started revision yet? See page 366.

21. The circuit diagram below is that of a circuit which Fatima set up.

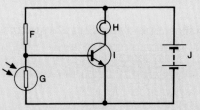

a) Name each of the five components F, G, H, I, J. [5]
b) Describe what the circuit is able to do. [2]
c) Devices based on this circuit can be used in the home or in factories. Describe **one** use of such a device. [1]
d) A burglar alarm can be built using this circuit with other pieces of apparatus. Describe how you could set up the burglar alarm near to a door so that the burglar's entry would be detected. Mention any other pieces of apparatus you would use. [3]
e) Suggest something which might be more useful than component H in your burglar alarm. [1] (LEAG)

22. This circuit is designed to switch on the beacon automatically when daylight fades.

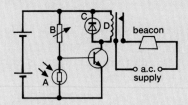

a) i) Give the full name of component A.
   ii) What is the abbreviation of its name?
   iii) What changes the value of this component? iv) Explain fully how its properties change. [4]
b) Name component B. [1]
c) Component D is the coil of a reed switch:
   i) Why has this coil no metal core?
   ii) What is the function of the coil? [2]
d) i) Explain how the transistor circuit works, mentioning the function of each component.
   ii) Explain how the transistor circuit controls the beacon. [4]

23. The diagram shows a combination of two logic gates, the two-input AND gate providing the input to the NOT gate.

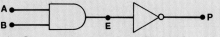

a) Copy and complete the truth table: [2]

A	B	E	P
0	0		
0	1		
1	0		
1	1		

b) The output of a logic gate can be displayed using an LED and a series resistor as shown:

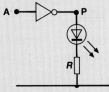

i) When the LED is ON, what is the logic state of the circuit at A?

When the LED is ON, the voltage drop across it is 2.0 V and the current through it is 10 mA. Calculate:

ii) the voltage drop across **R** if the voltage at P is 9.0 V,

iii) the resistance of **R**.

iv) Why must a resistor be included in the circuit? [8] (NEA)

24. Plants can be grown in a specially heated box with the correct amount of lighting. Sensors are included in the box and operate as shown below.

Copy and complete the diagram so that the buzzer sounds when the switch is on and when either the light level or the temperature level or both are low.

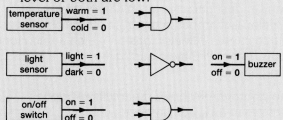

25. The diagram shows a circuit for a combination lock. When a switch is pressed, it changes its input from 0 to 1.

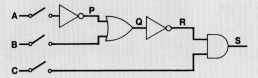

a) Draw up a truth table showing the state of P, Q, R, S for all combinations of A, B, C.

b) If S rises to logical 1, it opens the lock. What is the correct combination?

26. a) Write down the truth table for a NAND gate, shown below. A and B are the *two* inputs and X is the output.

b) A light detector gives a logic 1 when light hits it and a logic 0 when no light hits it. A temperature sensor gives a logic 1 when it is hot and a logic 0 when it is cold. A bell rings when the input is logic 1 and is silent when the input is logic 0.

The following circuit is set up:

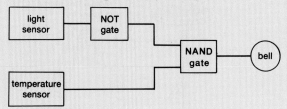

Copy and complete the truth table below to show the behaviour of the bell.

Light/Dark	Hot/Cold	Bell Ringing/Silent
Dark	Cold	
Dark	Hot	
Light	Cold	
Light	Hot	

c) Now design a circuit using **one** gate to ring the bell **only** when it is dark and cold. At all other times the bell is silent. Draw a circuit diagram showing clearly the gate you have chosen, and the connections between the sensors, gate and bell. (NI)

chapter 39

RADIOACTIVITY

In 1896, Henri Becquerel discovered almost by accident that some substances (like uranium and radium) can blacken a photographic film even in the dark. These substances are **radioactive**.

Radioactivity can be dangerous and experiments in this chapter will be teacher demonstrations only.

▶ Detecting radioactivity

1. Photographic film

The photograph shows how some film has been blackened by radioactivity except in the shadow of a key.

2. The gold-leaf electroscope

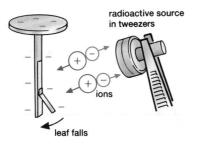

radioactive source
in tweezers

ions

leaf falls

Experiment 39.1
Charge up a dry gold-leaf electroscope as you did in experiment 30.6 (see page 241).

Does the leaf stay up for several minutes?
What happens if a radioactive source (e.g. radium) is brought near the leaf, using tweezers?

The leaf falls because the air is **ionised** by the radioactivity (see experiment 30.11 on page 242). A negative electroscope attracts the positive ions and so it is discharged. This **ionisation** is used in other detectors of radioactivity.

3. The cloud chamber

In a cloud in the sky, the water droplets tend to form on dust particles or on charged ions.
In a cloud chamber, a ray from a radioactive source causes a line of ions on which a thin cloud forms (rather like the cloud trail behind a high flying aircraft).

The photograph shows the tracks of 'alpha' particles in a cloud chamber.

4. The spark counter

Experiment 39.2
A spark counter consists of a fine wire stretched just below a piece of metal gauze.
A high voltage between the wire and the gauze is adjusted until it is almost, but not quite, sparking.
Using tweezers, a radioactive source (radium) is brought close to the gauze.

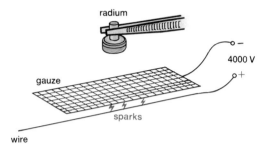

What happens? Why?

Rays from the radioactive source ionise the air so that it is a better conductor, and sparks are produced.

Do the sparks occur regularly or **randomly**?

Place a thick piece of paper between the source and the gauze. What happens? Why does it stop?

5. The Geiger–Müller tube (G–M tube)

This is a metal tube with a thin wire down the centre. It contains a gas at low pressure. It works on the same principle as a spark counter, but the voltage is lower so that no spark is formed. Instead, a pulse of current is produced. This is amplified and passed to either:

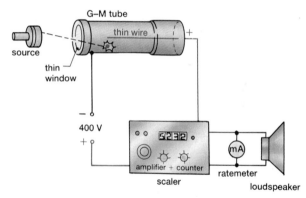

a **scaler**, which counts the pulses and shows the total number, **or**

a **ratemeter**, which shows the rate in 'counts per second'. In addition, a loudspeaker clicks each time a ray from the source passes through the tube.

Experiment 39.3
A radium source is placed near a G–M tube connected to a ratemeter and loudspeaker.

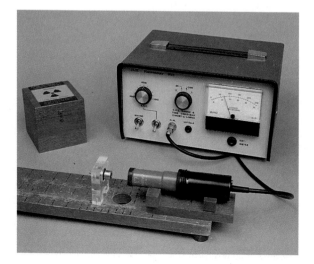

What do you hear as the source moves closer?

Place a thick piece of paper between the source and the tube.
Does the paper stop all the rays?

Because paper stopped the sparks in the last experiment, but not the clicks in this experiment, it seems that there are at least two kinds of radiation from radium.

In fact there are three kinds, called **alpha**, **beta** and **gamma** rays.

▶ Alpha, beta and gamma rays

The clicks you hear from the Geiger counter are
called the **background count**. It is due to natural
radioactivity, mainly from the rocks of the Earth
and cosmic rays from the Sun.

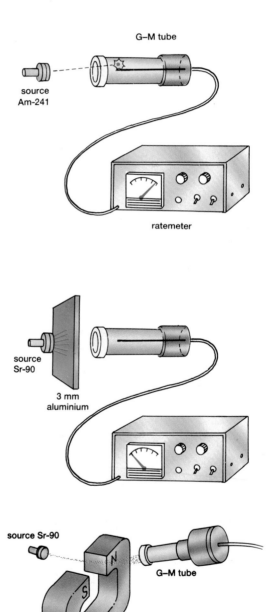

Now place a piece of paper between the source
and the tube. What happens to the count rate?

This radiation is called **alpha (α) radiation**.
It is easily stopped by paper or by your clothing.
It can be deflected by electric or magnetic fields.
It can be shown that **α-radiation is positively
charged particles moving at high speed**.

In fact, an alpha-particle is found to be the same
as the **nucleus** of a helium atom (see page 344).
It consists of 2 protons (+) and 2 neutrons.

Paper won't stop these rays. Try different
thicknesses of aluminium.

This radiation, from Strontium-90, is called
beta (β) radiation. It is more penetrating than
α-particles, but cannot penetrate through 3 mm
of aluminium.

Move the tube until the count-rate rises again:

The direction of the deflection shows that beta-
rays must be **negatively-charged particles** (using
Fleming's Left-hand Rule, see page 288).
In fact, **β-particles are electrons travelling at
very high speed**.

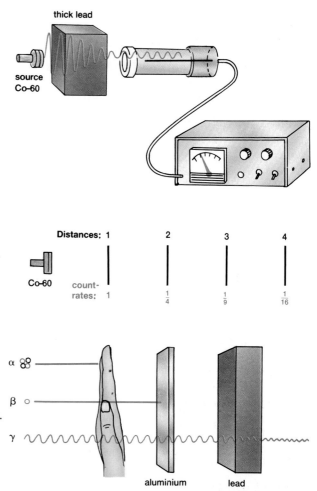

Experiment 39.8
Use the same apparatus, but with a cobalt-60 source. Do the same tests as before.

Paper and aluminium won't stop this radiation.

This radiation from cobalt-60 is called **gamma (γ) radiation**.
It is not affected by a magnetic field and is very penetrating. Although it is reduced, it is not stopped even by thick pieces of lead.

If the distance between the source and the G–M tube is increased, it is found that at twice the distance the count-rate is only one quarter. At three times the distance, the intensity of γ-rays is only one-ninth.
The intensity of light from a lamp changes in just the same way.

In fact, it can be shown that these **gamma-rays are really electromagnetic waves of very short wavelength** (see page 212).

Repeat all these experiments with a radium source. Do you find that it emits alpha- *and* beta- *and* gamma-rays?

Here is a **summary**:

Properties	Alpha-particle (α)	Beta-particle (β)	Gamma-ray (γ)
Nature:	Positive particle (helium nucleus) (about 7000 times the mass of an electron)	Negative electron e^-	Electromagnetic waves (see page 214). Very short wavelength
Affected by electric and magnetic fields?	Yes	Yes, bent strongly	No
Penetration:	Stopped by paper or skin (or 6 cm of air)	Stopped by 3 mm aluminium	Reduced but not stopped by lead
Causes ionisation?	Strongly	Weakly	Very weakly
Dangerous?	Yes	Yes	Yes
Speed:	10% speed of light	50% speed of light	Speed of light
Detectors:	Photographic film Cloud chamber Spark counter Gold-leaf electroscope Thin-window G–M tube	Photographic film Cloud chamber G–M tube	Photographic film Cloud chamber G–M tube

▷ Atomic structure

Experiments show that α-particles can pass through thin gold foil. This suggests that a gold atom might be almost entirely empty space!

To find more about the structure of an atom, Ernest Rutherford suggested an experiment in which α-particles were fired at thin gold foil. Detectors were used to find how the α-particles were scattered by the gold atoms:
It was found that some α-particles were scattered back **towards the source** – rather like firing a machine-gun at tissue paper and finding that some of the bullets bounce back!

In 1911, Rutherford showed that this could be explained if each atom has a tiny core or **nucleus** with a **positive charge**. A positive nucleus repels the positive α-particles so that they are scattered in different directions.

When Rutherford calculated the size of a nucleus, he found that it was very small even compared with a very small atom.
If you imagine magnifying an atom until it is about the size of your school hall, then the nucleus in the centre would be about the size of this full stop.

The simplest atom is the **hydrogen (H) atom**. The nucleus is a single positive charge called a **proton (p)**. There is a single negative **electron (e)** in an orbit round the nucleus.

Because the charges are equal but opposite, the atom is neutral. If the atom loses the electron, it becomes a charged atom or positive **ion**, H^+.

A slightly more complicated atom is the **helium (He) atom**. This has two protons in the nucleus and, to be neutral, two electrons outside. The helium nucleus also has two uncharged particles, called **neutrons (n)**.

A neutron and a proton are roughly equal in mass – each is about 1800 times the mass of an electron.

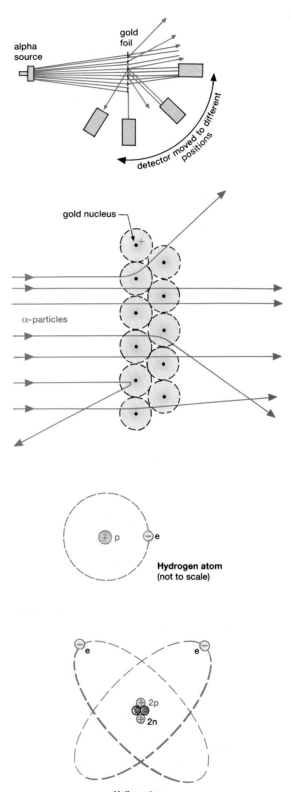

Hydrogen atom
(not to scale)

Helium atom

Proton number and nucleon number

Different elements have a different number of protons – for example, a *lithium* atom has 3 protons, a *beryllium* atom has 4 protons, a *boron* atom has 5 protons, a *carbon* atom has 6 protons and so on.
The number of protons in an atom is called its **atomic number**.

What is the atomic number of lithium?
A list of some elements in order of atomic number is shown at the bottom of the page.

Another important number is the **mass number**: this is the total number of *nucleons* (= protons + neutrons).
For example, the mass number of lithium is 7 (= 3 protons + 4 neutrons).
A shorthand way of describing a *nuclide* is shown here:

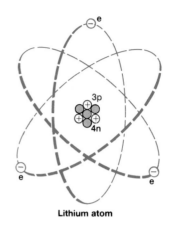

Lithium atom

Number of nucleons (protons + neutrons) **(mass number)**

$^{7}_{3}\textbf{Li}$

Number of protons **(atomic number)**

Isotopes

All the oxygen that you are breathing has an atomic number of 8. That is, it has 8 protons in the nucleus (and 8 electrons to make a neutral atom).

Most of the oxygen that you are breathing has also 8 neutrons (so the mass number = 8 + 8 = 16). This is $^{16}_{8}O$.
Look at the diagram of this atom:

Some of the oxygen you are breathing has got 9 neutrons (but still 8 protons, making a mass number of 17). This is $^{17}_{8}O$.
Draw a diagram of this atom.

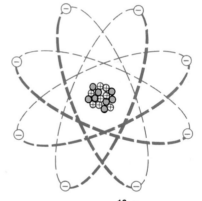

An atom of $^{16}_{8}O$

$^{16}_{8}O$ and $^{17}_{8}O$ are both atoms of oxygen.
They have the same chemical properties but have different masses. They are called **isotopes** of oxygen.

Oxygen has a third isotope, $^{18}_{8}O$. Draw a diagram of an atom of this isotope.

All elements have more than one isotope. Many of these isotopes are *unstable* – the nucleus breaks up into smaller parts.
When this happens α-particles, β-particles or γ-rays may be emitted *from the nucleus*. This is called *radioactive decay*.

Atomic number	Element + symbol		Commonest isotope
1	Hydrogen	H	$^{1}_{1}H$
2	Helium	He	$^{4}_{2}He$
3	Lithium	Li	$^{7}_{3}Li$
4	Beryllium	Be	$^{9}_{4}Be$
5	Boron	B	$^{11}_{5}B$
6	Carbon	C	$^{12}_{6}C$
7	Nitrogen	N	$^{14}_{7}N$
8	Oxygen	O	$^{16}_{8}O$
92	Uranium	U	$^{238}_{92}U$

▷ Radioactive decay

Plot a graph of the **number of dice-atoms surviving** against the **number of throws**.

Real atoms behave in a similar way: each atom disintegrates in a random, unpredictable way. A large number of atoms gives a smooth **radioactive decay curve**.

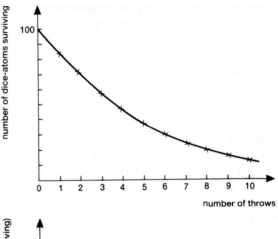

Half-life

The time taken for **half** the atoms to decay is called the **half-life** of the substance.

The graph shows a decay curve for a radioactive substance. On the graph you can see that after 1 half-life, half the atoms have disintegrated and half have survived.
After a further half-life, the activity has halved again so that only $\frac{1}{4}$ of the atoms survive. After 3 half-lives it has halved again and only $\frac{1}{8}$th survive.
What fraction survives after 4 half-lives?
What fraction survives after 5 half-lives?

The half-life of different substances varies widely – from fractions of a second up to millions of years.

The half-life of radium is 16 centuries.
This means that in a radium source in school (mass of radium only about five millionths of a gram), the atoms are disintegrating at the rate of 700 million in a physics lesson and yet it will do this for 16 centuries before half of the atoms have decayed!

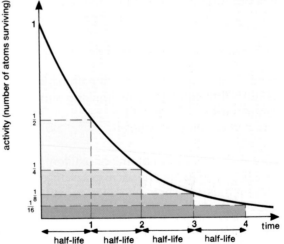

By mistake (in his lunch) Freddy Furze,
Ate radium and died with a curse,
But the point about Fred,
Is that now he is dead,
His half-life is sixteen hundred years!

Alpha-decay

An alpha-particle is a helium nucleus (see page 344). It is $^4_2$He.
It has 4 nucleons: 2 protons + 2 neutrons.

Radium-226 ($^{226}_{88}$Ra) decays by α-emission. When it loses the
α-particle, its mass number (226) must decrease by 4 (to 222).
Also, its atomic number (88) must decrease by 2, to become 86.
It has changed to a **different element**. It is now **radon** (Rn), a
radioactive gas.

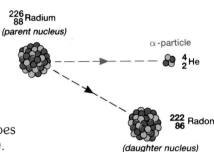

$^{226}_{88}$Radium
(parent nucleus)

α-particle
$^4_2$He

$^{222}_{86}$Radon

(daughter nucleus)

The nuclear equation is: $^{226}_{88}\text{Ra} \rightarrow \, ^{222}_{86}\text{Rn} \, + \, ^4_2\text{He} \nearrow$

Note that the top numbers balance on each side of the equation.
So do the bottom numbers.

Radon-222 is itself radioactive and decays by α-emission. What does
it become? Write down the equation (element 84 is polonium, Po).

Beta-decay

The $^{218}_{84}$Po from the last equation can decay by β-emission. This
is the emission of an electron *from the nucleus*. But there are no
electrons in the nucleus!
What happens is this: one of the neutrons changes into a proton
(which stays in the nucleus) *and* an electron (which is emitted as
a β-particle).
This means that the atomic number *in*creases by one, while the
total mass number stays the same.
The polonium changes into another element, called astatine (At).

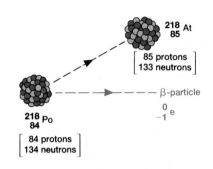

$^{218}_{85}$At

85 protons
133 neutrons

β-particle
$^0_{-1}$e

$^{218}_{84}$Po

84 protons
134 neutrons

The nuclear equation is: $^{218}_{84}\text{Po} \rightarrow \, ^{218}_{85}\text{At} \, + \, ^0_{-1}\text{e} \nearrow$

Notice again, the top numbers balance and so do the bottom ones.

Gamma-emission

When an α-particle or a β-particle is emitted, the nucleus
is usually left in an 'excited' state. It loses its surplus energy
by emitting a γ-ray.

'Splitting the atom' (transmutation)

In one of Rutherford's experiments, he fired α-particles
at the nuclei of nitrogen atoms.
He found that some of the nitrogen atoms turned into
oxygen atoms! The nuclear equation is:

$$^{14}_{7}\text{N} \; + \; ^4_2\text{He} \; \rightarrow \; ^{17}_{8}\text{O} \; + \; ^1_1\text{H}$$

nitrogen   alpha-   oxygen   hydrogen
nucleus   particle   nucleus   nucleus
                                          (proton)

Notice again how the numbers balance.
The photograph shows how this appears in a cloud chamber.

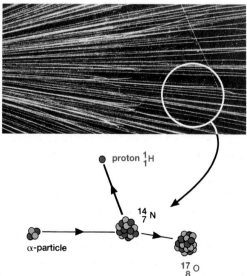

proton $^1_1$H

$^{14}_{7}$N

α-particle

$^{17}_{8}$O

▷ Uses

Radioactive isotopes, or *radio-isotopes*, can be made by bombarding substances inside a nuclear reactor. Handled with care, they have many uses.

1. Radioactive tracers

Radioactive fertiliser can be fed to plants and then traced through the plant using a Geiger counter. This method is used to develop better fertilisers.

Radioisotopes have important **medical uses**. For example, a doctor might suspect that a patient has a blocked kidney. To find out, the patient sits with a Geiger counter over each kidney:

A small amount of Iodine-123 is injected into the patient. Within 5 minutes both the kidneys should extract the iodine from the blood stream, and then within 20 minutes pass it with urine into the bladder. Which one of these kidneys is blocked?

The most common tracer is called Technetium-99. This is very useful, and safe, because:
- It emits only gamma-rays. The γ-rays can be detected outside the body by a 'gamma-camera'. It is also safer because γ-rays do not cause much *ionisation*. Ionisation can damage cells and cause cancer. (It is alpha-radiation, strongly ionising, that is the most dangerous *if it gets inside your body*, perhaps from a radioactive gas like radon.)
- It has a short half-life (of 6 hours). How much of it remains after one day?

Tracers are also used in **industry**. If radioactive pistons are fitted to a car, then a Geiger counter can be used to test the oil for traces of wear from the piston. Different piston designs can be tested.

Leaks from a pipeline carrying oil or gas can be traced by injecting a radioisotope into it. This saves digging it all up.
The right isotope is chosen so that:
- It has a half-life of only a few hours or days. This is so that it remains long enough to be detected but not so long that it remains a safety problem.
- It is a beta-emitter. Alpha-particles would be absorbed by the soil, whereas gamma-rays would pass through the metal pipe anyway.

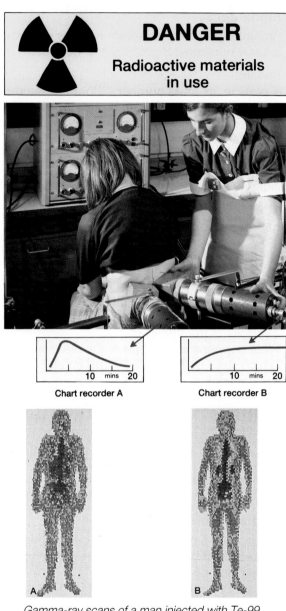

Chart recorder A   Chart recorder B

*Gamma-ray scans of a man injected with Te-99
A: initially   B: after 6 hours (one half-life)*

2. Sterilising

Gamma-rays can be used to kill bacteria, mould and insects in foods, even after the food has been packaged. This prolongs the shelf-life of the food, but it sometimes changes the taste.

Gamma-rays are also used to sterilise hospital equipment, especially plastic syringes that would be damaged by heating them.

Cancer cells in a patient's body can be killed by careful use of γ-rays (see the photo on page 214).

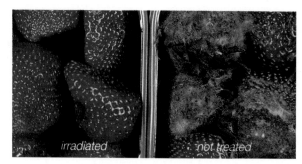
Strawberries after 7 days

3. Thickness control

In paper mills, the thickness of the paper can be controlled by measuring how much beta-radiation passes through the paper to a Geiger counter:

The counter controls the pressure of the rollers to give the correct thickness.
With paper, or plastic, or aluminium foil, β-rays are used. In a sheet-steel factory γ-rays are used. Why? Why should the source have a *long* half-life?

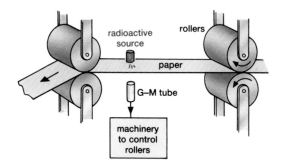

4. Smoke detection

Many homes are fitted with a smoke alarm. This contains a weak source made of Americium-241. This emits alpha-particles which ionise the air, so that it conducts electricity and a small current flows:

If smoke enters the alarm, it absorbs the α-particles, the current reduces, and the alarm sounds.
Am-241 has a half-life of 460 years. Is this helpful?

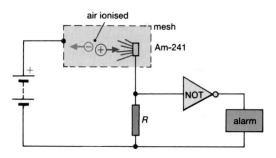

5. Checking welds

If a gamma source is placed on one side of the welded metal, and a photographic film on the other side, weak points or air bubbles will show up on the film (like an X-ray, see page 312).

6. Radioactive dating

The Uranium-238 in rocks decays steadily with a very long half-life of 4500 million years. It changes slowly into lead. By measuring how much of the uranium has changed into lead, it is possible to calculate the age of the rock.
The U-238 acts as a kind of clock.

With once-living material (like Egyptian mummies), Carbon-14 is used (see question 4 on page 353).

The Shroud of Turin.
Carbon-14 dating shows it is 600 years old.

► Nuclear energy

If an atom of Uranium-235 ($^{235}_{92}U$) is bombarded with slow-moving neutrons, it breaks up into two smaller atoms and three fast neutrons:
This process is called **nuclear fission**.

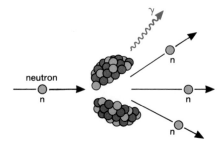

Mass and energy

The mass of the pieces after fission is found to be *less* than the mass of the pieces before the collision. The mass that is lost has been converted to energy. Albert Einstein showed how to calculate this:

$\begin{matrix}\text{Energy}\\\text{released}\end{matrix}$ (E) = $\begin{matrix}\text{mass}\\\text{lost}\end{matrix}$ (m) × velocity of light squared (c^2)	*or*	$E = mc^2$

The fission of 1 kg of Uranium-235 produces more energy than 2 million kg of coal.

When a Uranium-235 atom splits, three neutrons are emitted and these may hit three other uranium atoms so that they split and emit more neutrons which in turn hit other atoms and so on. This is called a **chain reaction**:

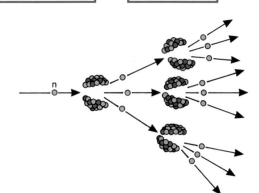

If the uranium is above a certain *critical size* (about the size of a tennis ball), then this chain reaction takes place very quickly, as in an atomic bomb.

Nuclear power station

A *nuclear reactor* is shown on the opposite page. The energy from the chain reaction makes the uranium **fuel rods** glow red-hot. The heat is carried away by a **coolant** (CO_2 gas, or water) which is pumped through the reactor.
In the **heat exchanger**, this heat boils water to make high-pressure steam, the same as in a coal-fired or oil-fired power station. The steam turns a turbine and a generator makes electricity.

- The chain reaction is controlled by movable **control rods** made of boron or cadmium. These are lowered to absorb neutrons, in order to reduce or stop the chain reaction completely.
- Graphite (or water) is included as a **moderator** to slow down the neutrons. This is because the fission of a uranium atom works more efficiently with slow neutrons.
- The whole reactor is shielded in steel and concrete to absorb the dangerous gamma-rays.

Mrs Messer: "What do nuclear scientists have for dinner?"
Professor Messer: "Fission chips."

The control room of a nuclear power station

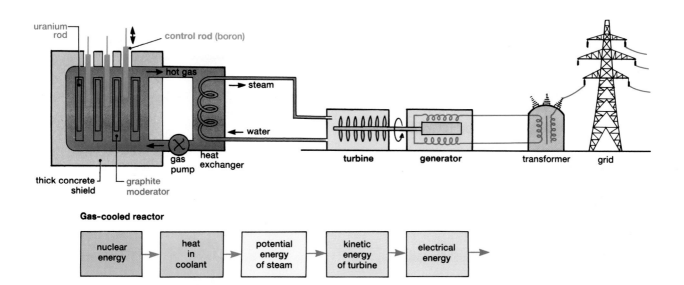

Gas-cooled reactor

nuclear energy → heat in coolant → potential energy of steam → kinetic energy of turbine → electrical energy

Nuclear fission raises strong feelings. This is partly because of thoughts of nuclear weapons, and uncertainty about the long-term effects of radiation. The world faces a severe energy crisis (see page 11) and decisions must be made.

Nuclear fusion is another way to get energy. In a hydrogen-bomb and in the Sun, hydrogen nuclei fuse together and release energy. This process has not yet been successfully developed for power stations (see page 11).

The nuclear debate. What do *you* think?

For:	Against:
• We need more nuclear power stations because of future energy demands, to keep our standards of living.	• We should save energy by better insulation of homes, and better use of the wasted heat from power stations.
• Nuclear power can save fossil fuels (coal, oil, gas), which are running out. They could be saved for the Third World, or not used at all (to reduce the Greenhouse Effect).	• The world's uranium is limited. We must develop renewable sources of energy, like wave power and solar power (see page 12). They do less damage to the environment.
• Nuclear power causes less damage to the environment than do coal-powered stations. Nuclear stations do not produce CO_2 and SO_2, and so do not make acid rain.	• Nuclear power stations produce waste which stays highly radioactive for thousands of years.
• The amount of nuclear waste is small. It can be safely stored by enclosing it in thick glass and burying it.	• Leaving waste is irresponsible – it pollutes the world for our grandchildren. The nuclear waste might leak out.
• Only 0.1% of our background radiation comes from the nuclear power industry (see page 352).	• No level of radioactivity is safe. Statistics show that children nearby are more likely to get leukaemia.
• Nuclear stations are very carefully designed to be safe.	• Chernobyl blew up, due to human error. A lot of people across Europe will die because of it. It is too risky.
• Risks due to the nuclear industry are less than other areas. Look at the figures: Disease: Heart disease 1 in 250, cancer 1 in 400 Accidents: On the road 1 in 8000, In the home 1 in 25,000 Radiation: Natural background 1 in 50,000, Nuclear industry 1 in 4 million	• Other risks are irrelevant. There is always a risk of a nuclear accident, with enormous consequences. Future generations would not forgive us.

▷ Radiation and You

Your body is receiving *background radiation* all the time. Most of this is due to natural sources:

The amount of radiation you receive depends on the number of X-rays you have, your job, and where you live (it is higher in Cornwall because of the rocks there).

The activity of a source is measured in *becquerels* (Bq), where:
1 becquerel = 1 nucleus decaying per second.
(Our bodies are slightly radioactive, with an activity of about 4000 Bq!)

More important is the amount of radiation that your cells *absorb* (the **dose**).
A dose is measured in *grays* (Gy), where:
**1 gray (1 Gy) = 1 joule of radiation energy
                absorbed in 1 kg of your body.**

However, alpha-particles do 20 times as much damage to cells as the same dose of betas, gammas, or X-rays. So it is more useful to talk of the **dose-equivalent**, measured in *sieverts* (Sv), where:

dose equivalent = absorbed dose × *Q*	
(in sieverts) (in grays)	
(*Q* = 20 for alphas; *Q* = 1 for β-, γ-, and X-rays)	

The average annual radiation dose in the UK is only 2.5 milli-sieverts (2.5 mSv). Fallout from the Chernobyl disaster raised this by 0.04 mSv.

Working with sources
A worker using sources should wear a film badge (a *dosemeter*) on his or her chest:

How would this show an exposure to beta-rays?

It is important to keep radiation from artificial sources as low as possible, and workers use thick shields and remote-handling equipment:

Waste from nuclear power stations is a problem because it is intensely radioactive and has a long half-life.
It is vital that the waste does not get into our food chain. It needs very careful long-term storage, deep underground.

Background radiation in Britain

mSv/year
- >0.4
- 0.3–0.4
- 0.25–0.3
- 0.2–0.25
- <0.2

Gamma-ray doses from the ground in Britain

film badge dosemeter

side view

lead

photographic film in light-proof jacket

aluminium

Summary

The properties and detectors of alpha-particles, beta-particles and gamma-rays are shown in the table on page 343.

An atom has a small, heavy, central nucleus containing protons (+) and neutrons (no charge). The number of protons is the atomic number. The number of nucleons (= protons + neutrons) is called the mass number.

In a neutral atom, the number of electrons in orbit = the number of protons.
If the numbers are not equal, then it is an ion.

An isotope has the same atomic number but a different mass number.

The half-life of a radio-isotope is the time taken for half the atoms to decay.

▶ Questions

1. Copy out and complete:
   a) An alpha . . . . is . . . .-charged and has the same charge and mass as the nucleus of a . . . . atom.
   b) Beta . . . . are . . . .-charged electrons travelling at very . . . . speed.
   c) Gamma rays are . . . . rays of very short . . . . .
   d) Beta . . . . are more penetrating than . . . . but less penetrating than . . . . .
   e) . . . . and . . . . can be deflected by electric and magnetic fields, but . . . . cannot.
   f) Alpha-, beta- and gamma-rays can be detected by . . . . film, a . . . . chamber, or a . . . . tube. A gold-leaf electroscope or a spark counter can be used to detect . . . . .

2. Copy out and complete:
   a) An atom has a small heavy . . . . containing . . . . (. . . .-charged) and . . . . (. . . .-charged).
   b) A . . . . and a . . . . are each about 1800 times the mass of an electron.
   c) The atomic number is the number of . . . . .
   d) The mass number is the number of . . . . .
   e) An isotope has the same number of . . . . but a different number of . . . . .
   f) The half-life of an element is the . . . . taken for . . . . of the atoms to decay. After . . . . half-lives, only one-eighth of the element remains.

3. Copy and complete the table:

	Mass	Charge
proton	1 unit	+ 1 unit
neutron		
electron		
α-particle		
β-particle		
γ-ray		

4. a) Most of the carbon in your body is $^{12}_{6}C$. Draw a diagram of this atom.
   b) Another isotope is $^{14}_{6}C$ with a **half-life** of 5700 years. What are its **atomic number** and **mass number**? Explain the words in italics and draw a diagram of this atom. What fraction of $^{14}_{6}C$ remains after 11 400 years?
   c) While animals or plants are living, the proportion of $^{14}_{6}C$ in them remains constant. But once they die, the Carbon-14 decays. Suppose a modern bone contains 80 units of C-14, and a fossil bone contains just 10 units. How old is the fossil?

5. Strontium-90 has a half-life of 28 years. It is a β-emitter and may be absorbed into human bone. How much time must pass before its activity falls to $\frac{1}{32}$ of its original value? Why would it be dangerous in our food chain?

6. Explain why workers in a nuclear power station wear badges containing photographic film.

7. Professor Messer fell asleep and dreamed that the nuclei of atoms could not be changed in any way. Why was his dream a nightmare?

Further questions on radioactivity

▷ Radioactivity

1. A certain radioactive source emits both β- and γ-radiation. Which of the following methods will **NOT** minimise the radiation hazard from such a source?

   A Always pick up the source using forceps.

   B Always carry out experiments using the source in a well ventilated laboratory.

   C Always store the source in a lead container when not in use.

   D Make sure that the source is out of the container for as short a time as possible.

   E Never point the source at a person's body.
   (NI)

2. a) Two students were set the task of measuring the half-life of a radioactive substance which emits beta particles only. They measured the count rate every 20 minutes and recorded their readings as shown below.

Counts/min	330	231	165	120	90	71	57
Time (min)	0	20	40	60	80	100	120

   They then discovered a count rate of 30/min was obtained even when there was no radioactive source near.

   i) What was the cause of the count rate of 30/min? [1]

   ii) Draw up a table showing how the count rate **due to the beta-source only** varies with time. [1]

   iii) Plot a graph of the beta count rate against time and use it to find the half-life of the source. [3]

   iv) What effect does the emission of a beta particle have on the structure of the nucleus? [1]

   b) Describe **one** medical or industrial use of a radioactive isotope. [2]
   (WJEC)

Further reading

Finding out about Nuclear Energy (CRAC)
Health Physics – McCormick (S. Thornes)
Chronology of Science & Discovery – Asimov

---

3. Nuclear power plants are not always big. Heart pacemakers can be powered from a small piece of plutonium. The isotope used is written as $^{238}_{94}$Pu.

   a) In a neutral atom of plutonium,

   i) how many protons are there?

   ii) how many neutrons are there?

   iii) how many electrons are there? [3]

   b) Make a **simple** sketch of a plutonium atom. Label the protons, neutrons and electrons. [4]
   (LEAG)

4. Radioactive iodine is used to treat tumours of the thyroid gland. It decays by emitting beta particles and gamma radiation. The beta decay process is represented by the following equation:

   $$^{131}_{53}I \rightarrow {}^{A}_{Z}Xe + {}^{0}_{-1}e$$

   a) What is:

   i) the nucleon number (mass number), **A**, of this nucleus of Xe? [1]

   ii) the proton number (atomic number), **Z**, of this nucleus of Xe? [1]

   iii) the number of neutrons in this nucleus of Xe? [1]

   b) A beta particle detector is set up, outside the body, close to the patient's thyroid gland to determine whether or not the gland is absorbing iodine correctly from the blood stream.
   Why can the beta particles **not** be detected by the detector? [1]

   c) The half-life of iodine-131 is 8 days. The total dose of iodine given to the patient initially emits 4×10^8 gamma rays per second.
   How many gamma rays does the total dose of iodine emit each second after 24 days? [3]

   d) For a particular patient, only 20% of the iodine-131 is absorbed by the thyroid gland. What is the maximum possible rate of emission of gamma rays from the thyroid gland after 24 days? [1]

   e) Give **two** reasons why the count rate recorded by the detector will differ from the above value. [2]
   (MEG)

5. a) The atomic number of americium-241 is 95. Complete the table to give the number of particles in an atom of americium-241. [3]

Particle	Number
Electrons	
Neutrons	
Protons	

b) Americium-241 decays by losing an alpha particle.
   i) Explain why smoke detectors containing americium-241 are not a danger to people.
   ii) Complete the equation showing this decay by giving values for **X** and **Y**. [4]

$$^{241}_{95}Am \rightarrow {}^{4}_{2}He + {}^{X}_{Y}Np + energy$$

c) The table shows how the activity of a sample of americium-241 changes over a long period of time.

Time (years)	Activity (counts per minute)
0	64
500	30
1000	14
1500	6
2000	2

   i) Use these numbers to draw a graph of activity against time for americium-241.
   ii) Use your graph to find the half-life of americium-241. [4]

d) Californium-241 is also radioactive and decays by losing an alpha particle. It has a half-life of 4 minutes.
   Suggest why it would be unsuitable for use in a smoke detector. [2] (LEAG)

6. a)

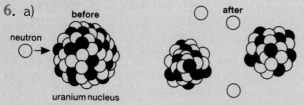

before   after

neutron

uranium nucleus

In a nuclear reactor, atoms of uranium-235 split into two nearly equal parts when they capture a neutron. An atom of uranium-235 has 92 electrons and 143 neutrons.

i) How many protons are there in one atom of uranium-235? [1]
ii) What is its proton number? [1]
iii) What is the nucleon number of the uranium-235 atom? [1]

b) It is proposed to build a nuclear power station at Lower Snodbury, Bumpshire. The Electricity Generating Board support the idea but the local residents oppose it. Suggest three advantages and three disadvantages of the scheme. Write a short paragraph on each. [6]

c) Radioactive isotopes have many practical uses. Describe one such use, giving details of how the isotope is used, the type of radiation it emits and an approximate value of its half-life (10 lines minimum). [5]
(LEAG)

7. The equation below represents part of the reaction in a nuclear reactor.

$$^{235}_{92}U + {}^{1}_{0}n \rightarrow {}^{236}_{92}U$$

a) Explain the significance of the numbers 235 and 92. [2]

b) $^{236}_{92}U$ atoms are unstable and disintegrate spontaneously into fragments approximately equal in size, together with two or three fast-moving neutrons and a large amount of energy.
What is this process called and what is the source of the energy? [2]

c) $^{235}_{92}U$ is much more likely to absorb slow-moving (thermal) neutrons than fast-moving neutrons. Describe how neutrons may be slowed in the reactor core. [4]

d) What is meant by a chain reaction? [4]

e) How is the rate of energy production in the reactor core controlled? [4]

f) The energy is produced in the form of heat in the reactor core. How is the heat removed from the core and how is it converted into electricity? [2]

g) When the fuel rods are withdrawn from the reactor core they are **radioactive**, containing **isotopes** with long **half-lives**. Explain the terms in italics. [8]

h) Outline three precautions which must be taken to ensure safe operation of the reactor. [6] (NEA)

The Nature of Science

Physics at work: Changing ideas

The study of science is based on the **scientific method**: an explanation (theory) is based on observations, and can be used to make predictions. This theory remains as the current theory until a new experimental fact contradicts it. Then the theory is modified, or changed to a new theory that fits the facts . . . until a new fact disproves this theory, and so on.
In this way, human understanding of our world has developed step by step. This is the nature of science.

Each of the boxes contains very brief outlines of the ways a scientific idea has changed.
Choose one of them and research it (working individually or sharing the work out among your group). You can use the library or any reference books. Key **names** and key *words* are shown here to help you. Then,
- write a detailed essay on your topic,   *and/or*,
- prepare a 5-minute illustrated talk. You can use posters, time-lines, drawings, models, photos, etc.
In each case concentrate on the *new ideas* that each scientist brought to human understanding at that time.

Ideas about Air and Air Pressure

1. In 1643 **Evangelista Torricelli** (at the suggestion of **Galileo Galilei**) investigated water pumps, a *mercury barometer* and the *vacuum* in it. **Otto von Guericke** experimented (1645) on sound in a vacuum, found the density of air, and did the *Magdeburg hemispheres* experiment (1654).
   Blaise Pascal argued that the mercury was held up by *atmospheric pressure* and persuaded his brother-in-law (François Périer) to climb a mountain with a barometer (1648).

2. **Robert Boyle** investigated the squeezing of air (1622, see opposite, 'Atoms'). **Joseph Black** found air is a mixture of gases (1754) and experiments by **Antoine Lavoisier** (1772), **Daniel Rutherford** (1772) and **Joseph Priestley** (1774) discovered its composition.

Ideas about Earth, Sun and Universe

1. *Astronomy* is an ancient science. It was investigated by the ancient Greeks, including **Thales** (585 BC), **Pythagoras** (500 BC), **Aristarchus** (280 BC), **Erastosthenes** (240 BC) and **Hipparchus** (150 BC). **Ptolemy** published detailed tables (AD 140), and his idea of the Earth being at the centre of the universe dominated for a long time. The Arabs perfected the *astrolabe* by AD 800.

2. In 1543 **Nicolaus Copernicus** argued that the Earth moved around the Sun, and observations by **Tycho Brahe** and **Johannes Kepler** confirmed this in detail (1609).
   Galileo Galilei built a telescope, observed the *Milky Way* (1609), *sunspots*, and *Jupiter's moons* (1610). He supported Copernicus and was punished (1633).

3. **Isaac Newton** published (1687) his *Laws of motion* and *gravitation*, explaining the movements of *planets*. **William Herschel** discovered *Uranus* (1781), *binary stars*, and showed (1783) that the Sun is moving through the universe. **John Adams** and **Jean Leverrier** predicted the existence of *Neptune* (1846).

4. **Albert Einstein's** theories predicted (1917) an *expanding universe*, confirmed (1929) by the work of **Edwin Hubble**. The '*Big Bang*' *theory* became more popular, and **Karl Jansky** began *radio astronomy* (1932). *Quasars* (1963) and *pulsars* (1967) were discovered. **Neil Armstrong** landed on the Moon (1969) and *Voyager-2* travelled through the *solar system*, eventually reaching Neptune in 1989.

5. Meanwhile **Alfred Wegener** proposed (1912) the idea of *continental drift*. Eventually, with evidence from *earthquakes*, *volcanoes*, and *mid-ocean ridges*, the ideas of *plate tectonics* were confirmed.

Ideas about Motion

Aristotle (350 BC) imagined heavier objects fell faster, and **Galileo Galilei** was the first to investigate motion (1589). **Robert Hooke** showed (1657) that a feather and a coin fall at the same rate in a vacuum. **Isaac Newton** published his three *Laws of motion* and his *Law of gravitation* in 1687. **Albert Einstein** in his *Theory of Relativity* (1905) showed that Newton's Laws are not completely true at very fast velocities.

Ideas about Light

Plato (400 BC) thought light went from the eye to the object, while **Pythagoras** (520 BC) thought that particles ('corpuscles') travelled to the eye. This *corpuscular theory* was supported by **Isaac Newton** after he investigated *refraction* and the *spectrum* (1666). **Olaus Römer** measured the *speed of light* in 1675. Meanwhile *diffraction* was discovered in 1660, and **Christiaan Huygens** suggested that light is a *wave motion* (1678). Eventually, in 1801, **Thomas Young** confirmed this, by *interference. Polarised light* was discovered in 1808, and showed that light is a *transverse* wave. In 1905, **Albert Einstein** explained the *photo-electric effect*, showing that light also behaves as particles, or *photons*.

Ideas about Electricity

In 1660, **Otto von Guericke** investigated *static electricity. Conductors* and *insulators* were discovered in 1729, and the two kinds of charge in 1733. **Benjamin Franklin** named them positive and negative 'fluids' (1747) and investigated *lightning* (1752). **Charles Coulomb** investigated electric forces (1785). **Luigi Galvani** investigated electricity in frogs' muscles (1790), and in 1800, **Alessandro Volta** constructed a battery, which was used by others to *electrolyse* water and copper sulphate (1800).

In 1820, **Hans Oersted** discovered *electromagnetism*, and **André Ampère** investigated it. **Michael Faraday** investigated the *motor effect* (1821), proposed the idea of *lines of force* (1821), and invented the *transformer* and *dynamo* (1831).
Experiments on 'cathode-rays' by many people, including **William Crookes** (1880), showed that electricity flows from the negative. In 1897, **Joseph Thomson** discovered the *electron* (see below).

Ideas about Atoms

1. **Democritus** (440 BC) suggested the idea of *atoms* but with no proof. Eventually **Robert Boyle**'s experiments (1662) supported the idea and **Daniel Bernouilli** proposed the *kinetic theory* (1738). **John Dalton** used *atomic weights* (1805) to support the idea, and **Thomas Graham** investigated *diffusion* (1831). In 1908, **Jean Perrin** used *Brownian movement* to find the size of an atom.

2. Meanwhile, **Dmitri Mendeleev** listed the *periodic table* (1869), and **Svante Arrhenius** used the idea of *ions* to suggest (1884) that atoms could be charged and are not solid balls. In 1897, **J. J. Thomson** investigated the particles ('electrons') in *cathode-rays*, and deduced that they are smaller than atoms. He suggested a 'current-bun' model of an atom. **Robert Millikan** measured the electronic charge (1911).

3. Meanwhile, **Henri Becquerel** discovered *radioactivity* (1896) and **Marie Curie (Maria Sklodowska)** discovered *radium* (1898). **Ernest Rutherford** discovered *alpha* and *beta* rays in 1897 and *gamma-rays* were discovered in 1900. Becquerel found (1900) that beta-rays are electrons. He suggested they came from atoms as they changed, and this was confirmed by Rutherford and **Frederick Soddy** (1902).

4. In 1911, Rutherford's *alpha-scattering experiments* supported the idea of an atomic *nucleus*, and in 1913 **Niels Bohr** suggested the idea of electrons in orbits. Soddy explained the idea of *isotopes* in 1913, and this was confirmed in 1932 when **James Chadwick** discovered *neutrons*.

Famous names

All the famous scientists were of course real people, and like all real people they had their hopes and their worries, and their different characters. Some of them were generous and kind; some were rude and unpleasant.
Here are brief biographies of six famous names that changed the world of science by their discoveries.

Isaac Newton
1642–1727

Isaac was born prematurely on Christmas Day 1642, during the English Civil War, in the year that Galileo died. His father died before he was born and his mother soon remarried but left Isaac with his grandmother for 9 years.
He was a very keen reader and went to Cambridge University when he was 19. Because of the plague, the university was closed in 1665–7 and Isaac went home for two years, doing his famous experiments on the spectrum (page 210), inventing the reflecting telescope (page 207), and observing the motions of the Moon and the planets. He developed his law of gravitation and his laws of motion (pages 85, 100, 140).
Isaac was an insecure man, arrogant and rude. He suffered two nervous breakdowns and often had flaming rows with other scientists, including Robert Hooke (page 82). He had no women friends and never married.

His great achievement was to show that mathematics can be used to state a set of laws which explain so much about our universe.

Galileo Galilei
1564–1642

Galileo was born in Italy on 15 February 1564 (the same year as Shakespeare). His father was a musician. Red-haired Galileo studied medicine at the University of Pisa, but soon became interested in doing experiments – an unusual idea at that time! He investigated the pendulum (page 115) and falling objects (page 138). Using a telescope, he was the first person to see Jupiter's moons and Saturn's rings (page 160). He observed the Moon, the phases of Venus, and sunspots rotating with the Sun. These observations convinced him that Nicolas Copernicus was right: that the Earth was not stationary at the centre of the universe, but that it was spinning on its axis and rotating round the Sun. This was against the views of the powerful Church at that time. At first Galileo agreed to keep quiet, but later he wrote a sarcastic book supporting Copernicus. The Church brought him to trial in 1633, with ten judges and the threat of torture even though he was ill and 69 years old. He eventually agreed to their opinions and was sentenced to house arrest for the last 9 years of his life.

Galileo was the founder of what we call the scientific method: he believed that theories could be tested or disproved by observations and experiment, not by opinions.

Albert Einstein   1879–1955

Albert was born on 14 March 1879 in Germany. At school he was lazy and slow in learning to read, but good at playing the violin. He failed some examinations but eventually got a degree at a polytechnic in Switzerland and then got a job in a patent office.
In 1905 he explained Brownian movement (page 16), published his theory of special relativity, and showed that $E = mc^2$ (page 350). He showed that Newton's Laws are not quite true for objects travelling very fast, at near the speed of light.
When the Nazis came to power in Germany Einstein left, to continue his work in America. He was a wise, cheerful and witty person, who spoke out against war and cruelty, and hated wearing socks!

His theory of relativity included and extended Newton's Laws, and changed the way that scientists look at space and time, energy and gravity.

Michael Faraday   1791–1867

Michael was born on 22 September 1791 near London, the son of a blacksmith from Yorkshire. His family was very poor and at age 13 Michael began work in a bookshop, and became a keen reader. He attended evening classes run by Sir Humphrey Davy and became his laboratory assistant.

Michael was probably the greatest experimental physicist the world has ever known. He discovered benzene, invented an electric motor (page 290), and later the dynamo (page 298) and transformer (page 302). He discovered electromagnetic induction (page 296) and the laws of electrolysis (page 270). He invented the idea of lines of flux (page 279).

He was a gentle, kind and generous man, and an enthusiastic lecturer.

He has been called 'the father of electricity'. When Michael started his experiments, electricity and magnetism were seen as toys. When he died, the laws of electromagnetism had been worked out and the technological basis of our electric society had been set up.

Further reading
The Ascent of Man – J. Bronowski
The Search for Solutions – H. Judson
The Physicists – C. P. Snow
A Brief History of Time – S. Hawking

Things to do
Find out the main historical events that happened during the lifetime of each of these scientists. Which of them do you think was the 'best' scientist? Can you justify your claim?

The Curie Family
Pierre   1859–1906
Marie   1867–1934
Irène   1897–1956

Maria Sklodowska was born on 7 November 1867 (the year Faraday died), in Poland where her father was a teacher. When she was 24 she went to university in Paris, France, where she met and married Pierre Curie.

Together, Marie and Pierre investigated radioactivity (discovered by their friend, Henri Becquerel, page 340). They discovered two new elements, radium (page 346) and polonium (named after her country). Pierre was run over and killed by a truck in 1906, but Marie carried on with the work and was the first person to receive two Nobel Prizes.

Their daughter Irène Curie married Frédéric Joliot, and together they found out how to make artificial radioactivity (which is now used widely to make radio-isotopes, page 348). They were awarded a Nobel Prize also.

Like her mother, Irène spoke out for women's rights, and like her mother she died of leukaemia, caused by working so long with radioactive substances.

Ernest Rutherford
1871–1937

Ernest was born on 30 August 1871 in New Zealand. His father was a Scottish immigrant, an odd-job man and farmer.

Ernest was a keen footballer, clever at school and went to university in England.

He was a great experimental physicist who rather distrusted theoretical physicists like Einstein. He had a good instinct for what should work.

He was a hard worker and covered his nervousness with a loud booming voice. He discovered alpha particles and beta particles (page 342), a radioactive series (page 347), and helped to invent the Geiger counter (page 341). He discovered protons and the transmutation of elements (page 347).

His greatest achievement was to show that atoms have a small nucleus surrounded by orbiting electrons (page 344).

Dates of some inventions and discoveries mentioned in this book

Invention or discovery	Inventor or discoverer	Date	Page number in this book
Wheel and axle	—	before 2000 BC	127
Pulley	—	about 700 BC	128
Screw	—	about 400 BC	126
Pinhole camera	—	about AD 1500	176
Compound microscope	Zacharias Janssen	1590	207
Air thermometer	Galileo Galilei	1597	30
Refracting telescope	Hans Lippershey	1608	206
Laws of refraction	Willebrord Snell	1621	187
Mercury barometer	Evangelista Torricelli	1644	95
Vacuum pump	Otto von Guericke	1654	95
Hooke's Law	Robert Hooke	1660	82
Boyle's Law	Robert Boyle	1662	35
Spectrum	Isaac Newton	1666	210
Reflecting telescope	Isaac Newton	1669	207
Pressure cooker	Denis Papin	1680	64
Law of gravity	Isaac Newton	1685	83
Laws of motion	Isaac Newton	1687	85, 140, 100
Steam engine	Thomas Newcomen	1705	69
Piano	Bartolomeo Cristofori	1709	229
Conductors and insulators	Stephen Gray	1729	241, 247
Celsius scale	Anders Celsius	1742	31
Capacitor	Pieter van Musschenbroek	1745	243
Lightning conductor	Benjamin Franklin	1752	243
Latent heat	Joseph Black	1760	59
High-pressure steam engine	James Watt	1765	69
Submarine	David Bushnell	1776	93
Hot-air balloon	Montgolfier brothers	1783	51
Law of force between charges	Charles Coulomb	1785	239
Parachute	Jean-Pierre Blanchard	1785	138
Gold-leaf electroscope	Abraham Bennet	1787	241
Hydraulic press	Joseph Bramah	1795	94
Aneroid barometer	A. A. S. Conté	1799	96

▷ Something to do

1. Plot a large labelled time-chart of inventions since 1500.

2. Each person in the group chooses one invention, and (after five minutes' preparation) has **one minute** to explain to the others how it made life and society different after that date.

Steam locomotive	Richard Trevithick	1804	69
Magnetic effect	Hans Øersted	1819	284
Electric motor	Michael Faraday	1821	288
Photography	Joseph Niépce	1826	200
Ohm's Law	Georg Ohm	1826	251
Brownian movement	Robert Brown	1827	16
Transformer and dynamo	Michael Faraday	1831	302, 298
Laws of electrolysis	Michael Faraday	1834	270
Semi-conductors	Munk Rosenschöld	1834	314
Refrigerator	Jacob Perkins	1834	65
Cranked bicycle	Kirkpatrick Macmillan	1839	127
Pneumatic tyre	Robert Thomson	1845	94
Cathode rays	(several people)	1859	309
Telephone	Alexander Graham Bell	1876	287
Four-cycle petrol engine	Nikolaus Otto	1876	70
Gramophone (record-player)	Thomas Edison	1877	306
Microphone	David Hughes	1878	297
Electric lamp	Thomas Edison	1879	263
Motor car	Carl Benz	1885	71
Ciné camera	William Friese-Greene	1889	203
Diesel engine	Rudolf Diesel	1892	71
X-rays	Wilhelm Röntgen	1895	312
Radioactivity	Henri Becquerel	1896	340
Electron	Joseph Thomson	1897	240, 308, 344
Radium	Marie & Pierre Curie	1898	341, 346
Loudspeaker	Horace Short	1900	289
Tape recorder	Valdemar Poulsen	1900	307
Vacuum cleaner	H. C. Booth	1901	97
Vacuum flask	James Dewar	1901	55
Diode valve	Ambrose Fleming	1904	308
$E = mc^2$	Albert Einstein	1905	350
G–M tube	Geiger, Müller, Rutherford	1908	341
Idea of nucleus	Ernest Rutherford	1911	344
Cloud chamber	Charles Wilson	1911	340
Sonar	Paul Langevin	1918	222
Television	John Logie Baird	1926	307
Jet engine	Frank Whittle	1930	72
Electrostatic generator	Robert Van de Graaff	1931	242
Radio telescope	Karl Jansky	1931	207
Neutron	James Chadwick	1932	344
Cats' eye reflectors	Percy Shaw	1934	190
Radar	Robert Watson-Watt	1935	215
Nuclear reactor	Enrico Fermi	1942	350
Transistor	Shockley, Bardeen & Brattain	1947	320
Hovercraft	Christopher Cockerell	1954	98
Integrated circuit	Jack Kilby	1958	324
LASER	Theodore Maiman	1960	195
Optical fibre	Kao, Newnes & Beales	1966	194

Doing your practical work

GCSE exams have at least 20% of the marks awarded for practical work. Often these marks will be awarded by your teacher while watching you do investigations in the laboratory. You may feel self-conscious when your teacher watches you doing practical work but it is important to overcome this feeling and get on with the work.

Your marks will be awarded for particular **skills**. The details vary from time to time but there are 6 basic skills. Your teacher will be looking to see how well you can do these 6 skills:

1. Follow instructions accurately and carefully.
2. Select and use the most suitable equipment.
3. Make observations accurately.
4. Design and carry out an investigation.
5. Record results in tables, graphs and charts.
6. Draw a conclusion, and communicate it to others.

When marking you on these skills, your teacher will usually have a detailed checklist of what to look for. For example, on skill 3, 'Making Observations', the checklist might be as in this table:

There will be a similar checklist for each of the other skills. Ask your teacher for more details.

Here are some more details on each of these 6 skills:

'Practical skills are important'

1. Following instructions
This sounds simple but it can be an easy way to lose marks! If you are given written instructions you should read them at least *twice*. Concentrate on each sentence and each word – sometimes even the commas are important. Follow the instructions step by step, doing exactly what they say, and tick off each step when you have done it.

2. Selecting and using the most suitable apparatus
It is important to choose the most suitable equipment – this usually means the most accurate equipment.
For example, to measure the volume of 100 cm^3 of water you should choose a measuring cylinder (see page 89) not a beaker with a 100 cm^3 mark on it. (Why?) To measure a smaller volume of water you should choose a narrower measuring cylinder. (Why?)

The table shows a list of equipment you should be familiar with:

Can you use each of them accurately?

You might also be expected to know about a joulemeter (page 42). ticker-timer (p. 133), CRO (p. 310), magnetic compass (p. 279), microscope (p. 207), G–M tube and ratemeter (p. 341).

Checklist for Skill 3: MAKING OBSERVATIONS	
Candidate's skill level	Marks awarded
The candidate reads the scales, but the observations are generally inaccurate. Makes frequent errors in estimation.	1
Some observations are inaccurate. Makes regular errors in reading scales. He/she has difficulty in handling units and forgets to check that the instrument reads zero before using it. Can make rough estimates with help.	2
Reads scales accurately and usually allows for zero-errors. He/she repeats readings only when instructed to. General observations tend to lack detail.	3
Uses instruments correctly but he/she sometimes need prompting to repeat results, especially unexpected ones. General observations are careful but lack fine detail.	4
Uses instruments correctly and checks results. Handles zero-errors and units easily. Makes detailed and systematic observations and checks unexpected ones.	5

metre rule	page 6
stopclock, stopwatch	7
thermometer	31
spring balance	83
measuring cylinder	89
balance (lever or electric)	89
ammeter	248
voltmeter	250

3. Making observations accurately

The checklist (opposite page) shows how this skill might be marked. You can see that accuracy is important as well as care in checking results. Remember to correct for **zero-errors** (for example, you should check that your spring balance or ammeter reads zero before you start to use it).

When reading a scale, make sure you look at right-angles to it, so that you read the correct number. When using a measuring cylinder remember to read the **bottom** of the meniscus (see page 19).

In some experiments you might be given 'extra' equipment which may not have an obvious use. What do you think each of the following might be useful for: plumb line, spirit level, set square, magnifying glass, blotting paper or cloth?

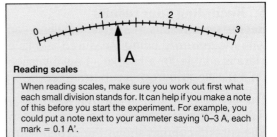

Reading scales

When reading scales, make sure you work out first what each small division stands for. It can help if you make a note of this before you start the experiment. For example, you could put a note next to your ammeter saying '0–3 A, each mark = 0.1 A'.

Reading scales can be tricky. Manufacturers usually put as many marks on the scale as the accuracy of it will allow. This means that you should only read to the nearest scale division (i.e. the nearest mark). For example, from the diagram above, you would write down 1.2 A.

However in some papers the examiners expect you to 'interpolate' and try to judge the fractions of a division. In this case you might write down 1.23 A. Your teacher will advise you which to do.

4. Designing and carrying out an investigation

As well as following instructions to do an experiment, often you may be asked to **design** an investigation to solve a problem. You may find it helpful to remember the seven steps in an investigation by remembering **HID-ERIC**:

The most important thing here is to devise a **fair test**. For example, look at experiment 8.10 on page 53. In this experiment it is important that we use two cans of the **same** size, and **same** shape, with the **same** amount of cold water, at the **same** temperature, placed the **same** distance from the **same** fire, with the temperatures taken at the **same** time. This is called 'controlling the variables' so that only **one** variable changes (the surface of the cans).

Skills 5 and 6 are discussed on the next page.

HID-ERIC: 7 steps of an investigation

H	HYPOTHESISE	Think about what you are going to investigate: What are the possibilities? What might happen?
I	IDEAS	What ideas can you suggest to tackle the investigation? Is each one a fair test?
D	DESIGN	Design and plan the investigation in detail. How will you measure the independent variable and the dependent variable (see the next page)? How can you control all the other variables?
E	EXPERIMENT	Do the experiment accurately and carefully, using suitable apparatus (see opposite page).
R	RESULTS	Record your results, using a table, graph, chart or other record (see the next page).
I	INTERPRET	Interpret your results, and 'draw a conclusion' about what they mean (see the next page).
C	COMMUNICATE	How will you communicate your results to other people? In writing? In a poster? In a talk?

Professor Messer's not too bright. It's up to you to put him right:

363

5. Recording your results

Record your results in a table, labelling each column with the **quantity** being measured and its **unit**.
Here is a table for a Hooke's Law investigation (p. 82) on a plastic strip from a supermarket shopping-bag:

Extension of a plastic strip

Load (in N)	Extension (in mm)
0	0
10	3
20	7
30	10

The first column shows the **independent variable** – this is what you change deliberately, step by step.
The second column shows the **dependent** variable – the size of this variable depends on the first one.
All other variables must be **controlled** (kept constant).

Often you will have to draw a line graph of your results. You can do this in 4 steps as shown here:

Four steps in drawing a graph:

1) Choose simple scales.
   For example, 1 large square = 1 newton, (or 2 N, 5 N, 10 N). *Never* choose an awkward scale like 1 square = 3 N or 7 N!
2) Plot the points and mark them neatly. Re-check each one.
3) If the points look as though they form a straight line, draw the *best* straight line though them with a ruler (and pencil). Check that it looks the *best* line.
4) If a point is clearly off the line you should always use your apparatus to repeat the measurements and check it.

Finding the gradient

This needs care, but can be done in 3 steps.

a) Draw a large right-angled triangle as shown in red:
b) Find the value of the two sides '**Y**' and '**X**', **in the units of the graph**. That is, you must find the size of **Y** and **X** using the scales on the graph (not just measuring with a ruler). For example, in the diagram, **Y** = 20 mm and **X** = 60 N. Can you see why?
c) Calculate the gradient (or 'slope') of the graph by dividing **Y/X**. Keep the units with the numbers, so that you find the unit of the gradient.
   For example, from the diagram:

$$\text{gradient} = \frac{Y}{X} = \frac{20 \text{ mm}}{60 \text{ N}} = \underline{0.33 \text{ mm/N}}$$

A calculator gave the result as 0.33333333. Why is it better to write it as 0.33?

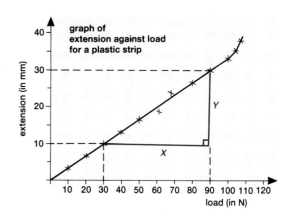

6. Drawing a conclusion

Every experiment has a 'conclusion'. This is a summary of what you found out (or sometimes what you didn't find!) Always look at your results or graph or chart and think hard to decide what you have discovered, or what reasonable or 'valid' deduction can be made from your results.

For example, from this Hooke's Law investigation (see also page 82) you might conclude:

a) The graph is a straight line through the origin (see also page 375). Therefore the extension is **proportional** to the load (up to the elastic limit).
b) The elastic limit is 100 N.
c) Below this value, the plastic extends by 0.33 mm per newton.

On the opposite page there are some practical questions for you to try.

Questions

1. In a school Science Club there are 18 pupils: 6 pupils are aged 14 years, 5 pupils aged 13, 4 aged 12, 2 aged 15 and 1 aged 16.
Plot a bar chart to show this information.

2. A waste bin is found to contain 50% paper, 25% glass, 20% plastic, 5% metal.
Plot a pie chart to show this information.

3. The table shows the results of weighing a number of 20p coins:

No. of coins	6	12	17	22	25	28	31	34	36	39
Mass (g)	29	59	83	108	127	137	152	167	177	187

   a) Plot a graph, discarding any doubtful results.
   b) Does it go through the origin?
   c) What is the mass of 20 coins?
   d) How many coins should make a mass of 127 g?
   e) What is the mass of 1 coin?
   f) What is the gradient of the graph?

4. The table shows the results of measuring the thickness of a number of sheets of paper:

No. of sheets	15	22	25	35	46	65	75	78	85
Thickness (mm)	1.7	2.4	2.7	4.4	5.1	7.2	8.1	8.6	9.4

   a) Plot a graph, discarding any doubtful results.
   b) Does it go through the origin?
   c) What is the correct thickness of 35 sheets?
   d) How many sheets would make 6.6 mm?
   e) What is the thickness of 1 sheet?
   f) What is the gradient of the graph?

5. *He's got his answers wrong again.*
   Take a look – can you explain?

For each of the following questions 6–18 use the ideas of HID-ERIC to plan an investigation. In each case you should:

a) Think of more than one method if possible and then select the best one.
b) Identify the variables, and ensure it is a fair test.
c) Sketch the apparatus you would use and add notes to explain the method.
d) Draw a table with headings and, if possible, say what graph you would draw.

After you have done this individually, you should discuss the investigation with a group of two or three others in your class.
Talk through all the details until you are all sure you have thought of everything. Use steps (c) and (d) to keep a record of the final plan.

6. Which make of paper towel is best for soaking up water?
7. Does a candle heat faster than a spirit burner?
8. Do the sleeves of different anoraks insulate equally well? (See page 57.)
9. When stretched, do copper wire and nylon fishing line behave the same way? (See p. 82.)
10. Who has the strongest hair in your class?
11. Which bounces better: a tennis or a golf ball?
12. Do pendulums with heavy bobs swing slower than pendulums with light bobs?
13. Which make of paper tissue is the strongest?
14. Does red paint absorb infra-red rays better than blue paint? (See page 53.)
15. What is the upper frequency limit of hearing for each member of your class? (See p. 224.)
16. How can you test the eyesight of your class?
17. Are the poles of a magnet of equal strength?
18. How would you put copper, iron, tin, constantan in order, by how well they conduct electricity?

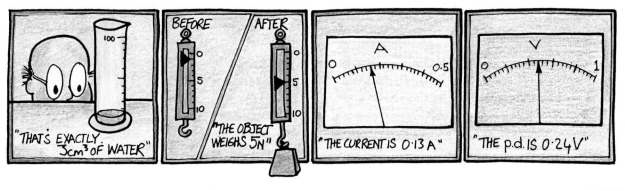

Revision techniques

Why should you revise?
You cannot expect to remember all the Physics that you have studied unless you revise. It is important to review all your course, so that you can answer the examination questions.

Where should you revise?
In a quiet room (perhaps a bedroom), with a table and a clock. The room should be comfortably warm and brightly lighted. A reading lamp on the table helps you to concentrate on your work and reduces eye-strain.

When should you revise?
Start your revision early each evening, before your brain gets tired.

How should you revise?
If you sit down to revise without thinking of a definite finishing time, you will find that your learning efficiency falls lower and lower and lower.
If you sit down to revise, saying to yourself that you will definitely stop work after 2 hours, then your learning efficiency falls at the beginning but **rises towards the end** as your brain realises it is coming to the end of the session (see the first graph).

We can use this U-shaped curve to help us work more efficiently by splitting a 2 hour session into 4 shorter sessions, each of about 25 minutes with a short, **planned** break between them.
The breaks **must** be planned beforehand so that the graph rises near the end of each short session.
The coloured area on the graph shows how much you gain:

For example, if you start your revision at 6.00 p.m., you should look at your clock or watch and say to yourself, 'I will work until 6.25 p.m. and then stop – not earlier and not later.'
At 6.25 p.m. you should leave the table for a relaxation break of 10 minutes (or less), returning by 6.35 p.m. when you should say to yourself, 'I will work until 7.00 p.m. and then stop – not earlier and not later.'

Continuing in this way is more efficient **and** causes less strain on you.

You get through more work **and** you feel less tired.

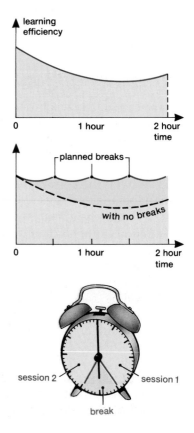

How often should you revise?

The diagram shows a graph of the amount of information that your memory can recall at different times after you have finished a revision session:

Surprisingly, the graph rises at the beginning. This is because your brain is still sorting out the information that you have been learning.
The graph soon falls rapidly so that after 1 day you may remember only about a quarter of what you had learned.

There are two ways of improving your recall and raising this graph.

● 1. If you briefly *revise the same work again after 10 minutes* (at the high point of the graph) then the graph falls much more slowly.
This fits in with your 10-minute break between revision sessions.
Using the example on the opposite page, when you return to your table at 6.35 p.m., the first thing you should do is *review*, briefly, the work you learned before 6.25 p.m.

The graph can be lifted again by briefly reviewing the work *after 1 day* and then again *after 1 week*. That is, on Tuesday night you should look through the work you learned on Monday night and the work you learned on the previous Tuesday night, so that it is fixed quite firmly in your long-term memory.

● 2. Another method of improving your memory is by taking care to try to *understand* all parts of your work. This makes all the graphs higher.
If you learn your work in a parrot-fashion (as you have to do with telephone numbers), all these graphs will be lower. On the occasions when you have to learn facts by heart, try to picture them as exaggerated, colourful images in your mind.

Remember: **the most important points about revision are that it must occur often and be repeated at the right intervals**.

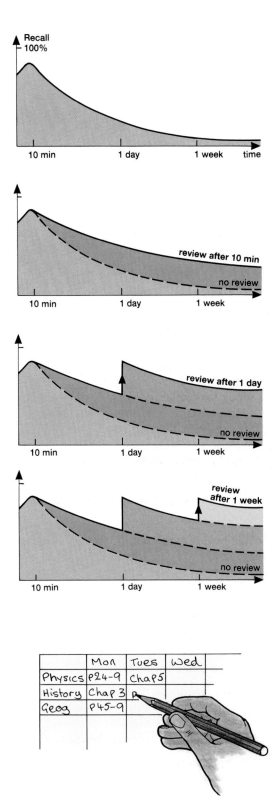

Suggestions for a revision programme

1. Beginning with Chapter 3 (on 'Molecules'), read the summary at the end of the chapter to gain some idea of the contents.
   Then read through the chapter looking at particular points in more detail, before reading the summary again.

2. Covering up the summary, check yourself against the fill-in-the-missing-word sentences at the end of the chapter.

   Remember to **re-read** the summary and to **review** each chapter after the correct revision intervals of 10 minutes, then 1 day, then 1 week (as explained on the previous page). Continue in this way with all the other chapters in the book.

3. While reading through the summaries and chapters like this, it is useful to collect together on a single sheet of paper all the formulae that you need to know. See also the checklist on the next page.

4. While you are going right through the book, doing this for every chapter:
   a) Attempt the questions on the 'Further Questions' pages. Your teacher will be able to tell you which are the most important ones for your syllabus.
   b) If there are multiple-choice questions in your examination, read the comments about them on page 371.

   There are some multiple-choice questions included on the 'Further Questions' pages, and there are 300 more, covering 30 topics, in a book called **Multiple Choice Physics for You.**

5. Read the section on 'Examination technique' on page 370, and check the dates of your exams. Have you enough time to complete your revision before then?

6. A few weeks before the examination, ask your teacher for copies of the examination papers from previous years. These 'past papers' will help you to see:
   – the particular style and timing of your examination
   – the way the questions are asked, and the amount of detail needed
   – which topics and questions are asked most often and which suit you personally.

When doing these past papers, try to get used to doing the questions **in the specified time**.

It may be possible for your teacher to read out to you the reports of examiners who have marked these papers in previous years.

Professor Messer:
If runners are defeated and cowboys are deranged, are examiners detested?

☑ Revision checklist: definitions, laws and formulas

In some examination papers you may be asked to *define* a certain quantity, or to *state* a law, or to *use* a formula.

The list below shows some that you might need. **Your teacher will be able to tell you which of them you should tick and then learn about.**

Heat
- Upper and lower fixed points   31
- Gas laws   35, 37, 38, 39
- Specific heat capacity   43
- U-value   49
- Specific latent heat of fusion   59, 60
- Specific latent heat of vaporisation   62
- Changing the melting point   61
- Changing the boiling point   63

Mechanics
- Force, weight   81, 83, 141
- Hooke's Law   82
- Newton's First Law   85
- Newton's Second Law   140
- Newton's Third Law   100
- Density   88
- Pressure   91
- Pressure in a liquid   92, 94
- Standard atmospheric pressure   95
- Scalar and vector quantities   102
- Parallelogram of forces   102
- Moment of a force   106
- Principle of moments   107
- Centre of mass (of gravity)   108
- Work and energy, joule   113, 115
- Principle of conservation of energy   114
- Gravitational potential energy   118
- Elastic potential energy   118
- Kinetic energy   119
- Power, watt   120
- Efficiency   124, 126, 129
- Average speed, velocity   132, 136
- Acceleration   132, 134, 138
- Constant acceleration equations   137
- Terminal velocity   138
- Momentum, impulse   146
- Principle of conservation of momentum   147

Waves and sound
- Wavelength, frequency, amplitude   167, 221
- Transverse and longitudinal waves   166
- Resonance   225
- Pitch, loudness, quality   226, 227

Light
- Law of reflection   179
- Real and virtual images   176, 181, 197, 198
- Focal length of a curved mirror   183
- Refractive index and Snell's Law   187, 188
- Critical angle   189
- Principal focus and focal length of a lens   196
- Magnification   197
- Spectrum and dispersion   210
- Electromagnetic spectrum   212–15, 233
- Primary colours of light   218

Electricity
- Law of electrostatics   239
- Electric field   242
- Capacitance   243, 323
- Resistors in series   248, 254, 259
- Resistors in parallel   249, 255, 259
- Ohm's Law and resistance   251
- Micro-, milli-, kilo-, mega-   7, 250
- E.m.f., internal resistance   250, 259
- Potential divider   256, 320
- Coulomb, ampere, volt   258
- Electric energy (in joules)   258, 259, 264
- Electric energy (in kW h) and cost   265
- Electric power   264
- Anode, cathode, electrolysis   269
- Law of magnetism   276
- Magnetic field and line of flux   279, 284
- Neutral point   279
- Right-hand Grip Rule   284
- Rule for polarity of a coil   285
- Fleming's Left-hand Rule   288
- Fleming's Right-hand Rule   296
- Faraday's Law and Lenz's Law   296, 297
- Effective value of a.c.   299
- Transformer   302
- Truth tables   326

Nuclear physics
- Alpha, beta and gamma rays   343, 347
- Atomic number, mass number, isotope   345
- Half-life   346
- Energy and mass   350

Examination technique

In the weeks before the examinations:

Attempt as many 'past papers' as you can so that you get used to the style of the questions and the timing of them.

Note which topics occur most often and revise them thoroughly, using the techniques explained on previous pages.
Read through your list of essential formulas as often as possible if you need to memorise them (see page 368).

Just before the examinations:

Collect together the equipment you will need:
- Two pens, in case one dries up.
- At least one sharpened pencil for drawing diagrams.
- A rubber and a ruler for diagrams.
  Diagrams usually look best if they are drawn in pencil and labelled in ink.
  Coloured pencils are usually not necessary but may sometimes make part of your diagram clearer (e.g. when drawing a spectrum).
- A watch for pacing yourself during the examination. The clock in the examination room may be difficult to see.
- For some examinations you may need special instruments e.g. compasses, a protractor or an electronic calculator.

Use your lists of essential formulas and definitions for last-minute revision.

It will help if you have previously collected all the information about the length and style of the examination papers (for all your subjects) as shown below:

Date, time and room	Subject and paper number	Length (hours)	Type of paper: – multiple-choice? – single word answers? – longer answers? – essays? – practical?	Sections?	Details of choice (if any)	Approximate time per question (minutes)
7th May 9.30 Hall	Physics 1	2	Multiple choice + long answers	2	A – all B – 3 out of 5	A – 2 min. B – 20 min.

In the examination room:

Read the introduction to the examination paper. It will often contain a phrase such as: 'Carelessness and untidy work **will** be penalised.' This means that you **must** take care to write and draw neatly or else you will lose marks.

Use the information on the front of the paper to calculate the length of time to be spent on each part of the paper. Then use your watch to pace yourself correctly.
Suppose for example that one section or a long question is supposed to occupy you for 20 minutes. If you finish in 15 minutes, this means that you know the work particularly well or, more likely, that you have not written in sufficient detail.
On the other hand, if you take 25 minutes or longer, it means that you may not have enough time to finish all the questions and so you will be at a disadvantage compared with other candidates who do finish.

If there is a choice of questions use your experience with past papers to choose a topic that suits you well.

When attempting a calculation, it usually helps to draw a diagram and then label it with all the available information. In this way you can see more easily what you know and which formulas are likely to be useful. (See also page 375.)

If you cannot do part of a question, leave a gap so that you can return to it later.
If you finish with time to spare, look over your answers and add to them or re-attempt them.

NATIONAL EXAMINING BOARD

Physics

23rd May
9·30 a.m.

Careless and untidy work will be penalised.

Section A — 1 hour
(30 compulsory questions)

Section B — 1 hour
Answer 3 questions (out of 5)

In what ways is your examination paper different from this one?

For multiple-choice questions:
- Read very carefully the coding instructions for A, B, C, D, E.
- Mark the answer sheet **exactly** as you are instructed.
- Take care to mark your answer (A, B, C, D, E) opposite the **correct** question number.
- Even if the answer looks obvious, you should look at all the alternatives before making a decision.
- If you do not know the correct answer and have to guess, then you can improve your chances by first eliminating as many wrong answers as possible.
- Ensure you give an answer to every question.

Careers

Have you thought what you want to do when you leave school?

After English and Mathematics, Physics is the most important qualification for a great many careers.

In the list below: ★ = Physics is essential
                    + = Physics is an advantage

Employers rate Physics qualifications very highly, particularly if you study it to A-level or higher.

And studying Physics can open the door to a surprising variety of jobs.

Aeronautical Engineer ★
Agricultural Scientist +
Air-traffic Controller +
Architect +
Army ★ or +
Astronomer ★
Audiologist ★
Automobile Engineer ★
Biomedical Engineer ★
Biophysicist ★
Building Technologist ★
Civil Engineer ★
Civil Service Scientific Officer +
Computer-aided Design +
Computer Programmer +
Dental Technician +
Dentist +
Doctor +
Draughtsperson +
Electrical Engineer ★
Electrician ★
Electronics Engineer ★
Environmental Health Officer +
Ergonomicist ★
Flight Engineer ★
Food Scientist +
Forensic Scientist ★
Geophysicist ★
Health & Safety Officer +
Industrial Designer +
Information Scientist +
Journalist (science) +
Laboratory Technician ★
Lighting Technologist ★

Marine Scientist +
Materials Scientist +
Mechanical Engineer ★
Medical Physicist/Technician +
Merchant Navy, deck,
    engineer, or radio officer ★
Metallurgist ★
Meteorologist ★
Mining Engineer ★
Motor Mechanic +
Nuclear Scientist ★
Optician ★
Patent Agent/Examiner ★
Pharmacist +
Physicist ★
Physiotherapist +
Pilot +
Production Engineer ★
Quantity Surveyor +
Radio and TV repair ★
Radiographer ★
Radio Studio Manager +
Recording Engineer ★
Royal Air Force ★ or +
Royal Navy ★ or +
Space Scientist ★
Structural Engineer ★
Systems Analyst +
Teacher (science) ★
Technical Writer +
Telecommunications (radio,
    telephone, satellite) ★
TV Camera Operator +
Veterinary Surgeon/Assistant +

a civil engineer

using computers

adjusting a jet engine

doing research

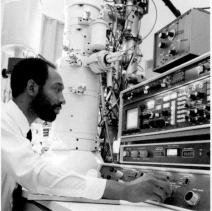

using an electron microscope

a technician

teamwork is important

using electronics

a quantity surveyor

Of course, for many of the careers listed or illustrated on these pages, you will need further study at school or in a College of Further Education. You should consult your careers teacher for more detailed information.

Also useful are the following books published by the Careers Research and Advisory Centre (CRAC), Bateman Street, Cambridge:

Decisions at 13/14+   **Decisions at 17/18+**
Decisions at 15/16+   **Your choice of A-levels**

Check your maths

Directly proportional

Look at the table. Do you see how **Y** and **X** are connected?
If **X** doubles, so does **Y**.   If **X** halves, so does **Y**, etc.

We say: **Y** is *directly proportional to* **X**
In symbols: $Y \propto X$

For an example, see Hooke's Law (page 82) or Ohm's Law (page 251).

Y	X
3	1
6	2
9	3
12	4

Making an equation

We can always change a proportionality into an equation by putting
in a constant, **k**:

$$Y = kX \qquad or \qquad \frac{Y}{X} = k$$

In the example in the table, $k = 3$.

Inversely proportional

Look at the table. Do you see how **P** and **V** are connected?
If **V** doubles, **P** halves.   If **V** halves, **P** doubles, etc.

We say: **P** is *inversely proportional to* **V**

In symbols: $P \propto \dfrac{1}{V} \qquad or \qquad \dfrac{1}{P} \propto V$

We make an equation in the same way as before:

$$P = k \times \frac{1}{V} \qquad or \qquad P = \frac{k}{V} \qquad or \qquad P \times V = k$$

In the example in the table, $k = 12$.

P	V
12	1
6	2
4	3
3	4

Changing the subject of an equation

As long as you do the same thing to **both** sides of an equation, it will still balance:

Example 1

From page 88: $D = \dfrac{M}{V}$

a) To change the subject to **M**:
   Multiply both sides
   by **V**, then cancel: $D \times V = \dfrac{M}{\cancel{V}} \times \cancel{V}$

$$\therefore \underline{M = D \times V}$$

b) To change the subject to **V**:
   First multiply by **V**: $D \times V = M$

   Then divide by **D**: $\dfrac{\cancel{D} \times V}{\cancel{D}} = \dfrac{M}{D}$

$$\therefore \underline{V = \frac{M}{D}}$$

Example 2

From page 137: $a = \dfrac{v - u}{t}$

To change the subject to **v**:

First multiply by **t**: $a \times t = \dfrac{v - u}{\cancel{t}} \times \cancel{t}$

$$\therefore at = v - u$$

Then *add u*: $at + u = v - \cancel{u} + \cancel{u}$

$$\therefore \underline{v = u + at}$$

Often it is easier if you put in your numbers
before changing the equation.

Graphs

a) *Directly proportional*
   For an equation like $Y = kX$ (see the opposite page), the graph is a **straight** line, **through the origin**.

   The gradient (or slope) $= k$ (see page 364).
   For an example, see Hooke's Law (page 82).

b) *Linear but not directly proportional*
   Consider an equation like $Y = kX + c$ or like $v = at + u$ (see example 2 opposite and page 137).
   For these equations the graph is a straight line but **not** through the origin.

c) *Inversely proportional*
   A graph of P against V (see table opposite) would give a curve.

   The equation is $P = k \times \dfrac{1}{V}$, so to get a straight line, we must plot P against $\dfrac{1}{V}$.

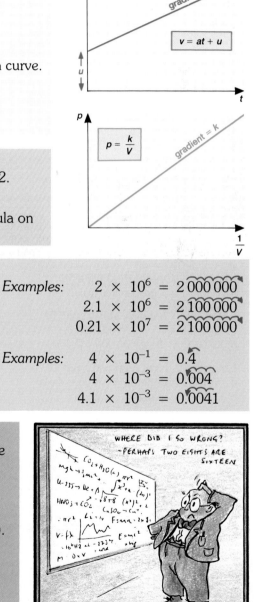

Dividing by fractions
Remember that dividing by $\frac{1}{2}$ is the same as multiplying by 2.
Dividing by $\frac{1}{3}$ is the same as multiplying by 3, etc.

Many people make mistakes with the parallel-resistor formula on page 255. Read the example on that page carefully.

Large and small numbers – indices
The small *index* tells us how many decimal places to move. *Examples:*

The + or − sign tells us which direction to move.

That is, for 10^n, move the decimal point n places to the right.

$$2 \times 10^6 = 2\,000\,000$$
$$2.1 \times 10^6 = 2\,100\,000$$
$$0.21 \times 10^7 = 2\,100\,000$$

For 10^{-n}, move the decimal point n places to the left. *Examples:*

$$4 \times 10^{-1} = 0.4$$
$$4 \times 10^{-3} = 0.004$$
$$4.1 \times 10^{-3} = 0.0041$$

Seven steps in solving a Physics problem
1. Draw a diagram showing all the *information* given in the question.
2. Decide which *formula* to use. (If you are not sure, write down the possibilities and then decide.)
3. Decide the *units* for the formula. If necessary, change your information to the correct units (e.g. mm to metres).
4. *Substitute* the numbers in the formula.
5. *Calculate* the answer.
   Check that your answer is a sensible size.
6. Decide the *units* for your answer.
7. Read the question to see if there is *another part* to do.

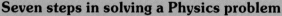

Answers

► Heat

Chapter 4 (Expansion) page 28
12. 0.004 m (4 mm); about 0.01 m (1 cm)

Chapter 5 (Thermometers) page 34
4. a) 283 K   b) 27 °C   c) 310 K
7. b) 40 °C   c) Between 2 hours and 4 hours
d) 4 hours   e) 9 hours   f) 0.5 h, 6.4 h   g) 5.5 h

Chapter 6 (Gas Laws) page 40
2. a) 300 K, 270 K, 423 K, 183 K   b) 100 °C, −73 °C
727 °C   **3.** 4 cm^3 **4.** 627 °C (900 K)
5. 3 atmospheres   **6.** 100 cm^3 **7.** 273 cm^3
8. 800 K (527 °C)   **9.** 980 K (707 °C) (assume the
volume is constant)

Chapter 7 (Measuring heat) page 45
2. Jack 4 MJ, Wife 18 MJ   **3.** Low specific heat
capacity, poisonous, expensive, heavy
4. a) 42 000 J   b) 8400 J   c) 880 J   d) 3800 J
5. 500 J/kg °C   **7.** a) 2.1 × 10^9 J (2100 MJ)
b) £63   c) 7.2 × 10^8 J (720 MJ)   d) 1.7 K (1.7 °C)

Chapter 8 (Conduction) page 58
3. Roof (cost regained in 2 years)

Chapter 9 (Changing state) page 67
7. a) 680 000 J   b) 170 000 J   c) 6 900 000 J
d) 230 000 J   **8.** 2 400 000 J/kg   **9.** 6 120 000 J

Chapter 10 (Heat engines) page 73
4. a) 1500 litres petrol, 1000 litres diesel
b) £600, £350 per year   c) 2 years

Further questions on Heat

Energy resources (page 74)
1. a) Coal, gas, oil   b) oil   c) i) 60 years   ii) 400 years
d) Oil finished; coal, water, nuclear increased

Molecules (page 74)
6. 5 × 10^{-9} m (5 nm), assuming the film is one
molecule thick; sprinkle powder first so oil shows area

Expansion (page 75)
8. a) Temperature, mercury, alcohol   b) 100 °C, 373 K
c) More sensitive, records minimum temperature; 310 K
d) See page 33

Gas Laws (page 75)
9. c) See page 375   d) 0.000 25 m^3 **10.** b) ii) 50 °C
iii) 2.08 kPa   iv) First is within the range of the graph;
second assumes it remains a gas and on the same
straight line

Specific heat capacity (page 76)
12. a) 940 J/kg K   b) iv) 18.4 minutes, 16 560 J
v) 920 J/kg K   vi) More heat lost to surroundings
13. a) 20 000 J   b) 4000 J/kg °C
14. a) i) 50 K (50 deg C)   ii) 8 400 000 J (8.4 MJ)
b) i) 3 600 000 J   ii) 2.33 units   iii) 19p (18.7p)
15. a) 800 J   b) 115 200 J   c) 144 s
16. a) See pages 16, 64, 50   b) i) 37 °C   c) See page
88; a bath measures roughly 0.4 m × 0.5 m × 2 m

Conduction, Convection, Radiation (page 77)
18. b) Door   c) Little temperature difference
d) 8020 kJ/h   e) 3   **19.** d) i) 30°   ii) 40° (add energy
in columns)   iii) 117 MJ   **20.** b) i) 20 K   iii) 1800 J
iv) 360 s (6 minutes)   **22.** a) i) walls (35%)   ii) 25%
iii) Hot air rises   iv) Carpet underlay   c) Walls (23%
saving)
23. a) i) See page 49   ii) 330 (328) J/s (330 W)
b) i) 110 kg   ii) 4.6 MJ

Change of state (page 79)
24. Melting point = 81 °C

► Mechanics

Chapter 11 (Pushes and pulls) page 87
2. 40 N (39.2 N)   **3.** a) 600 N (588 N)
b) Approx. 100 N   c) 60 kg   **4.** a) 10 cm   b) 15 cm
5. a) 50 mm   d) 70 mm length should be 66 mm
e) 5 N   f) 3.75 N   g) 86 mm
7. b) 6.6 cm should be 6.4 cm   c) 9.0 cm   d) 2600 N

Chapter 12 (Density) page 90
1. 5 g/cm^3 (5000 kg/m^3)   **2.** 8000 kg/m^3
3. 18 000 kg   **4.** a) B   b) C   c) A and C   d) D
5. 40 000 kg   **6.** 5 m^3 **7.** 3 g/cm^3 (3000 kg/m^3)
8. a) 2 m^3 b) 0.002 m^3 c) 5000 kg   d) 200 bricks
9. 130 kg   **11.** b) Density = 11 g/cm^3 (lead?)

Chapter 13 (Pressure) page 97
2. 50 N/m^2 **3.** 1 000 000 N (1 MN)
7. a) 100 N/cm^2 (1 000 000 N/m^2)   b) 2000 N

Chapter 14 (More about forces) page 103
3. b) 4.7 m/s   **7.** a) 14 kg   b) 2 N (in direction of 8 N)
8. a) 150 N at an angle of North 53° East
b) 150 N at an angle of South 53° West

Chapter 15 (Turning forces) page 112
3. 40 N m (4000 N cm)   **4.** a) 5 N   b) 4 N   **5.** c) 5 m

Chapter 16 (Work, energy and power) page 122
2. 10 J   **4.** 20 000 J (20 kJ)   **9.** a) 12 m
b) Longer   c) 24 m   d) Longer   e) At least 36 m
11. a) Straight line through origin (directly proportional, see page 375)   b) Non-linear (braking distance is proportional to square of speed, see page 119)   c) 76 m
12. 200 W   **13.** 100 W   **14.** 500 W
15. a) 7500 W   b) 7.5 kW
17. a) 100 J   b) 10 000 J (10 kJ)
18. a) 500 N   b) 2000 J   c) 2000 J
19. a) 450 000 J   b) 50 000 J   c) 400 000 J   d) 5000 N
20. 5 m

Chapter 17 (Machines) page 130
2. 25%, 35%, 10%, 20%   **3.** 2%, heat   **4.** 300 N
6. a) 600 J   b) 800 J   c) 75% (0.75)   **7.** 10 rev/s anticlockwise; 15 rev/s clockwise   **9.** i) 20 J, 30 J, 67% (0.67)   ii) 600 J, 800 J, 75% (0.75)

Chapter 18 (Velocity and acceleration) page 143
1. a) 20 m/s   b) 2 m/s^2 **3.** b) i) Zero   ii) 10 m/s^2
4. From rest, it accelerates uniformly for 4 s, at 2.5 m/s^2, up to 10 m/s. Travels at constant velocity of 10 m/s for 10 s. Then constant deceleration, of 2 m/s^2, for 5 s, to stop after a total time of 19 s.
5. a) 0.7 s   c) 4.7 m/s   **7.** a) 40 m/s   b) 200 m
9. b) 500 N   c) 50 kg   **10.** 7.7 s
11. a) 5 m   b) 2 s (1 s to go up, and 1 s down)
12. a) 20 m/s   b) 2 s   c) 10 m   **13.** 8.45 m
14. 5 m/s^2, 3 m/s^2
15. a) 3.0×10^7 N   b) 0.3×10^7 N (3×10^6 N)
c) 1 m/s   d) zero   **16.** a) 1000 N   b) 160 N

Chapter 19 (Momentum) page 149
3. a) 2 m/s   b) 64 J, 16 J   c) Heat, sound
4. 2 m/s   **5.** 40 m/s   **6.** 0.1 m/s
7. a) 32 500 000 kg m/s   b) 32 500 000 N (3.25×10^7 N)

Further questions on Mechanics

Hooke's Law (page 150)
1. c) i) 24 mm   ii) 5 N   **3.** c) i) See p. 362   ii) 27.5 mm

Pressure (page 151)
4. b) 500 N   c) 2000 N/cm^2 d) 1 N/cm^2 e) 1999
5. a) i) 12 N   ii) 20 cm^2 iii) 0.6 N/cm^2 iv) 0.3 N/cm^2
b) i) 160 cm^3 ii) 0.0075 kg/cm^3 (7500 kg/m^3)
6. b) 3000 N   c) 37.5 N/cm^2 (375 000 N/m^2 or 375 kPa)

Moments and machines (page 152)
13. c) i) 200 J   ii) 250 J   iii) 80% (0.8)   iv) Friction, weight of pulley block

Energy (page 152)
16. a) i) 231 MJ   ii) 269 MJ   iii) 185 MJ   b) gas

Energy and power (page 153)
18. c) 20% (0.20)   **19.** b) 4 m   c) 800 J
d) Gravitational potential energy   e) 840 J   f) Food
20. ii) 60 000 J (60 kJ)   iii) 240 000 J (240 kJ)
21. a) 14 400 J   b) 600 W   **22.** a) 3000 J   b) 600 W
23. a) 150 m   b) 120 000 J   c) 12 000 W (12 kW)
24. a) Zero, 200 kJ   b) 2000 N   c) 20 000 W (20 kW)
25. 1500 W, 600 joules/second (W)

Velocity and acceleration (page 154)
26. a) 0.2 s   b) 7.5 cm/s   c) 15 cm/s   d) Accelerating
f) Constant velocity   g) Zero   h) 0.6 s
27. b) Accelerating at 1 m/s; constant velocity = 15 m/s; decelerating at −0.75 m/s   c) 1 m/s^2 d) 112.5 m, 225 m, 337.5 m   **28.** a) 200 m   b) 11 s   c) 80 m
d) 10 s   e) 8 m/s   f) Constant   g) Zero
29. b) i) 36 s   ii) 750 m   iii) −2.5 m/s^2
iv) 250 000 N   c) Look at areas
30. a) 2 m/s^2 b) 10 N   c) 20 kg m/s   d) 40 J   e) 160 J

Force, acceleration, momentum (page 155)
31. a) i) Directly proportional   ii) 0.7 s   b) i) 89 m
ii) 21 m   iii) About 18 m/s   c) i) 120 000 J   iii) 30 m, 4000 N   d) See page 119
32. b) 1000 kg   c) i) 400 N   ii) 0.4 m/s^2 d) See p. 85, 148
33. a) 8 kg m/s   b) $2\frac{2}{3}$ m/s   c) 16 J   d) 10.7 J

▷ Waves: Light and Sound

Chapter 21 (Waves) page 171
3. 340 m/s   **4.** 300 000 000 m/s (3×10^8 m/s)
5. a) 2 cm   b) 3 Hz   **7.** b) 15 mm   c) 4 Hz

Chapter 22 (Light) page 177
4. 3.9×10^8 m (390 000 km) (see page 132)

Chapter 23 (Reflection) page 182
1. a) 60°   b) 60°   c) 60°   **3.** 2 m behind mirror, 5 cm wide; 1 m/s (2 m/s relative to him)

Chapter 25 (Refraction) page 193
11. a) 65°, 30°   b) 1.8   **13.** b) 15°   c) 23°   **14.** 4 m

Chapter 26 (Lenses) page 199
6. 7.5 cm from lens; 2 cm high; real, inverted, diminished
7. a) 100, 60, 40, 30, 24   c) 26 cm   d) 25 cm   e) 20 cm

Chapter 27 (Optical instruments) page 209
7. a) $\frac{1}{1000}$ s at f/5.6   b) $\frac{1}{1000}$ s at f/2
c) $\frac{1}{60}$ s at f/16   d) $\frac{1}{1000}$ s at f/4

Chapter 28 (Colour) page 219   **6.** 15 km   **7.** 0.3 s

Chapter 29 (Sound) page 231
2. 6800 m (6.8 km)   **3.** 3.4 m   **4.** a) 20 000 Hz, 20 Hz
b) 1.7 cm, 17 m   **5.** 1430 m   **7.** 680 m   **8.** 2250 m
9. 360 m/s   **13.** 104 dB, not safe

Further questions on Waves: Light and Sound

Waves (page 234)
3. c) i) 1.5 cm   ii) 15 cm/s

Refraction (page 235)
11. b) i) Real, inverted, diminished
ii) Focal length = 0.15 m (15 cm)

Sound (page 237)
22. b) 0.5 cm (0.005 m)   c) i) 0.2 m   ii) 0.0006 s
d) Cave walls appear closer, because echo quicker
25. a) 150 m   b) 75 m

▶ Electricity

Chapter 31 (Circuits) page 260
3. 3 A out,   b) 1 A in
5. a) 30 Ω b) 1500 Ω c) 1 Ω d) 5 Ω
8. a) 1 A, 0.5 A, 1 A   **9.** a) 10 V   b) 3 A   c) 10 Ω
10. a) 2 A   b) 2 A   c) 6 V   d) 14 V
11. a) 5 Ω b) 2 A   c) 4 V   d) 6 V   **12.** 12 V
13. a) 2 Ω b) 12 V   c) 12 V   d) 4 A   e) 2 A
14. 1000 A, 100 MJ (100 000 000 J)

Chapter 32 (Heating effect) page 268
2. 2 A
3. 4 A, 5 A; 1 A, 3 A; 480 W, 3 A; 4 A, 5 A; 2400 W, 13 A
5. a) 0.25 A   b) 960 Ω **6.** £1.20 (120 p)   **7.** 80 p

Chapter 33 (Chemical effect) page 271
4. 6 g   **5.** a) 0.33 g   b) 0.66 g

Further questions on Electricity (1)

Circuits (page 272)
3. a) 2 Ω b) 2 A   c) 10 V   d) 1.3 A
4. a) See page 257   c) i) 2.5 V   ii) 10 Ω
d) i) 1.2 Ω ii) 5.0 Ω iii) 0.8 A   iv) 0.96 V
v) 0.48 A   vi) 0.46 W
6. a) 2 A   b) 6 Ω c) 24 W   d) i) Increases
ii) Decreases
8. c) 0.064 K/min A^2 d) 0.051 K/min A^2 **9.** £4.20

Electric power (page 274)
11. b) i) S_3 ii) S_3 and S_1 c) i) 120 Ω ii) 4.5 A
iii) 7 A
12. b) i) 1.5p   ii) 2p   **13.** a) 48.0 W   b) 12.0 V
14. a) i) 4800 W   ii) 288 000 J   b) 252 000 J
c) 87.5% (0.875)

Chemical effect (page 275)
18. a) 970 g   b) 0.076 cm

Chapter 35 (Magnetic effect) page 295
13. b) i) 7.7 N   ii) 15.0 N
14. a) 30 J   b) 15 W   c) 20 W   d) 75% (0.75)

Chapter 36 (Electromagnetic induction) page 305
7. 1000 V a.c., 10 V a.c., 10, 1000
9. a) 10 : 1   b) 48 W   c) 48 W if 100% efficient   d) 0.2 A
10. a) i) 1000 A   ii) 1 A

Chapter 37 (Electron beams) page 313
5. a) 4 V (8 V peak to peak)   b) 20 ms (0.020 s)   c) 50 Hz

Further questions on Electricity (2)

Magnetism (page 335)
8. a) 80 J   b) 16 W   c) 40 W   d) 40% (0.40)

Electromagnetic induction (page 336)
12. b) A, 115 V; B, 230 V   **13.** c) i) 20 turns
ii) 0.1 A   d) i) Doubled to 40 turns   ii) 20 turns
14. c) i) 500 A   ii) 5000 V   iii) 2 500 000 W (2.5 MW)
iv) Heat   **15.** a) i) 500 turns, 0.25 A   b) See page 256

Electronics (page 337)
16. a) 1 V/cm (1 V cm^{-1})   b) 4 V   c) 0.5 ms/cm
(0.5 ms/cm^{-1})   d) 5 ms (0.005 s)   e) 200 Hz
19. c) i) About 350 Ω ii) 80 Ω
23. a) E: 0,0,0,1   P: 1,1,1,0   b) i) Low
(logical 0)   ii) 7.0 V   iii) 700 Ω iv) See page 316
25. A = 1; B = 0; C = 1

▶ Radioactivity

Chapter 39 (Radioactivity) page 353
4. b) 6, 14, $\frac{1}{4}$ c) About 17 100 years   **5.** 140 years

Further questions on radioactivity (page 354)
2. a) i) Background radiation   iii) 35 min   iv) See p. 347
3. a) i) 94   ii) 144   iii) 94   **4.** a) i) 131   ii) 54   iii) 77
c) 5×10^7 d) 1×10^7 per second
5. a) 95, 146, 95   b) ii) 237, 93   c) ii) 460 years (smaller
counts are less accurate)   **6.** a) i) 92   ii) 92   iii) 235

Graphs (page 365)
3. c) 98 g   d) 26   e) 4.9 g   f) 4.9 g/coin (0.20 coin/g)
4. c) 3.9 mm   d) 60   e) 0.11 mm   f) 0.11 mm/sheet

For help with your mathematics, see page 374

Dotty Definitions

kilogram	– dangerous record-player	principle	– royal tug of war
joule	– fight to the death	inclined plane	– there's writing on the walls of the aircraft
unit	– you do it with wool	velocity	– but we didn't lose the coffee
centigrade	– scale for perfumes	hertz	– painful
change of state	– emigration	amplitude	– well and truly eaten
insulate	– getting home after time	dispersion	– or that person
crankshaft	– lunatic coalmine	circuit	– for making a teacher?
newton	– up-to-date weight	anode, cathode	– female debtors
watt	– question	armature	– has strong teeth
power	– hit the lady	a.c./d.c.	– I view the ocean
sunspots	– his acne	dynamo	– eat briefly
maximum	– large mother		

Something to do

For each of these terms give the correct definition or write as much as you can to describe them.

E.g., kilogram: unit of mass; 1 kg = 1000 gram; here on Earth 1 kg weighs 9.8 newtons.

Wanted

A reward is offered for information leading to the arrest of Eddy Current, charged with assault and battery on a teenage coil named Milli Amp. He is also wanted in connection with the parallel theft of valuable joules from a bank volt.

Milli Amp tried to run but met series resistance and was overpowered. Later she was found by her friend Dinah Mo. After the couple had had a torque for a moment, Dinah said, "She almost diode but conducted herself well." Milli said, "Anode I would survive but it still hertz. It's enough to make a maltese cross."

Police say that the unrectified criminal escaped from a dry cell where he had been clamped in ions. First he had fused the electrolytes and then squeezed through a grid system. He was almost run to Earth in a magnetic field by a line of force, but he has been missing since Faraday.

Watt seems most likely is that he stole an a.c. motor. He may decide to switch it for a megacycle and return ohm by a short circuit.

How many puns can you find? (Over 30?)
Write a sentence to explain the correct meaning of each one.

Index

Illustration acknowledgements

5a Dr Mitsuo Ohtsuki/Science Photo Library 5b, 73, 148b, 163a NASA/Science Photo Library 10, 22b, 80c, 174, 176, 190, 198, 201d, 201e, 216, 340a Keith Johnson 11, 12a, 292 National Power plc 12b, 194, 213b, 333a British Telecommunications plc 13a Eric Thorburn/Scottish Power 13b Explorer/Robert Harding Picture Library 18a, 55, 184c, 186, 188, 208, 213a, 213c, 213d, 243, 271, 281, 302, 317a, 324a, 327, 332, 341, 346 Martyn Chillmaid 18b Stephen Dalton/Natural History Photographic Agency 19a, 214c, 309 Sally & Richard Greenhill 19b Marley Waterproofing Ltd 22a Dennis Hardley Photography 23 British Rail Research Unit 32 Chris Priest & Mark Clarke/Science Photo Library 48a Ake Lindau/Ardea, London 48b Electrolux Domestic Appliances Ltd 51 Quadrant Picture Library 52a Moulinex Swan Holdings Ltd 52b, 69b, 80a, 111a, 118b, 127, 130, 140, 144a, 144b, 145a, 145d, 147 Colorsport 53a Robert Harding Picture Library 53b Spanish Tourist Office 53c, 111b, 126, 330a Barnaby's Picture Library 56a Martin Bond/Science Photo Library 56b, 215a Dr R. Clark & M. Goff/Science Photo Library 56c, 311a, 312a Science Photo Library 57a Bryan & Cherry Alexander 57b Lewis Woolf Griptight Ltd 57c Dr R. Clark/Science Photo Library 57d British Aerospace Ltd 59 Spectrum Colour Library 68a Genesis Space Photo Library 68b Allsport/Pascal Rondlau 68c, 72, 191b Adrian Meredith Photography 68d ScotRail 69a Millbrook House Collection 71, 113 Leyland DAF 80b, 118a Allsport/Steve Powell 80d Allsport/David Leah 85a Transport and Road Research Laboratory 85b Volvo Concessionaires Ltd 86, 124 Allsport/Mike Powell 88 A Shell Photograph 94 JCB Ltd 98 Hoverspeed Ltd 99 L.C. Marigo/Bruce Coleman Ltd 111c London Buses Ltd 111d Allsport/Bob Martin 119a Lee Lyon/Survival Anglia 119b, 166, 201a Tony Stone Worldwide 129 Grove Worldwide 132, 135, 138a, 145c Allsport/Vandystadt 137 Gunter Ziessler/Bruce Coleman 138b Allsport/Simon Ward 145b Zefa/Globus Brothers 148a Securon Ltd 156 European Space Agency/Science Photo Library 159, 160, 162, 175a, 373h NASA 163b Sally Bensusen/Science Photo Library 164 U.S. Naval Observatory/Science Photo Library 169 Airfotos Ltd 170 W. Llowarch/Clarendon Press 173, 229 London Features International Ltd 175b George East/Science Photo Library 179 Allsport/Simon Bruty 182 Chris Johnson 184a Amstrad plc 184b Ian Beames/Ardea, London 184d Volumatic Ltd 191a Petit Format/Nestle/Science Photo Library 195, 214b Philippe Plailly/Science Photo Library 201b Allsport/Pascal Rondeau 201c Royal Albert Hall 202 David Parker/Science Photo Library 203 Joel Finler Collection 207 The Lovell Telescope, Jodrell Bank Science Centre 210 Gordon Garradd/Science Photo Library 211 Fred Burrell/Science Photo Library 212a U.S. Dept of Energy/Science Photo Library 212b, 212d, 233a, 233b National Medical Slide Bank 212c Jamaica Tourist Board 214a, 348a, 352 AEA Technology 215b Agema Infrared Systems/Science Photo Library 225 Memtek International Ltd 227 Yamaha–Kemble Music (UK) Ltd 230 Castle Associates 232a John Mason/Ardea, London 232b Dr Allan Dodds, Blind Mobility Research Unit, Nottingham 232c St. Bartholomew's Hospital, London 233c Zefa 233d, 233e National Gallery, London 233f, 233g Birmingham Forensic Science Laboratories 238a, 265, 267 Electricity Association 238b, 286 Boxmag-Rapid Ltd 238c Terry Why/Barnaby's Picture Library 238d Steve Percival/Science Photo Library 238e ICL 242 Ontario Science Centre 248 Condor PR/Omega Lighting Ltd 250, 258 Jane Howard PR Ltd/Ever Ready 253a, 256, 317b, 317c, 324b R.S. Components Ltd 253b–d Fisons Scientific Equipment 280 Moorfields Eye Hospital, London 282 Elcometer Instruments Ltd 289 Redferns 291 Robert Bosch Ltd 297 Eugene Adebari/Rex Features London 300 GEC Alsthom Turbine Generators Ltd 311b Racal Marine Group 312b Stammers/Thompson/Science Photo Library 312c Associated Press 324c, 373f GEC Ferranti 330b Charles Falco/Science Photo Library 333b PLI/Science Photo Library 333c, 333d National Remote Sensing Centre 340b, 347 Science Museum Library, London 348b Elscint/Science Photo Library 349a Leatherhead Food Research Association 349b G. Tortoli/Ancient Art & Architecture Collection 350 CEGB 359a, 359d, 359f National Portrait Gallery, London 359b Scala 359c Camera Press 359e Vivien Fifield Picture Library 373a Maggie Murray/Format 373b, 373i Brenda Prince/Format 373c, 373e, 373g Rolls Royce plc 373d Warren Spring Laboratory

Thanks to Chris Johnson for checking the answers section.
Acknowledgement is also made to the following Examining Groups for permission to reprint questions from their examination papers.

LEAG   London and East Anglian Group
MEG   Midland Examining Group
NEA   Northern Examining Association
NI   Northern Ireland Schools Examinations Council

SEG   Southern Examining Group
WAEC   West African Examinations Council
WJEC   Welsh Joint Education Committee

Other books by Keith Johnson

Multiple Choice Physics for You
This book provides carefully constructed and tested questions which can be used in a number of ways to supplement the course.
- It contains 300 questions grouped into 30 key topics which occur in all Physics courses.
- The questions are for use during a course and may be used for classwork or homework, testing or revision, or for practice for examinations.
- Diagrams have been widely used to help the student to understand the questions and appreciate the principles involved.
- Simple numbers have been used in calculations to keep the arithmetic straightforward and to highlight the principles involved.

Timetabling
This book is a complete and practical guide for those staff responsible for, or interested in, timetabling in schools and colleges.

TimeTabler is a comprehensive computer program for PC, Nimbus or BBC computers, which will do the actual scheduling of a school or college timetable. Friendly, fast and easy-to-use, this interactive program allows the user to sit at the keyboard controls and 'drive' quickly through the timetable.

Have you any comments to make?
Have you got any good cartoons or rhymes?
If you have, why not write to:
Keith Johnson
c/o Stanley Thornes (Publishers) Ltd
Old Station Drive
Leckhampton,
Cheltenham GL53 0DN